Signal Analysis in Pharmacovigilance

This book provides detailed concepts and information on principles and processes of signal analysis in pharmacovigilance along with case studies. It covers the fundamental concepts and principles of pharmacovigilance, emphasizing the need for robust signal detection and analysis methods. The book reviews the diverse array of databases and tools employed for signal detection, including electronic health records (EHRs), social media mining, claims data, and distributed data networks. In turn, the book discusses the application of molecular dynamics, molecular docking, and the use of the FDA Adverse Event Reporting System (FAERS) database in signal analysis.

- Explores the essential role played by signal analysis in the field of pharmacovigilance
- Examines various tools and software commonly used in signal analysis
- Reviews the process of data extraction, highlighting key considerations and best practices
- Provides insights into the specific regulations and guidelines that govern signal analysis in pharmacovigilance
- Entails the use of network pharmacology integrated with molecular docking and molecular dynamic techniques in the validation of identified signal
- Highlights case studies focusing on the practical application of signal analysis techniques in pharmacovigilance

Toward the end, the book explores the identification, validation, and assessment of signals associated with vaccines. This book is useful for graduate, post-graduate students of pharmaceutical sciences, and scientists in pharmacology research and drug development.

Signal Analysis in Pharmacovigilance
Principles and Processes

Edited by
Anoop Kumar

CRC Press
Taylor & Francis Group
Boca Raton London New York

CRC Press is an imprint of the
Taylor & Francis Group, an **informa** business

Designed cover image: Shutterstock

First edition published 2025
by CRC Press
2385 NW Executive Center Drive, Suite 320, Boca Raton, FL 33431

and by CRC Press
4 Park Square, Milton Park, Abingdon, Oxon, OX14 4RN

CRC Press is an imprint of Taylor & Francis Group, LLC

ISBN: 9781032629704 (hbk)
ISBN: 9781032629988 (pbk)
ISBN: 9781032629940 (ebk)

DOI: 10.1201/9781032629940

Typeset in Times
by KnowledgeWorks Global Ltd.

Contents

Editor ix
Contributors xi

Chapter 1   Introduction to Pharmacovigilance: Ensuring Patient Safety 1

Sachin Kumar and Vikas Maharshi

Chapter 2   Drug Safety Signal Detection and Management in Pharmacovigilance....................22

Mira Desai

Chapter 3   Databases and Tools for Signal Detection of Drugs
in Post-Marketing Surveillance...32

Jeesa George, Prizvan Lawrence Dsouza, Yalamanchili Jahnavi,
Hemendra Singh, and Pramod Kumar A

Chapter 4   Role of Electronic Medical Records, Insurance Claims Data,
Distributed Data Networks and Social Media in Signal Analysis 44

Vandana Roy

Chapter 5   Uncovering False Alarms: Examining Biases and Confounders in
Disproportionality Analysis 73

Lipin Lukose, Mansi Pawar, Aina M Shaju, Gouri Nair,
and Subeesh K Viswam

Chapter 6   Role of Medical Coding in Signal Detection 95

Dipika Bansal, Beema T Yoosuf, and Muhammed Favas KT

Chapter 7   Regulatory Aspects in Signal Detection and Assessment.. 111

James Buchanan

Chapter 8   Disproportionality Methods for Signal Detection in Pharmacovigilance................ 124

Krishna Undela and Christy Thomas

Chapter 9   Bayesian Methods of Signal Detection 135

Dipika Bansal, Muhammed Favas KT, and Beema T Yoosuf

Chapter 10 Subgroup and Sensitivity Analysis.. 148

Faiza Javed and Anoop Kumar

Chapter 11 Causality Assessment of ICSRs 158

Vivekanandan Kalaiselvan and Rishi Kumar

Chapter 12 Signal Analysis of Drug–Drug Interactions Using Spontaneous
Reporting Method 168

Rima Singh, Ruchika Sharma, Deepti Pandita, and Anoop Kumar

Chapter 13 Use of Network Pharmacological Approaches in Signal Analysis 175

Vipin Bhati, Ruchika Sharma, Deepti Pandita, and Anoop Kumar

Chapter 14 Docking Studies in Signal Analysis: An Overview 181

Kumari Kala Shah, Ruchika Sharma, S Latha, and Anoop Kumar

Chapter 15 Molecular Dynamics Simulations in Signal Validation 184

Debanjan Dey and Anoop Kumar

Chapter 16 Identification and Validation of Novel Signals of Drugs Using the FAERS
Database: A Case Study 198

*Sweta Roy, Vipin Bhati, Ruchika Sharma, Deepti Pandita,
and Anoop Kumar*

Chapter 17 Molecular Docking and Molecular Dynamics
in Signal Analysis: A Case Study.. 210

Sweta Roy, Ruchika Sharma, and Anoop Kumar

Chapter 18 Available Databases and Tools for Signal Detection of Vaccines........................... 226

Sachin Kumar and Vikas Maharshi

Chapter 19 Identification and Validation of Signals Associated with Vaccines........................ 241

Juny Sebastian

Chapter 20 Neurological and Behaviour Alterations with COVID-19 Vaccines Using the
Vaccine Adverse Event Reporting System (VAERS): A Retrospective Study 253

Mohd Amir, Saquib Haider and Anoop Kumar

Chapter 21 Pharmacovigilance-Based Drug Repurposing via Inverse Signals......................... 261

Simran Ohra and Anoop Kumar

Chapter 22 Prospective Role of Preclinical Experimentation in Validating
Safety Signals 269

Chandragouda Raosaheb Patil, Chandrakant Gawli, and Shvetank Bhatt

Chapter 23 Role of Artificial Intelligence in Signal Detection 279

*Pramod Kumar A, Jeesa George, Rethesh Kiran, Priyanka,
and Prizvan Lawrence Dsouza*

Index 287

Editor

Dr. Anoop Kumar is currently working as an Assistant Professor in Department of Pharmacology, Delhi Pharmaceutical Sciences and Research University (DPSRU), Govt. of NCT of Delhi, New Delhi, India. He earlier served as an Assistant Professor in the Department of Pharmacology and Toxicology at the National Institute of Pharmaceutical Education and Research (NIPER), Raebareli. He has also worked as an Associate Professor and Head of the Department of Pharmacology at ISF College of Pharmacy, Moga, Punjab; and as a Research Scientist at Translational Health Science Institute (THSTI), Faridabad; and the Medical Affairs and Clinical Research Department of Sun Pharmaceutical Limited, Gurugram, India.

He has authored more than 100 research and review articles as well as 20 book chapters in international journals and publishers of repute. His current h-index is 25, i10 index 59 with citations of 2340. He is the author of three books and has filed five Indian patents. He has received research grants from the Indian Government (DST SERB, ICMR, National Innovation Foundation (NIF)-India and IIT Delhi). He was a DST INSPIRE Fellow (2012–2016), as well as recipient of the Distinguished and Notable Researcher Award (2021–2022). He is included in the top 2% list of scientists released by Stanford. Currently, he is working as member secretary of the Human Ethics Committee approved by the Central Drugs Standard Control Organisation (CDSCO), Ministry of Health & Family Welfare, Government of India and Treasurer of the International Society for Pharmacoeconomics and Outcomes Research (ISPOR) India Chapter. He is an expert member of board of studies of various reputed universities. He is a lifetime member of various professional societies such as the Indian Pharmacological Society (IPS), Association of Pharmaceutical Teachers of India (APTI), and Society for Alternatives to Animal Experiments (SAAE), India. He has worked as a member secretary and animal house incharge approved by the Committee for Control and Supervision of Experiments on Animals (CCSEA), Government of India. His research interests are drug repurposing using computational and experimental studies, meta-analysis, signal analysis in pharmacovigilance, and pharmacoeconomic studies.

Contributors

Pramod Kumar A
M S Ramaiah University of Applied Sciences
Bengaluru, India

Mohd Amir
Delhi Pharmaceutical Sciences and Research
    University (DPSRU)
New Delhi, India

Dipika Bansal
National Institute of Pharmaceutical Education
    and Research (NIPER)
Mohali, India

Shvetank Bhatt
School of Health Sciences and Technology
Dr. Vishwanath Karad MIT World Peace
    University
Pune, India

Vipin Bhati
Delhi Pharmaceutical Sciences and Research
    University (DPSRU)
New Delhi, India

James Buchanan
Covilance, LLC
Belmont, California, USA

Mira Desai
SAL Institute of Medical Sciences
and
Signal Review Panel
Pharmacovigilance Programme of India
Ahmedabad, Gujarat, India

Debanjan Dey
Delhi Pharmaceutical Sciences and Research
    University (DPSRU)
New Delhi, India

Prizvan Lawrence Dsouza
M S Ramaiah University of Applied Sciences
Bengaluru, India

Chandrakant Gawli
R. C. Patel Institute of Pharmaceutical
    Education and Research
Dhule, India

Jeesa George
M S Ramaiah University of Applied Sciences
Bengaluru, India

Saquib Haider
Delhi Pharmaceutical Sciences and Research
    University (DPSRU)
New Delhi, India

Yalamanchili Jahnavi
Thomas J. Long School of Pharmacy
University of the Pacific
Stockton, California, USA

Faiza Javed
Delhi Pharmaceutical Sciences and Research
    University (DPSRU)
Delhi, India

Anoop Kumar
Delhi Pharmaceutical Sciences and Research
    University (DPSRU)
New Delhi, India

Vivekanandan Kalaiselvan
Indian Pharmacopoeia Commission
Ghaziabad, India

Rethesh Kiran
M S Ramaiah University of Applied Sciences
Bengaluru, India

Muhammed Favas KT
National Institute of Pharmaceutical Education
    and Research (NIPER)
Mohali, India

Rishi Kumar
Indian Pharmacopoeia Commission
Ghaziabad, India

Sachin Kumar
Delhi Pharmaceutical Sciences and Research
    University (DPSRU)
New Delhi, India

Lipin Lukose
Independent Researcher
Baltimore, Maryland, USA

S Latha
Delhi Pharmaceutical Sciences and Research
    University (DPSRU)
New Delhi, India

Vikas Maharshi
All India Institute of Medical Sciences
    (AIIMS)
Patna, India

Gouri Nair
M S Ramaiah University of Applied Sciences
Bengaluru, India

Simran Ohra
Delhi Pharmaceutical Sciences and Research
    University (DPSRU)
New Delhi, India

Chandragouda Raosaheb Patil
R. C. Patel Institute of Pharmaceutical
    Education and Research
Dhule, India

Deepti Pandita
Delhi Pharmaceutical Sciences and Research
    University (DPSRU)
New Delhi, India

Mansi Pawar
Independent Researcher
Hyderabad, India

Priyanka
JSS College of Pharmacy
JSS Academy of Higher Education and
    Research
Mysuru, India

Sweta Roy
Delhi Pharmaceutical Sciences and Research
    University (DPSRU)
New Delhi, India

Vandana Roy
Maulana Azad Medical College
New Delhi, India

Aina M Shaju
Population Health Sciences Institute
Newcastle University
Newcastle upon Tyne, UK

Hemendra Singh
Ramaiah Medical College
M S Ramaiah University of Applied Sciences
Bengaluru, India

Juny Sebastian
College of Pharmacy
Gulf Medical University
Ajman, United Arab Emirates

Kumari Kala Shah
Delhi Pharmaceutical Sciences and Research
    University (DPSRU)
New Delhi, India

Rima Singh
Delhi Pharmaceutical Sciences and Research
    University (DPSRU)
New Delhi, India

Ruchika Sharma
Delhi Pharmaceutical Sciences and Research
    University (DPSRU)
New Delhi, India

Subeesh K Viswam
Independent Researcher
Bangalore, India

Christy Thomas
Department of Pharmacy Practice
National Institute of Pharmaceutical Education
    and Research (NIPER)
Guwahati, India

Krishna Undela
National Institute of Pharmaceutical Education
    and Research (NIPER)
Guwahati, India

Beema T Yoosuf
National Institute of Pharmaceutical Education
    and Research (NIPER)
Mohali, India

1 Introduction to Pharmacovigilance

Ensuring Patient Safety

Sachin Kumar and Vikas Maharshi

1.1 INTRODUCTION

Pharmacovigilance (PV) is described by the World Health Organization (WHO) as "the science and activity relating to the detection, assessment, understanding and prevention of adverse effects or any other drug-related problem" [1]. Drug safety and public health are both dependent on PV. It entails keeping track of, evaluating, and avoiding drug side effects. Healthcare practitioners can choose drugs wisely by methodically gathering and examining data on side effects. This aids in the identification of potential safety issues, enabling prompt regulatory actions like label modifications or recalls. The safety of patients exposed to pharmaceuticals, biologicals, vaccines, blood products, and medical equipment is covered by PV. It also helps to reduce harm, increase therapeutic efficacy, and protect patients' well-being, making it an essential part of contemporary healthcare systems [2].

1.1.1 THE IMPACT OF PHARMACOVIGILANCE ON ADVERSE DRUG REACTIONS

More than 100,000 people died in 1994 as a result of undesirable medication reactions. Additionally, from 1999 to 2008, 26,399 fatalities and 0.9% of emergency hospital admissions were attributable to adverse drug responses, indicating that these reactions are to blame for significant proportion of morbidity and mortality among patients [3]. Despite different prevention measures, research indicates that between 5% and 10% of patients may experience an adverse drug reaction (ADR) upon admission or during hospitalization. PV has a profound effect on the frequency of occurrence of ADRs. PV has considerably decreased the frequency and severity of ADRs through systematic data collection, analysis, proactive risk assessment and taking appropriate measures. PV has made quick action and risk reduction possible via early ADR detection, frequently in the post-marketing phase. Continuous real-world data monitoring allows for the detection of new safety signals, which in turn leads to regulatory steps like updated drug labeling, dosage changes, or even the recall of dangerous drugs. ADRs have been seen to occur less frequently as a result of improved drug labeling and more public awareness, which have given patients and healthcare professionals the power to make wise decisions [4–7].

PV focused its efforts on managing the drug adverse effects through its global central database, which compiles reports of drug side effects from all around the world, in order to combat this phenomenon. The safety profile of the medications, patient care, and safety are all significantly improved by this position, which also supports the efforts of national drug regulatory agencies [8].

1.1.2 IMPORTANCE OF PHARMACOVIGILANCE

There are still a lot of unanswered questions regarding the safety profile of new pharmaceuticals when they are first brought to the market. These drugs are used by a variety of patients for a variety of illnesses. These patients may also be taking other medications, and they must adhere to various

DOI: 10.1201/9781032629940-1

customs and dietary restrictions, all of which could have a negative impact on how well the treatments work. Additionally, different formulations and components within the same drug are possible. While taking medications with conventional and herbal treatments, ADRs may also occur and need to be watched for using PV. A certain medication's ADRs may occasionally exclusively happen in one nation or region. PV is a crucial monitoring system for the safety profile of medications in a nation, with the cooperation of doctors, chemists, nurses, and other health professionals of the country. It aims to prevent any unnecessary physical, mental, and financial suffering of patients [9]. Besides patient safety, PV has an important role in the following areas.

1.1.2.1 Regulatory Clinical Trials

The number of clinical studies has significantly increased during the last few years, and this phenomenon has two sides. Although the development of new medicines is important for many patients, an arbitrary rise in the number of clinical trials may have an impact on how ethics committees and regulators operate, which could result in unethical patient behavior, inadequate reporting of side effects, and subpar patient monitoring throughout the course of clinical trials [1]. PV participates in clinical trial regulations to ensure correct monitoring, reporting, and evaluation of clinical data in order to address these issues. Along with standardizing and managing the international ADR reporting process between pharmaceutical makers and regulatory bodies, the Council for International Organization of Medical Sciences (CIOMS) working groups on PV share the same responsibilities [10].

1.1.2.2 Monitoring Post-Marketing Safety

After being released, 10% of recently approved medications are removed from the market. A new medicine's safety profile is not assured by its release onto the market after receiving clearance. Pre-market analysis cannot adequately determine a drug's safety profile. The results of the drug safety profile may be impacted by the small number of participants, short research length, and exclusion of particular patient populations, such as the elderly and children. Additionally, medication manufacturers put more effort into demonstrating the effectiveness of their products than their safety [10, 11]. Additionally, a drug's safety is not shown by the fact that a significant majority of patients have used it for a number of years. Food and Drug Administration (FDA)-approved in 1999, rofecoxib (Vioxx) was taken off the market in 2004. Despite the fact that Vioxx has throughout the years been shown to be an effective non-steroidal anti-inflammatory drug (NSAID) with a low risk of gastrointestinal bleeding, it has also caused many consumers to suffer catastrophic cardiovascular events [12]. Therefore, it is essential for both new and old medications to have PV as part of their monitoring for any adverse drug events. In this situation, the community and healthcare professionals play a crucial role in PV by reporting any adverse medication events [11].

1.1.2.3 Surveillance of Herbal and Traditional Medicines

Natural chemicals and supplements are increasingly used in herbal medicine to treat a variety of disorders since they are considered safe and free from side effects, in contrast to allopathic medications or chemically synthesized treatments. This is not entirely true, as there have been several reports of the negative effects of using herbal or conventional treatments. Herbal medications may cause these adverse reactions because they contain unpurified plant parts or extracts. To monitor their safety profile, Pharmacovigilance Programs also include traditional and herbal medications under the umbrella [13].

1.1.2.4 Pharmacovigilance for Vaccines

Vaccine PV is described as the science and activities linked to the detection, assessment, understanding, prevention, and communication of adverse events associated with immunization, or any other vaccine, or difficulties related to immunization. Strengthening PV is critical in every country because it enables healthcare professionals to avoid immunization difficulties and protect people's health from adverse events during immunization. Success of immunization system lies in having minimal vaccine-related morbidity and mortality, along with its efficacy in preventing diseases. Vaccines are biological products meant to prevent infectious diseases, but they can occasionally induce untoward occurrences called adverse events following immunization (AEFI). The detection of AEFI is a critical step in the prevention of issues in the immunization system [14].

1.1.3 ROLE OF PHARMACOVIGILANCE CENTERS IN EDUCATING AND TRAINING HEALTHCARE PROVIDERS

The contribution of healthcare professionals to promoting PV globally cannot be understated. The healthcare provider needs to be quite knowledgeable in order to recognize an adverse medication event properly. National Pharmacovigilance Centers are crucial in preparing medical staff to identify and report any adverse drug occurrences [1]. These facilities have received the WHO's approval, as have the nations that participate in the WHO Program for International Drug Monitoring. Any adverse event suspected to be due to a drug can be reported to the National Pharmacovigilance Centre by the public and/or healthcare professionals [15].

1.1.4 EVOLUTION OF PHARMACOVIGILANCE

PV has a history of roughly 170 years; however, it was not known as a separate entity in the beginning. To assess pharmacological risk/benefit ratios, improve patient safety, and improve quality of life, it is a deliberate action with major social and economic ramifications in the field of professional healthcare. To fully grasp the historical evolution, we describe the PV milestones up to the present day. The initial reports, which were essentially letters or cautions from clinicians to the editors of important and well-respected scientific publications, all the way up to the current, ultra-structured electronic registries of today, represent these achievements. The historical stages also assist us in determining the issues that PV will face in the coming years as well as why PV assisted us in achieving such significant results for human health and pharmacology itself.

Significant improvements in patient safety and regulatory supervision have characterized the dynamic process of PV development. PV, which was initially developed in reaction to horrible incidents like the Thalidomide catastrophe, has changed from a mostly reactive approach to a proactive and systematic method for regulating the safety profile of drugs. ADRs now have a deeper understanding of their possible effects, which has enhanced risk assessment, mitigation tactics, and ultimately better informed healthcare decisions. PV is a field that is constantly changing, but its most important function is still protecting the public's health and promoting confidence in medical treatments.

On January 29, 1848, 169 years ago, Hannah Greener, a young child from North England, died as a result of receiving chloroform anesthesia before having an infected toenail removed. This is the first incident of PV that has been documented. Sir James Simpson had discovered and developed the use of chloroform, a strong and secure anesthesia. The goal of the investigation into Hannah's death was to shed light on the circumstances of her departure, but it proved difficult to determine the precise cause of her death. It is likely that either pulmonary aspiration or a lethal arrhythmia were to blame for her demise [16].

Lancet established a commission to look into this matter in response to rising mortality and growing worries about the safety of anesthesia expressed by both medical professionals and the general public. This panel advised English physicians, especially those practicing in colonies, to keep records of any fatalities connected to anesthesia. *Lancet* subsequently published the findings of this investigation in 1893 [17].

Medication purity and adulteration-free requirements were first established with the passage of the US Federal Food and Drug Act on June 30, 1906. The propagation of bogus pharmacological or therapeutic indications was also forbidden by this organization in 1911 [17]. Tragically, in 1937, the use of the diethyl glycol-based sulfanilamide elixir in the United States resulted in the deaths of 107 people. These fatalities were blamed on the producers' ignorance of the solvent's toxicity at the time [16, 18, 19]. In 1938, the Federal Food, Drug, and Cosmetic Act was passed as part of an effort to update the public health system.

In fact, the revolutionary approach brought the capacity to conduct factory inspections and laid the groundwork for the need for safety profiles of medications be demonstrated before their clearance for market distribution [20]. Douthwaite proposed the theory that the occurrence of melena, a type of gastrointestinal bleeding, may be related to acetylsalicylic acid (ASA) in 1938 [21]. A variety of findings were obtained from the investigation of the toxicity of ASA on the digestive system. At this time, ASA is not recommended for people with gastrointestinal ulcers since data suggest that it may cause gastrointestinal problems [22].

A considerable improvement in European PV was motivated by the thalidomide catastrophe in 1961. Australian physician Dr. McBride made a potential connection between thalidomide use and congenital defects in neonates in a letter that was published in *Lancet*. He noticed that when thalidomide was given during pregnancy, the frequency of congenital abnormalities in infants (1.5%) increased by as much as 20% [23]. Parallel to this, Dr. Lenz discovered a connection between thalidomide and abnormalities at a German pediatric conference. His discoveries were subsequently reported in the German newspaper *Welt am Sonntag* [24]. The correlation between thalidomide use during pregnancy and the development of congenital abnormalities in infants was confirmed by retrospective research carried out in 1973 [25].

Due to Dr. Kelsey's strong safety concerns about thalidomide use during pregnancy, the United States was spared the tragedy of thalidomide [26]. The thalidomide tragedy raised a number of questions and important issues, such as the validity of animal testing, the operation of the industrial sector, and the requirement for post-marketing drug monitoring. The PV system has drastically changed as a result of this tragedy, with the methodical, coordinated, and controlled reporting of ADRs. This contained all the necessary components to encourage spontaneous reporting and establish a connection between the medicine and the adverse event [26].

The "Yellow Card" (YC) method was created in the United Kingdom in 1964 to make it easier to gather spontaneous complaints of drug toxicity. This was accomplished by using a particular form called the YC [27]. A crucial amendment requiring the submission of safety and efficacy data for drugs prior to premarketing authorization was passed in the United States in 1962. This mandate includes the inclusion of the outcomes of teratogenicity tests performed on three different animal species in safety data [18]. The EC Directive 65/65 was introduced to address these worries in 1965 in reaction to the fatal effects of thalidomide in Europe [28]. In addition, the Boston Collaborative Drug Surveillance Program pilot project was launched in 1966.

This organization, a pioneer in the area, set the standard for performing epidemiological studies to calculate possible drug side effects through in-hospital monitoring. Additionally, it was crucial in developing and putting into practice drug epidemiology approaches [29]. A total of 11 nations participated in the founding of the WHO Programme for International Drug Monitoring in 1968: Australia, the United States, the United Kingdom, Canada, Germany, Sweden, Ireland, New Zealand, Denmark, and the Netherlands. Later, in 1975, Italy joined this effort [30]. Numerous studies were carried out between 1968 and 1982 to investigate proven adverse medication responses [3].

The International Society of Pharmacovigilance (IsoP) superseded the European Society of Pharmacovigilance (ESoP) in 1992. This organization's main goals included developing PV and all matters concerning the safe and efficient use of medicines [31]. The European Medicines Agency (EMA) was founded in the year 1995 [32]. Funding for EudraVigilance began in 2001, and it now serves as the official database for tracking and analyzing data on reported side effects associated with drugs that have acquired marketing authorization or are being studied in European clinical trials [33]. The environment of European PV underwent significant changes in 2012 as a result of the passage of the new law (Directive 2010/84/EU) [33].

Marketing authorizations will have improved access to the EudraVigilance database in order to successfully carry out their PV duties. In November 2017, the updated EudraVigilance format was released. According to Commission Implementing Regulation (European Union [EU]) No. 520/20121 [32], these requirements include continual evaluation of EudraVigilance data and the dissemination of validated signals to the Agency and national regulatory bodies. The timeline of the historical evolution of PV is depicted in Figure 1.1.

1.1.5 ROLE OF ARTIFICIAL INTELLIGENCE IN PHARMACOVIGILANCE

A growing component of PV is artificial intelligence (AI), a highly interdisciplinary discipline [34]. Large volumes of data may be processed and analyzed using AI, and it can be used to facilitate treating a variety of diseases. Although additional research is required to determine whether this process improvement has an impact on the caliber of safety assessments, automation and machine learning models can improve PV procedures and offer a more effective way to examine safety-related data. As a result of its utility in predicting side effects and ADRs, it is anticipated that usage will rise in the near future [35].

FIGURE 1.1 Timelines of the historical evolution of pharmacovigilance.

1.1.6 Role of Pharmacogenetics in Pharmacovigilance

PV relies on pharmacogenetics to customize drug therapy to a person's genetic profile and reduce adverse responses [36]. For instance, information about genetic variations like *HLA-B*15:02* help prescribe carbamazepine safely for the treatment of epilepsy [37]. The optimal dose of warfarin is determined by the genotypes of CYP2C9 and VKORC1 [38], while CYP2D6 testing helps prevent codeine-related toxicity [39]. Pharmacogenetics is an essential tool for tracking and enhancing the effectiveness of medication because of these tailored approaches that improve drug's safety profile and efficacy.

1.1.7 Pharmacovigilance Regulations in Different Countries

It is clear from examining how PV trends have changed over the past ten years that there has been a definite transition from a mostly reactive to a more proactive PV approach with a strong emphasis on risk management and communication tactics. Although there are numerous areas of divergence and dispute, a direct comparison of specific PV rules, systems, and processes throughout the four regions (the United States, United Kingdom, Canada, and India) reveals major differences. For instance, there are significant differences among the specific data gathered by foreign regulatory bodies, with different categories of safety data being collected/reported in different forms and at varying frequencies, as well as adverse event reporting.

Additionally, risk management is frequently carried out inconsistently. For instance, while specific elements of the ICH good clinical practice recommendations are included in various sections of regional legislation, they are not always integrated across the four regions or presented in the same manner. Table 1.1 shows a comparison of PV laws across different countries, which amplifies the various, significant inconsistencies among the legislation in these four regions.

1.1.8 Adverse Drug Reactions

The WHO defines an ADR as an unexpected and unpleasant effect that occurs in patients after taking drugs for the prevention, diagnosis, or treatment of a disease at doses that are typically used [30]. ADRs significantly increase mortality, morbidity, longer hospital admissions, and medical costs, posing a substantial burden to the healthcare system [41, 42]. According to a meta-analysis [43], 8.7% of ADR-related hospitalizations in patients over 60 were computed correctly. To lessen their impact, ADRs must be studied in terms of early diagnosis and prevention, and motivating healthcare professionals to report ADRs is also essential [44]. Overall, 60% of ADRs can be avoided, according to a WHO assessment [45]. Drug–drug interactions (DDIs) are a major contributor to avoidable ADRs. Polypharmacy has become commonplace as a result of the rise in multimorbidity among patients and the complexity of therapeutic drugs [46, 47].

An adverse event, on the other hand, is a negative outcome following drug exposure that isn't always brought on by the drug [48]. As few patients have been exposed to a medicine, little is known about its safety profile when it is marketed [48, 49]. Since detection and diagnosis frequently depend on clinical expertise, medication safety evaluation should be seen as an essential component of routine clinical practice.

1.1.8.1 Significance of Adverse Drug Reactions

An important clinical problem is the occurrence of ADRs, which account for 2%–6% of all hospital admissions [50–52]. Adverse medication events lengthen hospital stays and raise hospital expenses, according to recent surveys conducted in the United States [52, 53].

TABLE 1.1

Comparison of Pharmacovigilance Regulations in Different Countries

Parameters	United States	India	United Kingdom	Canada
Regulatory authority	Food and Drug Administration	Central Drugs Standard Control Organization	Medicines and Healthcare Products Regulatory Agency (MHRA)	Health Canada
Responsible body for regulatory pharmacovigilance	Center for Biologics Evaluation and Research (CBER) and Center for Drug Evaluation and Research (CDER)	National Coordination Centre— Pharmacovigilance Programme of India (NCC-PvPI), IPC	Commission on Human Medicines (CHM)	Marketed health products, directorate of the health products and food branch
Guidelines/Rules	Guidance for Industrial GVP and Pharmacoepidemiologic Assessment in 21 CFR 314.80 and 314.98	Drug and Cosmetics Act (1940) and Rules (1945)	Directive 2001/20/ EC, Directive 2001/83/EC, and Regulation (EC) No. 726/2004—all contain articles 106.	GVP guidelines (GUI-0102)
Reporting process	Through MedWatch form and online through FAERS	ADR reporting form, VigiFlow, mobile application, or email	Via email, the Yellow Card portal, the online reporting form, or another channel	Canada Vigilance Program (MedEffect Canada) either online, by fax/mail or through telephone at Canada Vigilance Regional Office
Pharmacovigilance inspection	Via PADE inspections	Not mentioned	Via risk assessment strategy	GVP inspection program Inspection strategy for GVP for drugs (POL-0041)
Pharmacovigilance audit	Via post approval audit inspections	Not mentioned	In accordance with EU GvP guidelines	GVP inspection program Inspection strategy for GVP for drugs (POL-0041)
Risk management system	Guidance on risk management is provided under guidance for industry GVP and pharmacoepidemiologic evaluation	Cited in the recommendation document for reporting spontaneous adverse medication reactions	Follows risk management plan as per EMA guidance	Submission of risk management strategies and follow-up commitments is mentioned in the advice document
Reporting timelines for SAE	15 calendar days after the event	Within 24 hours of occurrence (in clinical trials)	Within 15 calendar days, reporting by QPPV	15 calendar days after an ADR occurs
Database	FAERS database	WHO ICSR Database (VigiBase)	Yellow Card database	Canada Vigilance Adverse Reaction Online Database

(Continued)

TABLE 1.1 *(Continued)*

Comparison of Pharmacovigilance Regulations in Different Countries

Parameters	United States	India	United Kingdom	Canada
ADR reporting form	Forms 3500, 3500A, and 3500B	Healthcare professionals' suspected ADR reporting form and patients' medication side effect reporting form	Yellow Card for reporting an adverse drug reactions from the MHRA	Industry-suspected adverse medication reaction to commercialized goods Consumer reporting form for possible drug reactions
PSUR submission	For drug products, to CDER and for biological products, to CBER	To DCG (I) and PvPI	To PSUR repository	Therapeutic Products Directorate of Health Canada, Submission and Information Policy Division
Data lock point for PSUR	70/90 days	30 days of the last reporting period	6 months after the commission date	70/90 days
Toll-free/Helpline number	1-800-332-1088	1800-180-3024	0-808-100-3352	1-866-337-7705
Connection with UMC	FAERS data are communicated to WHO UMC	Via VigiFlow, the ICSRs are directly reported to the UMC database	Following a causation analysis, Yellow Card reports are submitted to UMC	Via MedEffect program

Source: Reference 14.

Abbreviations: ADR, Adverse drug reactions; CDSCO, Central Drugs Standard Control Organization; CHM, Commission on Human Medicines; DCG, Drugs Controller General; DCG(I), Drugs Controller General of India; EMA, European Medicines Agency; EU, European Union; FAERS, FDA Adverse Event Reporting System; GVP, Good pharmacovigilance practices; ICSRs, Individual Case Safety Report; IPC, Indian Pharmacopoeia Commission; MHRA, Medicines and Healthcare Products Regulatory Agency; PADE, Post-marketing adverse drug experience; PSUR, Periodic Safety Update Report; PvPI, Pharmacovigilance Program of India; QPPV, Qualified Personnel for Pharmacovigilance; UMC, Uppsala Monitoring Centre; WHO, World Health Organization.

1.1.8.2   Types of Adverse Drug Reactions

Adverse medication reactions are divided into six categories (each with a mnemonic): dose-related (Augmented), non-dose-related (Bizarre), dose- and time-related (Chronic), time-related (Delayed), withdrawal (End of use), and failure of therapy (Failure) [54]. Type A (pharmacological) and type B (idiosyncratic) adverse medication reactions are two different categories [41]. The pharmacological effects of a medicine are amplified in type A reactions, which are dosage-dependent and reversible by dose decrease or withdrawal and are frequently may be discovered in preclinical or clinical trials. On the other hand, type B side effects are distinctive and cannot be foreseen purely based on the drug's known pharmacology. Type B ADRs are associated with drug- and patient-specific features (idiosyncratic), environmental hazards, and an unclear relationship to increasing dose. Post-marketing testing frequently reveals rare type B reactions. Type "A" and type "B" reactions are described below in more detail.

1.1.8.2.1 Pharmacological Adverse Drug Reactions

Over 80% of all ADRs are of type A, which are more frequent than type B reactions [50]. Thus, the primary pharmacological side effects of beta-blockers are bradycardia and hypotension, while bronchospasm is a secondary side effect.

The recent incident with the investigational hepatitis B medication fialuridine highlights the need for continued progress in developing appropriate in vivo and bridging in vitro testing regimes. These frameworks are designed to foresee secondary drug-related adverse reactions in human subjects. Five of the 15 patients who took part in a phase II trial in June 1993 passed away, and 2 more needed emergency liver transplants because of liver and kidney failure [55, 56]. Four separate animal species had not previously seen this outcome. Fialuridine and its metabolites may inhibit mitochondrial DNA polymerase, which may be the cause of the toxicity, according to the findings of in vitro tests performed on cultivated hepatoblasts [55, 57].

DDIs, variations in pharmacokinetics or pharmacodynamics, dosage differences, changes in pharmaceutical drug formulation, and anomalous pharmacokinetic or pharmacodynamic responses all increase the risk of pharmacological adverse events. Lower doses of some drugs, such as captopril, were later found to be both safe and efficacious. Initially, higher doses of these drugs were associated with an intolerable level of toxicity. Type A reactions are more common in the elderly and people with illnesses like renal failure that reduce a drug's efficacy. Additionally, the probability of harmful interactions increases with the number of prescriptions written, as demonstrated by the fact that 50% of interactions take place when five medications are taken at once [55, 58].

1.1.8.2.2 Idiosyncratic Adverse Drug Reactions

An unpleasant reaction that is unusual among patients taking a particular prescription and has nothing to do with the medication's intended therapeutic effect is referred to as an "idiosyncratic drug reaction" (IDR), a term that has been used in a variety of situations but lacks a clear definition. IDRs are unpredictable and frequently lethal, despite not being the most frequent type of ADR. Although almost any organ can be impacted by IDRs, the liver, blood cells, and skin are the most frequently affected. While medications can have an impact on many organs, sometimes concurrently, others can only have an effect on one organ. While some IDRs share certain traits with the majority of other IDRs and can be caused by different medications, each medicine can cause a somewhat varied spectrum of IDRs. The majority of IDRs seem to be immunologically mediated; conclusive proof is frequently absent, and it is undoubtedly unknown exactly how a drug causes an immune response [59].

IDRs also generally don't seem to grow in risk with dose, which is another defining feature of them [59]. IDRs are not dose-independent, despite the fact that some people have claimed this to be the case. The highest incidence of an IDR typically occurs at a dose below what is regarded as therapeutic. As a result, the risk does not rise above the level required for effective therapy. But occasionally, due to chance variables, the two dose-response curves coincide, which increases the likelihood of IDR within the therapeutic range [60]. Idiosyncratic refers to something unique to a person, and generally speaking, it is impossible to forecast who would experience an IDR from a certain medication. Major histocompatibility complex and HLA gene have always been associated with the discovery of significant genetic components [61]. Abacavir hypersensitivity reactions are the exception, when 50% or more of HLA-B*57:01 recipients of abacavir would develop an IDR [62].

1.1.9 CURRENT METHODS OF PHARMACOVIGILANCE

The only focus of PV, a subfield of pharmacoepidemiology, is the epidemiological examination of adverse effects or pharmacological occurrences. The term "events" in this context refers to happenings that are noted in the patient's notes during the course of a drug monitoring period. These events may be brought on by the illness or infection for which the medication is being taken, a different infection or illness that is also present at the same time, an ADR that is being watched, the activity of a medication that is being taken concurrently, or any of these factors. Pharmacoepidemiological methods are used as tool, in PV for generating initial suspicions (hypothesis-generating methods) or testing hypotheses (hypothesis-testing methods) about changing in adverse effect profiles of medicines. The details of these two methods are provided below.

1.1.9.1 Hypothesis-Generating Methods

1.1.9.1.1 *Spontaneous ADR Reporting*

ADR forms are given to doctors on which they can report any suspected ADRs to a central body. These forms may be used by patients and other healthcare providers in some nations. Since 1964, the "Yellow Card" has served this function in the United Kingdom. To promote reporting, the MedWatch form is utilized and widely distributed to health professionals in the United States. The key benefit of spontaneous reporting is that it is cost-effective and may be used to identify extremely rare ADRs for all medications during their entire lifetime.

The information may only reflect the reporter's suspicions, but a doctor or other healthcare provider treating an actual patient offers their insight. However, the program is incredibly helpful, and giving medical professionals a way to report their suspicions is crucial. Numerous unexpected and dangerous ADRs have been found and verified thanks to spontaneous reporting. Many commercially available pharmaceuticals have been removed as a result of these results, or new information has been made available to help consumers use the products more safely.

Underreporting of ADRs is a frequent problem in PV [63–65]. This is due to the fact that most countries, including India, adopt a spontaneous or voluntary method of ADR reporting. Patient-related causes of underreporting include failure to notice ADR or inability to link the ADR to a medication. The most common doctor-related causes include guilt, fear of lawsuit, ignorance, laziness, higher patient load, inadequate risk perception of newly marketed medications, apprehension, insufficient training to recognize ADRs, and a lack of information about the PV program [66].

1.1.9.1.2 *Prescription-Event Monitoring*

The idea of using event monitoring to find pharmacological side effects was initially put forth 25 years ago. Prescription-event monitoring (PEM), created by the Drug Safety Research Unit, is the first systematic post-marketing surveillance strategy to use event tracking on a wide scale in the United Kingdom. From photocopies of National Health Service prescriptions that are centrally processed in England, PEM can identify patients who have been prescribed a specific drug and their doctors. Every patient's general practitioner receives a personalized follow-up questionnaire ("green form") in the mail, typically on the first anniversary of the initial prescription, asking for details about the patient, particularly any "events" that may have occurred since starting treatment with the medication.

This crucial strategy, used notably in the United Kingdom, makes use of several National Health Service (NHS) components. When a prescription is written by a general practitioner inside the NHS, the Prescription Pricing Authority (PPA) receives it after it is filled. The PPA facilitates the process of obtaining confidential duplicates of specific prescriptions for recently approved drugs that are currently being evaluated by the Drug Safety Research Unit.

PEM's drawback is that there is no way to measure compliance (even though information is only gathered on prescriptions that have been filled) [67].

1.1.9.2 Hypothesis-Testing Methods

1.1.9.2.1 *Case-Control Studies*

These studies compare individuals who have a condition against healthy controls who are susceptible to it. This approach analyzes the exposure rates in the cases and controls, with statistical adjustments made for potential confounding variables. Much attention must be given in the design, just as with any official epidemiological or clinical investigation. It is crucial to use caution while creating examples to make sure that they accurately depict the desired result. For example, it's crucial to distinguish Stevens-Johnson syndrome from other rashes. The meticulous selection of a suitable control group that accurately represents the population from which the cases originated is equally crucial.

The level of bias in a study can be reduced through careful design; appropriate control in the analysis is also crucial. Major medication safety concerns have received a significant body of information from case-control studies. Aspirin and Reye's syndrome have been linked, according to research [68], and vaginal cancer has been linked to the use of the hormone diethylstilbestrol (DES) during pregnancy in women's offspring [69]. These are only two notable examples. Furthermore, case-control research showed that prenatal vitamin supplementation has a protective impact against the emergence of neural tube abnormalities [70].

The final findings of this research give an estimate of the likelihood that the outcome related to the investigated exposure would occur, given as an odds ratio. The absolute risk can only be ascertained under highly unique conditions. Compared to case-control studies, cohort studies are less effective since they necessitate considerable data collection just for the relevant cases and controls. Case-control studies can frequently be included in larger clinical trials or earlier cohort studies.

Recall bias is the most frequently mentioned drawback of case-control research [71]. In a case-control study, recall bias refers to the higher likelihood that participants who experienced the result would remember and report exposure as compared to participants who did not.

1.1.9.2.2 Cohort Studies

Research design must consider cohort studies. The word "cohort" comes from the Latin "cohors", which means "a group of soldiers". This kind of study is non-experimental or observational. The word "cohort" refers to a group of individuals who have been included in a study as a result of an occurrence that meets the criteria established by the researcher. For instance, a group of people who were born in Mumbai in 1980. This group is referred to as a "birth cohort". The cohort will also include smokers as an example. Alternative names for these studies include "prospective studies" and "longitudinal studies".

The participants in a cohort study are not initially having the desired outcome. In choosing them, consideration is given to the person's exposure status. The outcome of interest is then monitored throughout time to gauge its likelihood of occurring. Cohort studies include the Framingham cohort study, the HIV cohort study in Switzerland, and the Psoriasis and Depression cohort study in Denmark, to name a few. These studies could be either prospective, retrospective, or a hybrid of both. The timing between exposure and outcome is well defined in a cohort design since people who enter the cohort study do not yet have an outcome.

Compared to a prospective cohort study, a retrospective cohort study can be conducted quickly and is less expensive. In a cohort study, participant follow-up is crucial, and losses are a significant source of bias in this kind of research. The cumulative incidence and incidence rate are calculated using the results of these research studies. The longitudinal nature of the data is one of a cohort study's key advantages. Both time-dependent and time-independent variables will be present in the data.

A prospective cohort design has a number of significant drawbacks, including the fact that it is costly and time-consuming. For instance, it may take many years of follow-up to determine the incidence of cardiovascular disease in psoriasis patients [72].

1.1.9.2.3 Randomized Controlled Trials

Prospective studies called randomized controlled trials (RCTs) are used to assess how well a novel intervention or treatment works. Randomization lessens bias and offers a rigorous technique to explore cause-effect links between an intervention and outcome, even if no study is likely to be able to prove causality on its own. This is so that any differences in results can be attributed to the research intervention. Randomization balances participant characteristics (both observed and unobserved) between the groups. With any other study design, this is not feasible.

Researchers must carefully choose the population, the interventions to be compared, and the outcomes of interest while designing an RCT. Once they are specified, a power calculation is done to determine the number of participants required to accurately assess the existence of such a link. Following participant recruitment, they are randomly allocated to either the intervention group or the control group [73]. It is crucial to make sure that the participant's group assignment is secret at the time of recruiting; this is known as concealment. Automated randomization systems, such as computer-generated ones, are frequently used to assure this. RCTs frequently use blinding to further reduce bias by keeping participants' treatment information secret from investigators, nurses, and researchers. The lack of generalizability or low external validity of RCTs is a limitation in community health research [74].

1.1.10 Roles and Responsibilities in Pharmacovigilance Systems

Stakeholders in different PV systems have varying nomenclature according to region; however, as mentioned previously, the basic functioning remains the same. Stakeholders directly involved in the usage of drugs (e.g., prescribers, dispensers, patients, patient's attendants, and those involved in drug administration/care providers) should identify and suspect an adverse event to be related to drug usage when appropriate, and they should report it to the nearest ADR monitoring/reporting center/manufacturer through a suitable mode (filled forms or telephonically). It is also the responsibility of informer to provide as much right and detailed information as possible.

Marketing authorization holder should also retrieve relevant safety information from various sources, including but not limited to literature, patients, healthcare providers, media reports, their PV persons/employees, contractual partners, medical inquiries. Marketing authorization holders are required to validate the received information and submit a safety report of their medicinal product in a specified format to the drug regulatory authority of the region in a timely fashion (this may vary from country/region/zone). Marketing authorization holder is also required to implement the suggestions received from the regulatory authority based on the received event information.

At the next level, responsibilities of personnel reporting the received event information online are to accurately enter the information into the software, do the causality assessment, follow-up on the case till there is a complete resolution of the adverse event, report the suspected ADR cases in a timely manner, and maintain the records of reported and received cases.

Responsibilities of stakeholders at the next level are to perform a causality assessment, to inform regulatory authorities, establish ADR monitoring/reporting centers, provide manpower to the centers for reporting, organize programs for regular training and sensitization of various stakeholders, formulate and avail tools (forms, apps, toll-free helpline, etc.) for reporting (in English and other languages) and update them from time to time, management of database and analysis, signal detection, disseminate the relevant information to other stakeholders, expand the coverage of reporting by collaborating with other national/international health programs.

Regulatory authorities also play an important role in PV. Based on the information received, regulatory authorities may take appropriate regulatory action, like the withdrawal of a batch/lot, change in drug label (package insert), and drug recall/revoke [75, 76].

1.1.11 Causality Assessments

In context with PV, a causality assessment is an evaluation of the likelihood of a drug being a cause of an adverse event [77, 78]. Causality assessments assist in the evaluation of the safety profile of available medicines [77]. Broadly, the methods of causality assessments can be categorized under three heads, *viz.* "Expert judgment/Global introspection", "Algorithmic approaches", and "Bayesian approaches". Here we provide a brief overview of these three types of approaches, with details of the most commonly practiced methods from among these [77, 78].

1.1.11.1 Expert Judgment/Global Introspection

This approach involves an assessment of causality by an expert or a group of experts, considering all the available data regarding a particular adverse event. Judgment here is made regarding the causal relation between drug and event based on the prior knowledge and experience of expert(s). Examples of such methods are the "Swedish method" and the WHO-UMC (World Health Organization–Uppsala Monitoring Center causality assessment scale). The WHO-UMC scale considers four parameters for evaluating the causality, namely (a) temporal relation between drug administration and appearance of adverse events, (b) other factors (drugs and/or diseases) that can explain the causation of event of interest, (c) effect of withdrawal of suspected drug(s), also known as dechallenge, and (d) effect of re-administration of the drug, called rechallenge. There are six levels of causal association (between suspected drugs and adverse events) in the WHO-UMC scale (Certain, Probable, Possible, Unlikely, Unclassified, and Unclassifiable) [77–79]. The causality assessment according to the WHO-UMC scale is mentioned in Table 1.2. Arbitrary criterion for causality assessment according to the WHO-UMC scale is as follows.

1.1.11.2 Algorithmic Approaches

Algorithms are step-by-step instructions in the form of flow charts or questionnaires to guide the ranking of a causal association between a drug and the adverse event. These methods provide consistent and reproducible results because of systematic approach. A well-known example of an algorithmic approach is the Naranjo et al. method (simply called the Naranjo scale).

The Naranjo scale consists of a questionnaire with 10 items. Each item has three options, "Yes", "No", and "Don't know" [77, 78, 80]. All the items with scoring for their options are mentioned in Table 1.3.

Other examples of causality assessment scales under the head "Algorithmic approaches" are the Dangaumou French method, the Kramer et al. method, the balanced assessment method, the summary time plot, the Ciba-Geigy method, the Loupi method, the Roussel Uclaf method, the Maria and Victorino scale, and the drug interaction probability scale [77, 78].

1.1.11.3 Bayesian/Probabilistic Approaches

These approaches use epidemiological information (prior probability) to determine posterior probability (which combines epidemiological information with the evidence for the case of interest). These approaches do not put any bounds on the amount of information assessed to determine the causality. These approaches also allow for the assessment of multiple causes simultaneously. Examples of causality assessment methods under the Bayesian approach are the "Australian method", the "Bayesian Adverse Reaction Diagnostic Instrument", and the MacBARDI spreadsheet [77, 78].

TABLE 1.2
Causality Assessment According to the WHO-UMC Scale

Levels of Causal Association	Temporal Relationship	Other Plausible Drugs/ Diseases Ruled Out	Dechallenge	Rechallenge
Certain	+	+	+	+
Probable	+	+	+	−
Possible	+	−	−	−
Unlikely	−	−	−	−
Unclassified	Adequate information is not there; additional information is required and is being sought			
Unclassifiable	Adequate information is not there; additional information is required but cannot be sought			

TABLE 1.3
Naranjo Scale and Scoring System [61]

Sl. No.	Items	Scores for Options		
		Yes	No	Don't know
1.	Are there previous conclusive reports on this reaction?	+1	0	0
2.	Did adverse event appear after the suspected drug was administered?	+2	−1	0
3.	Did the adverse reaction improve when the drug was discontinued or a specific antagonist was administered?	+1	0	0
4.	Did the adverse reaction reappear when the drug was re-administered?	+2	−1	0
5.	Are there alternative causes (other than the suspected drug) that could, on their own, have caused the reaction?	−1	+2	0
6.	Did the reaction reappear when a placebo was given?	−1	+1	0
7.	Was the drug detected in the blood (or other fluids) in concentrations known to be toxic?	+1	0	0
8.	Was the reaction more severe when the dose was increased or less severe when the dose was decreased?	+1	0	0
9.	Did the patient have a similar reaction to the same or similar drugs in any previous exposure?	+1	0	0
10.	Was the adverse event confirmed by any objective evidence?	+1	0	0

Note: Causality based on the sum of scores of all 10 items is as follows: Total score: 0 (Doubtful); 1–4 (Possible); 5–8 (Probable); ≥9 (Definite).

1.1.12 PHARMACOVIGILANCE SYSTEMS AND DATABASES

Most countries developed PV systems after the 1960s thalidomide disaster, and many of them joined the WHO's program of international drug monitoring. The number of countries establishing their own PV system and joining the program is increasing gradually [81]. Examples of major PV systems are "Yellow Card scheme" (in the United Kingdom), "MedWatch" in the US and "VigiFlow" (by the WHO), the Japanese adverse event reporting system, and EudraVigilance (in the European economic area) [82–85].

The basic principles of the functioning of these PV systems remain the same. These systems largely rely on voluntary reporting (online, offline, or telephonically) of adverse events suspected to have been caused by drug(s), by healthcare professionals and/or drug users to the designated centers with or without comment on causality. Thereafter, the strength of causal association is assessed by authorities, and a level of causality is designated for each reported case, which may require communication with the reporter to retrieve more information and for the follow-up of the case [82, 86]. Adverse events with a designated causality are called adverse drug reactions (ADRs) [87]. Based on the level of causality of ADRs, the seriousness, the commonness of the usage of the culprit drug, the number of cases received of such a reaction, and the novelty of the reaction, regulatory authorities of that region may take appropriate action. Based on these factors, actions taken by the regulatory authorities may include, but are not limited to, generating a signal and disseminating the information to the stakeholders (manufacturers, prescribers, dispensers, drug administrators, drug users) in the form of newsletters, circulars, orders, etc. [82, 88]. Regulatory action in this regard may lead to signal generation, withdrawal of a particular lot/batch of that drug or whole drug recall from the market (there are examples of drug recalls after their marketing, e.g., Rofecoxib because of adverse cardiovascular events, Terfenadine because of fatal cardiac arrhythmias, and Tegaserod because of the risk of stroke) [89], change in label (package insert) of the respective drug with the addition/deletion of some boxed warning, an indication, and dosage (Figure 1.2).

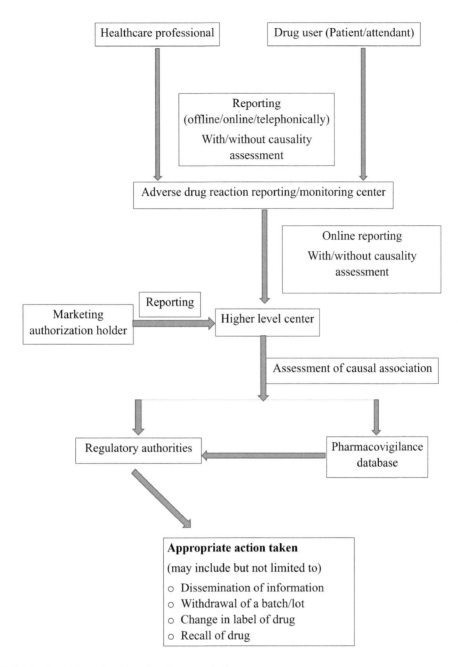

FIGURE 1.2 Basic functionality of a pharmacovigilance system.

All the reports following causality assessment are saved and stored in an online platform called a database. Various databases are there for different PV systems, e.g., "VigiBase" for the events reported to VigiFlow, EudraVigilance in the European economic area, the FDA adverse event reporting system in the United States, and Japanese adverse event reporting system [90, 91]. Among these, VigiBase is the largest global database of reported potential adverse effects of drugs, containing >30 million such reports. VigiBase is also linked to various drug nomenclature systems like WHO-Drug, WHO ICD, WHO-ARD, MedDRA, which enable entry, retrieval, and analysis of information in structured and systematic way [91].

1.1.13 Challenges and Future Prospects

PV systems and ADR reports received are still disproportionate worldwide. One of the biggest challenges for the current PV system is the lack of reporting of suspected ADRs by healthcare professionals and/or drug users. Multiple reasons could be there for non-/underreporting, e.g., over-burdened outpatient clinics, especially in the developing world, and doubts about reporting (what, when, where, why, and how to report) because of a lack of awareness about PV and its importance among drug users. Many healthcare professionals, especially those from specialties other than phar-macology, consider this a useless activity because of a lack of sensitization and motivation. Many times, reporting does not reflect the true reporting; rather, falsified information is reported just to meet the target set by the superior authority on the personnel working under them.

Scales used for causality assessment have a major subjectivity component in them, which may make signal detection difficult, and the problem is exaggerated by underreporting. Establishing adequate infrastructure, providing sufficient manpower, and training stakeholders is also not free of cost and is a big challenge, especially in developing world.

PV system in low-middle-income countries may be provided with adequate budgetary sup-port, and those countries should also have robust systems to effectively utilize that support. Timely sensitization and awareness activities must be organized for healthcare professionals, drug users, information collectors, and reporters to improve PV coverage and reduce the burden of underreporting.

Some appropriate incentives may also be offered to healthcare professionals directly involved in patient care and drug users (as a patient-centric approach) for reporting suspected ADRs, balancing undue coercion and quality of reporting, to further improve the coverage.

PV officials may also need to collaborate with other sectors, including information technology and engineering, to make use of AI in making more robust and efficient tools that can facilitate convenient reporting of events, data storage, retrieval, analysis, interpretation, and decision making, but simultaneously they should be able to maintain privacy and safeguard against cyber-attacks.

Moreover, the introduction of novel therapeutics like gene therapies, personalized medicines, and biologics also contributes to complexity in PV. Assessment of their safety may not be addressed by traditional methods and may need advanced tools.

1.1.14 Conclusion

In conclusion, the field of PV is an indispensable pillar of the modern healthcare system. Its fun-damental purpose is to safeguard public health by meticulously monitoring the safety of pharma-ceutical products throughout their lifecycle. PV ensures that the advantages of medicines continue to outweigh any potential risks by methodically gathering, analyzing, and evaluating adverse drug responses.

This chapter's introduction to PV has shown how important it is. By discovering unexpected dangers, assisting with regulatory choices, and enabling the improvement of pharmacological regi-mens, it acts as a vital link between pharmaceutical innovation and patient well-being.

PV is a globally interconnected field that may collaborate across borders and promote openness and trust while disseminating vital safety information quickly. PV is still flexible and robust in a time when healthcare concerns are always changing, from the advent of novel therapies to the dif-ficulties of adverse event reporting in a digital age. In addition, as technology develops, PV is enter-ing a revolutionary period. A new era of proactive and individualized drug safety is about to dawn thanks to advancements in AI, empirical data, and patient-centered methodologies.

PV is essentially a promise made to patients that their health and well-being are of the utmost priority. It is not merely a medical specialty. To make sure that the medications we use are not only efficient but also as safe as possible, healthcare professionals, pharmaceutical corporations, regula-tors, and society at large have made a commitment. In the chapters that follow, we will delve deeper into the intricacies of PV, exploring its methodologies, challenges, and future prospects, all in pursuit of the ultimate goal: a safer and healthier world for all.

REFERENCES

1. World Health Organization. Safety alerts for medicines and vaccines. 2024. Available from https://www.who.int/teams/regulation-prequalification/regulation-and-safety/pharmacovigilance/safety-alerts (Last accessed on 10.06.2024).
2. World Health Organization. The importance of pharmacovigilance. 2002. Available from https://apps.who.int/iris/handle/10665/42493 (Last accessed on 30.08.2023).
3. Kumar, Atul. "Pharmacovigilance: Importance, concepts, and processes." *Am Journal of Health-System Pharmacy* 74, no. 8 (2012): 606–612.
4. Bates, David W., Lucian L. Leape, and Stephen Petrycki. "Incidence and preventability of adverse drug events in hospitalized adults." *Journal of General Internal Medicine* 8 (1993): 289–294.
5. Lazarou, Jason, Bruce H. Pomeranz, and Paul N. Corey. "Incidence of adverse drug reactions in hospitalized patients: A meta-analysis of prospective studies." *Survey of Anesthesiology* 43, no. 1 (1999): 53–54.
6. Pirmohamed, Munir, Sally James, Shaun Meakin, Chris Green, Andrew K. Scott, Thomas J. Walley, Keith Farrar, B. Kevin Park, and Alasdair M. Breckenridge. "Adverse drug reactions as cause of admission to hospital: Prospective analysis of 18 820 patients." *BMJ* 329, no. 7456 (2004): 15–19.
7. Davies, Emma C., Christopher F. Green, Stephen Taylor, Paula R. Williamson, David R. Mottram, and Munir Pirmohamed. "Adverse drug reactions in hospital in-patients: A prospective analysis of 3695 patient-episodes." *PLoS One* 4, no. 2 (2009): e4439.
8. Sahu, Ram Kumar, Rajni Yadav, Pushpa Prasad, Amit Roy, and Shashikant Chandrakar. "Adverse drug reactions monitoring: Prospects and impending challenges for pharmacovigilance." *Springerplus* 3, no. 1 (2014): 1–9.
9. Mccormak, M. H., N. Arthurs, and J. Feely. "The spontaneous reporting of adverse drug reaction by nurses." *Br J Clin Pharmacol.* 1995 Aug; 40(2): 173–175.
10. Cioms, C. H. Geneva. Benefit-risk balance for marketed drugs. Evaluating safety signals: Report of CIOMS working group IV CIOMS, Geneva. 1998. Available from: http://www.cioms.ch/publications/g4-benefit-risk.pdf [Last accessed on 25.09.2023]
11. Najafi, S. "Importance of pharmacovigilance and the role of healthcare professionals." *Journal of Pharmacovigilance* 6 (2018): 252. doi: 10.4172/2329-6887.1000252.
12. Krumholz, Harlan M., Joseph S. Ross, Amos H. Presler, and David S. Egilman. "What have we learnt from Vioxx?" *BMJ* 334, no. 7585 (2007): 120–123.
13. Arya, Prabha, Kamana Singh, Divya Sharma, Mahaveer Dhobi, Kamal Kumar Gupta, Indrakant K. Singh, Veeranoot Nissapatorn, Jyoti Kayesth, and Sunil Kayesth. "Herbal and traditional medicines pharmacovigilance for holistic treatment." *Indian Journal of Natural Products and Resources* 14 (2023): 13–21. doi: 10.56042/ijnpr.v14i1.1137.
14. Kuçuku, Merita. "Role of pharmacovigilance on vaccines control." *Journal of Rural Medicine* 7, no. 1 (2012): 42–45.
15. Reumerman, Michael, J. Tichelaar, B. Piersma, M. C. Richir, and M. A. Van Agtmael. "Urgent need to modernize pharmacovigilance education in healthcare curricula: Review of the literature." *European Journal of Clinical Pharmacology* 74 (2018): 1235–1248.
16. Routledge, Philip. "150 years of pharmacovigilance." *Lancet* 351, no. 9110 (1998): 1200–1201.
17. Fornasier, Giulia, Sara Francescon, Roberto Leone, and Paolo Baldo. "An historical overview over pharmacovigilance." *International Journal of Clinical Pharmacy* 40 (2018): 744–747.
18. Woolf, Alan D. "The Haitian diethylene glycol poisoning tragedy: A dark wood revisited." *JAMA* 279, no. 15 (1998): 1215–1216.
19. FDA Consumer Magazine. (1981). Available from https://www.fda.gov/downloads/AboutFDA/WhatWeDo/History/Origin/ucm125604.doc (Last accessed on 25.09.2023).
20. Food and Drug Administration (FDA). History. (2023). Available from https://www.fda.gov/AboutFDA/WhatWeDo/History/ (Last accessed on 25.09.2023).
21. Douthwaite, A. H. "Recent advances in medical diagnosis and treatment." *British Medical Journal* 1, no. 4038 (1938): 1143.
22. Levy, Micha. "The epidemiological evaluation of major upper gastrointestinal bleeding in relation to aspirin use." In *Epidemiological concepts in clinical pharmacology*, pp. 100–104. Berlin, Heidelberg: Springer Berlin Heidelberg, 1987.

23. McBride, William Griffith. "Thalidomide and congenital abnormalities." *Lancet* 2, no. 1358 (1961): 90927–8.
24. Lenz, Widukind, and K. Knapp. "Foetal malformations due to thalidomide." In *Problems of birth defects: From Hippocrates to thalidomide and after*, pp. 200–206. Dordrecht: Springer Netherlands, 1962.
25. Kajii, T., M. Kida, and K. Takahashi. "The effect of thalidomide intake during 113 human pregnancies." *Teratology* 8, no. 2 (1973): 163–166.
26. International Council for Harmonisation of Technical Requirements of Pharmaceuticals for Human Use guideline. (2003). Available from https://database.ich.org/sites/default/files/E2D_Guideline.pdf (Last accessed on 10.06.2024).
27. Medicines and Healthcare Products Regulatory Agency. Yellow Card: making medicines and devices safer. (2024).Available from https://yellowcard.mhra.gov.uk (Last accessed on 10.06.2024).
28. European Coalition on Homeopathic & Anthroposophic Medicinal Products. Council Directive 65/65/EEC. (1965). Available from http://www.echamp.eu/eu-legislation-and-regulation-documents/directive_65-65-eec__-__consolidated_version.pdf (Last accessed on 25.09.2023).
29. Boston Collaborative Drug Surveillance Program. (2024). Available from http://www.bu.edu/bcdsp/ (Last accessed on 10.06.2024).
30. World Health Organization. Pharmacovigilance. (2024). Available from http://www.who.int/medicines/areas/quality_safety/safety_efficacy/pharmvigi/en/ (Last accessed on 10.06.2024).
31. ISoP - ESOP/ISoP History. (2024). Available from http://isoponline.org/about-isop/esopisop-history/ (Last accessed on 10.06.2024).
32. European Medicines Agency. History. (2023). Available from http://www.ema.europa.eu/ema/index.jsp?curl=pages/about_us/general/general_content_000628.jsp (Last accessed on 25.09.2023).
33. European Medicines Agency. EudraVigilance history. (2023). Available from http://www.ema.europa.eu/ema/index.jsp?curl=pages/regulation/general/general_content_000633.jsp (Last accessed on 25.09.2023).
34. Hauben, Manfred, and Craig G. Hartford. "Artificial intelligence in pharmacovigilance: Scoping points to consider." *Clinical Therapeutics* 43, no. 2 (2021): 372–379.
35. Salas, Maribel, Jan Petracek, Priyanka Yalamanchili, Omar Aimer, Dinesh Kasthuril, Sameer Dhingra, Toluwalope Junaid, and Tina Bostic. "The use of artificial intelligence in pharmacovigilance: A systematic review of the literature." *Pharmaceutical medicine* 36, no. 5 (2022): 295–306.
36. Oates, J. T., and D. Lopez. "Pharmacogenetics: An important part of drug development with a focus on its application." *International Journal of Biomedical Investigation* 1, no. 2 (2018).
37. Dean, L. "Carbamazepine therapy and *HLA* genotype." 2015 Oct 14 [updated 2018 Aug 1]. In: Pratt, V. M., S. A. Scott, M. Pirmohamed, B. Esquivel, B. L. Kattman, and A. J. Malheiro, editors. *Medical genetics summaries* [Internet]. Bethesda (MD): National Center for Biotechnology Information (US); 2012–. PMID: 28520367.
38. Dean, L. "Warfarin therapy and VKORC1 and CYP genotype." 2012 Mar 8 [updated 2018 Jun 11]. In: Pratt, V. M., S. A. Scott, M. Pirmohamed, B. Esquivel, B. L. Kattman, and A. J. Malheiro, editors. *Medical genetics summaries* [Internet]. Bethesda (MD): National Center for Biotechnology Information (US); 2012–. PMID: 28520347.
39. Dean, L., and M. Kane. "Codeine therapy and CYP2D6 genotype." 2012 Sep 20 [updated 2021 Mar 30]. In: Pratt, V. M., S. A. Scott, M. Pirmohamed, B. Esquivel, B. L. Kattman, and A. J. Malheiro, editors. *Medical genetics summaries* [Internet]. Bethesda (MD): National Center for Biotechnology Information (US); 2012–. PMID: 28520350.
40. Hans, M., and S. K. Gupta. "Comparative evaluation of pharmacovigilance regulation of the United States, United Kingdom, Canada, India and the need for global harmonized practices." *Perspectives in Clinical Research* 9 (2018): 170–174.
41. Khan, Lateef Mohiuddin. "Comparative epidemiology of hospital-acquired adverse drug reactions in adults and children and their impact on cost and hospital stay—A systematic review." *European Journal of Clinical Pharmacology* 69 (2013): 1985–1996.
42. Angamo, Mulugeta Tarekegn, Leanne Chalmers, Colin M. Curtain, and Luke RE Bereznicki. "Adverse-drug-reaction-related hospitalisations in developed and developing countries: A review of prevalence and contributing factors." *Drug Safety* 39 (2016): 847–857.
43. Oscanoa, T. J., F. Lizaraso, and Alfonso Carvajal. "Hospital admissions due to adverse drug reactions in the elderly. A meta-analysis." *European Journal of Clinical Pharmacology* 73 (2017): 759–770.
44. Arulappen, Ann L., Monica Danial, and Syed A. S. Sulaiman. "Evaluation of reported adverse drug reactions in antibiotic usage: A retrospective study from a tertiary care hospital, Malaysia." *Frontiers in Pharmacology* 9 (2018): 809.

45. Lau, Phyllis M., Kay Stewart, and Michael J. Dooley. "Comment: Hospital admissions resulting from preventable adverse drug reactions." *Annals of Pharmacotherapy* 37, no. 2 (2003): 303–304.
46. Obreli-Neto, Paulo Roque, Alessandro Nobili, André de Oliveira Baldoni, Camilo Molino Guidoni, Divaldo Pereira de Lyra Júnior, Diogo Pilger, Juliano Duzanski et al. "Adverse drug reactions caused by drug–drug interactions in elderly outpatients: A prospective cohort study." *European Journal of Clinical Pharmacology* 68 (2012): 1667–1676.
47. Scondotto, Giulia, Fanny Pojero, Sebastiano Pollina Addario, Mauro Ferrante, Maurizio Pastorello, Michele Visconti, Salvatore Scondotto, and Alessandra Casuccio. "The impact of polypharmacy and drug interactions among the elderly population in Western Sicily, Italy." *Aging Clinical and Experimental Research* 30 (2018): 81–87.
48. Asscher, A. W., G. D. Parr, and V. B. Whitmarsh. "Towards the safer use of medicines." *BMJ* 311, no. 7011 (1995): 1003–1005.
49. Rawlins, Michael D. "Pharmacovigilance: Paradise lost, regained or postponed?: The William withering lecture 1994." *Journal of the Royal College of Physicians of London* 29, no. 1 (1995): 41.
50. Einarson, Thomas R. "Drug-related hospital admissions." *Annals of Pharmacotherapy* 27, no. 7–8 (1993): 832–840.
51. Bates, D. W., D. J. Cullen, N. Laird, L. A. Petersen, S. D. Small, D. Servi, et al. Incidence of adverse drug events and potential adverse drug events—Implications for prevention. *JAMA* 274 (1995): 29–34.
52. Bates, D. W., N. Spell, D. J. Cuilen, E. Burdick, N. Laird, L. A. Petersen, et al. The costs of adverse drug events in hospitalized patients. *JAMA* 277 (1997): 307–311.
53. Classen, D. C., S. L. Pestotnik, R. S. Evans, J. F. Lloyd, and J. P. Burke. "Adverse drug events in hospitalized patients. Excess length of stay, extra costs, and attributable mortality." *JAMA* 277 (1997): 301–306.
54. Edwards, I. Ralph, and Jeffrey K. Aronson. "Adverse drug reactions: Definitions, diagnosis, and management." *Lancet* 356, no. 9237 (2000): 1255–1259.
55. Rawlins, M. D., and J. W. Thompson. Mechanisms of adverse drug reactions. In: Davies DM, editor. *Textbook of adverse drug reactions.* Oxford: Oxford University Press; 1991. pp. 18–45.
56. McKenzie, R., M. W. Fried, R. Sallie, H. Conjeevaram, A. M. Dibisceglie, and Y. Park, et al. Hepatic-failure and lactic-acidosis due to fialuridine (fiau), an investigational nucleoside analog for chronic hepatitis B. *New England Journal of Medicine* 333 (1995): 1099–1105.
57. Lewis, W., E. S. Levine, B. Griniuviene, K. O. Tankersley, J. M. Colacino, J.-P. Sommadossi, et al. Fialuridine and its metabolites inhibit DNA polymerase gamma at sites of multiple adjacent analog incorporation, decrease mtDNA abundance, and cause mitochondrial structural defects in cultured hepatoblasts. *Proceedings of the National Academy of Sciences of the United States of America* 93 (1996): 3592–3597.
58. Atkin, P. A., and G. M. Shenfield. "Medication-related adverse reactions and the elderly: A literature review." *Adverse Drug Reactions and Toxicological Reviews* 14 (1995): 175–191.
59. Uetrecht, J. "Idiosyncratic drug reactions: Current understanding." *Annual Review of Pharmacology and Toxicology* 47 (2007): 513–539.
60. Cameron, H. A., and L. E. Ramsay. "The lupus syndrome induced by hydralazine: A common complication with low dose treatment." *British Medical Journal (Clinical Research Ed)* 289 (1984): 410–412.
61. Daly, A. K. "Using genome-wide association studies to identify genes important in serious adverse drug reactions." *Annual Review of Pharmacology and Toxicology* 52 (2012): 21–35.
62. Mallal, S., E. Phillips, G. Carosi, J. M. Molina, C. Workman, J. Tomazic, et al. HLA-B*5701 screening for hypersensitivity to abacavir. *New England Journal of Medicine* 358 (2008):568–579. doi: 10.1056/NEJMoa0706135.
63. Pushkin, R., L. Frassetto, C. Tsourounis, E. S. Segal, and S. Kim. "Improving the reporting of adverse drug reactions in the hospital setting." *Postgraduate Medicine* 122 (2010): 154–164.
64. Irujo, M., G. Beitia, M. Bes-Rastrollo, A. Figueiras, S. Hernández-Díaz, and B. Lasheras. "Factors that influence under-reporting of suspected adverse drug reactions among community pharmacists in a Spanish region." *Drug Safety* 30 (2007): 1073–1082.
65. Hazell, L., and S. A. Shakir. "Under-reporting of adverse drug reactions: A systematic review." *Drug Safety* 29 (2006): 385–396.
66. Khan, S. A., C. Goyal, N. Chandel, and M. Rafi. "Knowledge, attitudes, and practice of doctors to adverse drug reaction reporting in a teaching hospital in India: An observational study." *Journal of Natural Science, Biology and Medicine* 4 (2013): 191–196.
67. Mann, R. D. Prescription-event monitoring–recent progress and future horizons. *British Journal of Clinical Pharmacology* 46 (1998): 195–201. doi: 10.1046/j.1365-2125.1998.00774.x.

68. Hurwitz, E. S., M. J. Barrett, D. Bregman, W. J. Gunn, P. Pinsky, L. B. Schonberger, et al. Public Health Service study of Reye's syndrome and medications. Report of the Main Study. *JAMA* 257 (1987): 1905–1911. Erratum in: JAMA 1987;26;257:3366.

69. Herbst, A. L., S. J. Robboy, R. E. Scully, and D. C. Poskanzer. "Clear-cell adenocarcinoma of the vagina and cervix in girls: Analysis of 170 registry cases." *American Journal of Obstetrics and Gynecology* 119 (1974): 713–724.

70. Werler, M. M., S. Shapiro, and A. A. Mitchell. "Periconceptional folic acid exposure and risk of occurrent neural tube defects." *JAMA* 269 (1993): 1257–1261.

71. Sedgwick, P. "Bias in observational study designs: Case-control studies." *BMJ* 350 (2015): h560.

72. Setia, M. S. Methodology series module 1: Cohort studies. *Indian Journal of Dermatology* 61 (2016): 21–25. doi: 10.4103/0019-5154.174011.

73. Bonnie, Sibbald, and Roland Martin. "Understanding controlled trials: Why are randomised controlled trials important?" *BMJ* 316 (1998): 201.

74. Kostis, J. B., and J. M. Dobrzynski. Limitations of randomized clinical trials. *The American Journal of Cardiology* 129 (2020): 109–115. doi: 10.1016/j.amjcard.2020.05.011.

75. Indian Pharmacopoeia Commission. Pharmacovigilance Guidance Document for Marketing Authorization Holders of Pharmaceutical Products. (2018). Available from https://www.ipc.gov.in/PvPI/pub/Guidance%20Document%20for%20Marketing%20Authorization%20Holders.pdf (Last accessed on 28.08.2023).

76. Suke, S. G., P. Kosta, and H. Negi. Role of pharmacovigilance in India: An overview. *Online Journal of Public Health Informatics* 7 (2015): e223. doi: 10.5210/ojphi.v7i2.5595.

77. Hire, R. C., P. J. Kinage, and N. N. Gaikwad. Causality assessment in pharmacovigilance: A step toward quality care. *Scholars Journal of Applied Medical Sciences* 1 (2013): 386–392. doi: 10.36347/sjams.2013.v01i05.008.

78. Pande, S. Causality or relatedness assessment in adverse drug reaction and its relevance in dermatology. *Indian Journal of Dermatology* 63 (2018): 18–21. doi: 10.4103/ijd.IJD_579_17.

79. World Health Organization: Uppsala Monitoring Centre. The use of the WHO-UMC system for standardised case causality assessment. (2018) Available from https://who-umc.org/media/164200/who-umc-causality-assessment_new-logo.pdf (Last accessed on 28.08.2023).

80. Naranjo, C. A., U. Busto, E. M. Sellers, P. Sandor, I. Ruiz, E. A. Roberts, E. Janecek, C. Domecq, and D. J. Greenblatt. A method for estimating the probability of adverse drug reactions. *Clinical Pharmacology & Therapeutics* 30 (1981): 239–245. doi: 10.1038/clpt.1981.154.

81. Garashi, H. Y., D. T. Steinke, and E. I. Schafheutle. A systematic review of pharmacovigilance systems in developing countries using the WHO pharmacovigilance indicators. *Therapeutic Innovation & Regulatory Science* 56 (2022): 717–743. doi: 10.1007/s43441-022-00415-y.

82. Maharshi, V, and P. Nagar. Comparison of online reporting systems and their compatibility check with respective adverse drug reaction reporting forms. *Indian Journal of Pharmacology* 49 (2017): 374–382. doi: 10.4103/ijp.IJP_733_16.

83. European Medicines Agency. EudraVigilance. (2024). Available from https://www.ema.europa.eu/en/human-regulatory/research-development/pharmacovigilance/eudravigilance (Last accessed on 10.06.2024).

84. US Food & Drug Administration. MedWatch: The FDA safety information and adverse event reporting program. (2024). Available from https://www.fda.gov/safety/medwatch-fda-safety-information-and-adverse-event-reporting-program (Last accessed on 10.06.2024).

85. Pharmaceuticals and medical Devices Agency. PMDA Medical Safety Information. (2024). Available from https://www.pmda.go.jp/english/safety/info-services/safety-information/0001.html (Last accessed on 10.06.2024).

86. Medicines and Healthcare Products Regulatory Agency. The Yellow Card scheme: guidance for healthcare professionals, patients and the public. (2021). Available from https://www.gov.uk/guidance/the-yellow-card-scheme-guidance-for-healthcare-professionals#full-publication-update-history (Last accessed 28.08.2023).

87. Coleman, J. J., and S. K. Pontefract. Adverse drug reactions. *Clinical Medicine (London)* 16 (2016): 481–485. doi: 10.7861/clinmedicine.16-5-481.

88. Medicines and Healthcare Products Regulatory Agency. Pharmacovigilance- how the MHRA monitors the safety of medicines. Available from https://assets.publishing.service.gov.uk/media/5feefb56d3bf7f089a791a2c/Pharmacovigilance___how_the_MHRA_monitors_the_safety_of_medicines.pdf (Last accessed on 10.06.2024).

89. Gossell-Williams, M., and S. A. Adebayo. The PharmWatch programme: Challenges to engaging the community pharmacists in Jamaica. *Pharmacy Practice (Granada)* 6 (2008): 187–190. doi: 10.4321/s1886-36552008000400003.
90. Pozsgai, K., G. Szűcs, A. Kőnig-Péter, O. Balázs, P. Vajda, L. Botz, et al. Analysis of pharmacovigilance databases for spontaneous reports of adverse drug reactions related to substandard and falsified medical products: A descriptive study. *Frontiers in Pharmacology* 13 (2022): 964399. doi: 10.3389/fphar.2022.964399.
91. Uppsala Monitoring Centre. About VigiBase: The unique global resource at the heart of the drive for safer use of medicines. Available from https://who-umc.org/vigibase/ (Last accessed on 28.08.2023).

2 Drug Safety Signal Detection and Management in Pharmacovigilance

Mira Desai

2.1 INTRODUCTION

Although the strategies in randomized controlled clinical trials (RCTs) are to identify the safety and the risk associated with the use of medicines, it does not reflect the way the medicines are used in real-life practice. The population exposed during post-marketing period varies vastly than those studied during the drug development process. The conventional clinical trials are done with strict inclusion-exclusion criteria, for a limited duration, on restricted sample size and may lack sensitivity or statistical power to detect rare and latent adverse drug reactions (ADRs) [1]. Thus, the complete safety profile of medicines cannot be established in conventional clinical trials [1]. In addition, post-marketing phase can also generate information on the benefits of the product that helps to assess the benefit-risk profile of the drug. The limitations of RCTs impose pharmacovigilance (PV) obligations on marketing authorization holders and regulatory authorities. Therefore, the post-marketing safety monitoring of medical products through PV system is essential and continues throughout the marketed life cycle of medicines [2].

The essence of PV is to collect, analyze and interpret data relevant to patient safety, suspect a causal link between a drug and ADR and generate evidence-based hypothesis that is not documented or known previously. Hypothesis can be generated by identification of unexpected serious or non-serious adverse events (AEs) or change in its severity, increase in reporting frequency or experience in special population such as pediatric, elderly, and hepato-renal compromised patients. Nevertheless, hypothesis does not establish any causal relationship between the drug and the event, albeit advocates further investigation. PV professional's role is to investigate the hypothesis and to prove or disprove the causal relationship between the drug and AE. The investigative activities move from evidence-gathering task to information analysis known as signal detection. The objective of signal detection is to discover potential drug safety alerts within shortest possible time after the medicine has been marketed, to distinguish serious signals at risk for public health and to detect the class effect, if any. Thus, the drug safety signal detection in PV requires monitoring of individual case safety reports (ICSRs) and other data sources to identify AEs that are worthy of further exploration to ensure the safety of medical products and patients.

2.1.1 What Is a Signal

2.1.1.1 Definition

A "signal" is reported information on a possible causal relationship between an AE and a drug, the relationship being unknown or incompletely documented previously [3, 4]. As per the Council for International Organizations of Medical Sciences (CIOMS), information that arises from one or multiple sources (including observations and experiments), which suggest a new potentially causal association, or a new aspect of a known association, between an intervention and an event or set of related events, either adverse or beneficial, is judged to be of sufficient likelihood to justify verificatory action [5]. As per both definitions, a causal association between an AE and drug needs to be verified.

DOI: 10.1201/9781032629940-2

Signal detection can be a new safety information that is unidentified, unlabeled or change in frequency, severity or information on risk factors of an existing signal. Thus, a signal is essentially a hypothesis of a risk associated with the use of a medicine based on the supporting data and arguments derived from one or more of many possible sources. However, it is only a preliminary indication that needs to be confirmed and validated. It can be stated that the signal detection process is the set of activities performed to determine whether, based on an examination of ICSRs, accumulated data from various active surveillance systems or clinical studies, published biomedical literature or other data sources, there are new aspects of a known association between an active substance or a medicinal product and an event, either adverse or beneficial that justifies verificatory action [6]. The signal management process shall cover all steps from signal detection to validation, confirmation and recommendations for subsequent regulatory action.

2.1.1.2 Sources of Signals

Multiple sources to identify medicine safety signals are explored, although each has its own advantages and limitations. Ideally, all sources of safety information should be considered together for a single active substance. However, the quality and completeness of all the essential data is the key for meaningful signal detection and assessment. Data sources used for signal detection in post-authorization drug safety periods are mentioned in Table 2.1.

2.1.2 Spontaneous Reports

Worldwide spontaneous reporting system (SRS) is the cornerstone for continuous monitoring of pharmaceutical and biological products during the post-marketing period. The suspected ADRs are reported by means of ICSRs and submitted to national PV coordinating centers and databases. The PV data collected in ICSRs is structured and systematic as per international safety reporting guidance issued by the International Conference on Harmonization to establish causal relationships and identify early safety signals from the real world [7]. The SRS has been valuable, flagging the early

---

TABLE 2.1

Data Sources to Support Signal Detection in Post-Authorization Drug Safety Period

No.	Data	Source
1.	Post-marketing PV data	• National pharmacovigilance database – VigiBase • Regulatory databases, e.g., FDA AERS, EudraVigilance • Post-marketing reports • Clinical trial serious adverse event reports
2	Electronic data	• Medical records of the patients • Post-marketing non-interventional studies, e.g., post-authorization safety studies • Administrative files • Health insurance claims databases • Diseases and patient registries by hospitals and professional organizations • Periodic safety update reports submitted to regulatory authorities by marketing authorization • Prescription event monitoring databases • The media – the internet (including company-sponsored websites, internet forums, social media) • Newspapers
3.	Biomedical literature	• Published case reports, case series • Pharmacoepidemiologic studies • Clinical trials • Observational studies

detection of patient safety issues either due to a medical product or its clinical uses. For example, initial reports of an association of myopathy and rhabdomyolysis with cerivastatin resulted in regulatory actions in different countries based on ICSRs [8].

VigiBase is the unique WHO global database for spontaneous reports. It is one of the largest databases in the world for ADR reports of medicinal products and has increased exponentially over the years [9, 10]. Additionally, U.S. Food and Drug Administration (FDA) adverse event reporting (FAERS), the European EudraVigilance and the vaccine adverse event reporting (VAERS) are the most prominent ICSR management systems. The real-world data collected by SRSs facilitates continuous monitoring of pharmaceutical products with all the essential elements required to establish causal relationships and support early safety signal detection. Despite its benefits, due to the voluntary reporting of ICSRs by healthcare professionals, underreporting, selective reporting and inability to quantify and calculate real incidence rates are some of the well-known limitations [2, 11].

2.1.3 Electronic Health Records (EHRs)

EHRs consist of medical records of patients, including discharge summaries, laboratory findings, administrative files, insurance and reimbursement claims, and so on. The limitation of SRSs such as underreporting and selective reporting can be overcome by EHRs; however, the data is unstructured and poorly maintained medical records, their quality, completeness and retrieval could be an arduous task. Furthermore, there can be privacy and ethical concern and ownership issues to access the data. While disease-specific registries based on facts and evidence can be a good source for generating quality data, provided it is well maintained and organized with predetermined scientific and clinical objectives. In context with healthcare system in India, it requires substantial efforts for quality and completeness [12].

2.1.4 Biomedical Literature

This includes comprehensive screening and review of published medical literature such as case reports, case series, pharmacoepidemiologic studies, clinical trials and observational studies for important PV data. In addition, periodic safety update reports submitted to regulatory authorities by marketing authorization holders have been reviewed for benefit and risk associated with use of medicines. Nevertheless, lack of essential information for ICSR reporting, unstructured data for meaningful assessment, right searching strategy and duplicate publications can be challenging. Unbiased effective information-extraction techniques and considerable time and effort are vital for monitoring medical literature accurately and efficiently. Thus, it has been reported that a considerable number of published literatures are not retrieved and preferred for signal detection [13].

2.1.5 Signal Detection Methods

Signal detection methods are 'qualitative' as per case-by-case assessment of safety reports and 'quantitative' by means of data mining tools using real-world databases. The qualitative and quantitative assessment is further subjected to validation and confirmation by clinical evaluation and subject expert judgment.

2.1.5.1 Qualitative Methods

This is a traditional approach, also known as case-by-case analysis of ICSRs, case series, aggregate datasets (periodic safety update reports) and other sources (e.g., published biomedical literature, health authorities, media, internet, social media). Each D-AE combination case is reviewed and thoroughly assessed by medically qualified persons to establish evidence to confirm or reject causal associations. This is followed by systematic evaluation of multiple case reports of D-AE combinations, wherein the cumulative data (number of reports in database in the given time period), frequency trends over time and frequency rates to specific time period, system organ classification and Medical Dictionary for Regulatory Activities (MedDRA) coding with combined retrospective analysis using computerized tools are undertaken. In addition, scientific literature, characteristic of patient population exposed, pharmacological plausibility are evaluated in detail. However, manual assessment of individual reports is time consuming and practically not suitable for large database analyses. This has resulted in the development and application of statistical data mining tools for drug safety surveillance.

2.1.5.2 Quantitative Methods

The quantitative methods offer unique prospects that leverage to complement or augment existing qualitative approaches. These methods can also detect potential signals for further investigation that are not readily recognizable or apparently evident on a single case report. These are systematic examinations of the reported AEs using statistical or mathematical tools that are quick and detect signals earlier. However, they are primarily useful for large datasets and not suitable for solo ICSR. A recommended minimum database size is 5,000 reports, as a suitable lower limit to avoid excessive rates of false-positive associations [14]. These methods have been valuable to establish the relationship between pioglitazone and bladder cancer, rofecoxib exposure and thrombotic ADRs [15].

Various statistical algorithms are applied to a database to identify drug-AE pairs (or frequent combinations of a drug and an event) that occur with disproportionately high frequency in large spontaneous report databases. Typical methodologies include the disproportionality analysis (DPA) (statistics) or the Empirical Bayesian Geometric Mean (EBGM). The system automatically generates statistical values or scores that indicate potential safety issues, strength of the association between drug and AE and is time saving. The higher the score, the stronger the statistical association. However, statistical associations may not necessarily always designate causal relationships and signals, albeit, indicate further investigation [16].

2.1.6 MEASURE OF DISPROPORTIONALITY

DPA is the main driving force behind most computerized PV methods for SRS. DPA methodologies use frequency analysis of 2×2 contingency tables to estimate surrogate measures of statistical associations between specific drug-event combinations mentioned in spontaneous reports, as mentioned in Table 2.2.

The objective of DPAs is to detect the 'unexpectedly' frequently reported AE relative to a background of other reports. Among all reported suspected AE reports *'whether the occurrence of drug-event combination purely by chance?'* or *'Is drug-event combination disproportional?'* The essence of disproportionality is that what is observed is different from what is expected.

TABLE 2.2

2×2 Contingency Table of Adverse Event Report Data for Disproportionality Analysis

	No. of AE Reports of Interest (Reaction of Interest)	Reports for All Other Events in the Database (All Other Reactions)	Total
Reports for drug of interest	A	B	A + B
Reports for all other drugs	C	D	C + D
Total	A + C	B + D	A + B + C + D

$$PRR = \frac{A/A \pm B}{C/C + D} \geq 2$$

In addition, it also quantifies the degree to which a drug-event combination co-occurs 'disproportionally' compared to what would be expected if there were no association [17].

The major use of this method is to confirm or refute a potential association based on a pharmacological hypothesis between a specific drug and ADR. However, the statistics of disproportionate reporting alone do not imply causal relationship.

Commonly used statistical measures of association are as follows:

1. *Information component (IC)*: Bayesian measure assesses the strength of the quantitative dependency between a specific drug and specific ADR. The UMC regularly scans the WHO database for potential signals using this automated data mining program. A positive IC value indicates that D-AE combination is reported more frequently than expected based on all reports in the database. The values can be plotted as graphs to identify any trend over a period of time. A positive signal will have an IC value that becomes more significant over time as more cases are included. If IC025 value is positive, suggesting an association between D and AE however does not imply causality and clinical assessment remains essential to identify signal.

2. *Proportional reporting ratio (PRR)*: It is a simple measure that indicates how many times higher the risk is for an ADR in the drug-exposed category as compared to the drug-unexposed category. It is easiest to interpret. If the PRR for a particular D-AE combination is significantly high and is not a recognized reaction, it may represent a signal.

3. *Reporting odds ratio (ROR)*: The odds of a certain event occurring with a medicinal product are compared to the odds of the same event occurring with all other medicinal products in the database. A signal is considered when the lower limit of the 95% confidence interval (CI) of the ROR is greater than one.

4. *Chi-square test*: It allows the comparison of two attributes (drug exposure and ADR occurrence) in a sample of data to determine if there is any relationship between them.

Automated methods are validated to generate drug safety signals that complement and strengthen a signal identified by qualitative assessment. They may identify signals that were missed during manual assessment of the safety reports. However, these automated methods simply calculate indices based on a 2×2 contingency table to determine safety signals as shown in Table 2.2. It is based on the contrast between observed and expected numbers of reports, for any given combination of drugs and an AE. All signals identified by statistical methods (Bayesian Confidence Propagation Neural Network [BCPNN] or PRR) require subsequent clinical evaluation.

2.1.7 SIGNAL DETECTION AND MANAGEMENT PROCESSES

Drug safety signal detection is both a cyclical and dynamic process. The practice of signal detection is essential for marketed drugs wherein the drug-AE safety concern may be new, unknown or not recognized risk during premarketing phase. It may be a subtle pattern of already known ADRs, albeit out of proportion than expected, significant enough to raise an alert and relevant for healthcare professionals for a wise therapeutic decision.

VigiBase, one of the largest global databases of potential AEs of medicinal products, is periodically screened by VigiLyze for unlisted drug-ADRs combination by strength of evidence. ADRs, which are serious, associated with new drugs, previously unknown, potential risk to public health are prioritized and verified by initial manual review and statistical screening. The initial manual assessment of D-ADR hypothesis includes the number of AEs/ADRs reported in national database of Pharmacovigilance Programme of India (PvPI) and global database, excludes other likely causes and determines disproportionality by VigiLyze. A human assessor reviews line listings or individual ICSR form for a given period and decides to proceed further for in-depth assessment by multidisciplinary signal review panel. Subsequently, each D-ADR case is evaluated in depth by PV professionals and medical subject experts, considering factors like patient characteristics, temporal time relationship, case narratives, pharmacological plausibility, confounding factors, dechallenge-rechallenge information, outcome, class effect, related similar events, potential public health impact of an ADR and maturity of the product, i.e., number of years on the market. Furthermore, published biomedical literature (peer reviewed and indexed in major database) is reviewed and cross-checked whether ADR has been mentioned in the product information leaflets, package insert and or summary of product characteristic of well-regulated countries. In addition, DPA and IC values determined by statistical methods are aligned with the qualitative assessment report. A combined approach of clinical evaluation along with statistical methods has been optimal to dismiss or formulate a safety signal detection that requires further monitoring and recommendations to regulatory authorities for subsequent action, if any (Figure 2.1).

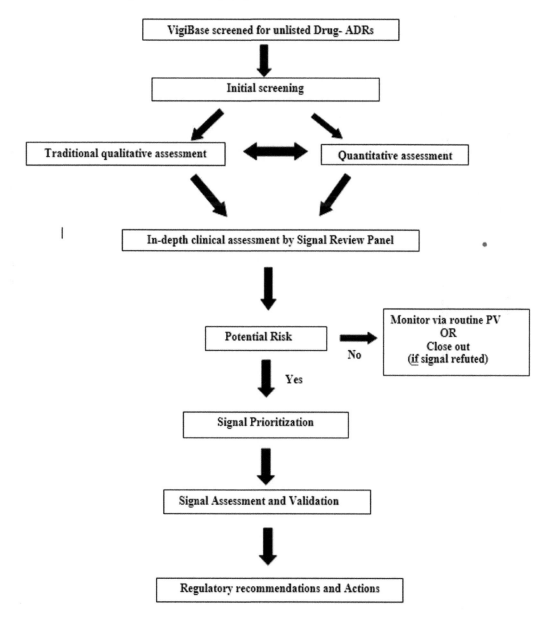

FIGURE 2.1   Schematic presentation of a signal detection plan and process.

2.1.8 Challenges and Favorable Situations in Signal Detection Processes

Establishing a relationship between a drug and AE is not a simple straightforward process. Practically, the signal detection process encounters limitations and challenges related to data, pharmacological and medical aspects. The quality of information, incomplete data, missing essential elements of ICSRs and incomplete or vague case narratives impacts the evaluation of the case. In addition, AEs that cannot be explained or are plausible by pharmacological action (kinetic or dynamic) or the absence of supporting evidence (laboratory findings) and are unexpected may be difficult to detect. For example, coughs associated with captopril, tendinitis due to fluoroquinolone, and so on. Similarly, when the AE manifests only in a small number of patients and the temporal time relationship or dose relationship is uncertain or the phenomenon is trivial or may be coincidental, it is challenging to attribute the relationship with the causal drug. For example, myocarditis, hypothyroidism and fibromyalgia manifestations may be difficult to distinguish as causal drug-induced or a coincidental disease symptom in individual patient. Moreover, when the AE is difficult to reproduce as an experimental effect and is non-specific and/or subjective in nature such as memory disturbances, headache, and fatigue with no evidence, makes the signal detection process arduous. Furthermore, when the frequency of AE is low, the background frequency of AE is high (cough, diarrhea, constipation, fall, skin rashes, hepatitis) or low frequency of drug exposure makes the connection tough to prove or refute. On the other hand, favorable situations for signal detection processes include clear temporal time relationships, AEs observed in patients of similar age, disease and medicine use, known to be drug induced (e.g., anaphylaxis, Stevens-Johnson syndrome, fixed drug eruptions), high frequency of exposure to a drug, positive rechallenges and plausible pharmacopathological mechanisms.

Furthermore, the number of case reports required to raise a suspicion for a potential signal has always been questioned. It has been recommended that for a meaningful assessment at least three cases need to be reported, which provides a 95% chance of identifying a specific event that has an incidence of $\geq$ 1:3000, i.e., an uncommon or rare event [11]. However, a single case report with all essential information is sufficient to generate a signal, depending on the seriousness of the event and quality of the information [3]. In fact, a strong signal will have several ICSRs with a 'certain' or 'probable' causality category.

2.1.9 Signal Prioritization

Signal prioritization is a critical part of the signal management process. Subsequent to qualitative and quantitative assessment, high priority signals are identified for further evaluation. The criteria for prioritization include D-AE combinations that are serious, caused by a new medical product, likely to have substantial impact on public health with quality of the evidence for a causal relationship and significantly increased IC values, for example, Bradford Hill criteria [18].

2.1.10 Signal Assessment

Once a signal is prioritized, other data sources are systematically assessed to determine sufficient evidence of causality. The data sources include ICSRs that trigger the signal and other ICSRs with similar event terms identified by using Standardized MedDRA Queries (SMQs). In order to retrieve and review similar cases of interest within the database, the use of SMQs is recommended. Additionally, scientific literature, clinical trial, pre-clinical and epidemiological data sources are reviewed [19, 20].

Signals can be assessed on the strength of the ICSRs triggering the signal and the caseload (number/volume of cases). The data may be stratified according to age, gender, ethnicity, concomitant medication or disease to categorize populations at highest risk for the AE and overcome confounding factors [21]. Signal validation depends on the number and quality of case reports, the nature of the reaction, type of drug and the population exposure [21].

TABLE 2.3

Action Plan to Gather Information to Strengthen the Signal

Clinical Studies

- Targeted safety studies, targeted spontaneous reporting
- Comparative observational studies (cross-sectional study/survey, case-control study, cohort study, epidemiological studies; retrospective or prospective)
- Enhanced monitoring or follow-up techniques
- Active surveillance schemes (sentinel sites, drug event monitoring, registries)

2.1.11   DECISION MAKING FOLLOWING AN ASSESSMENT

2.1.11.1   Close Signal

If there is no evidence of a causal relationship, the signal is refuted and no further action is required. However, it needs to be documented, and the signal is closed. Nonetheless, if further evidence becomes available, the signal can be re-assessed.

2.1.11.2   Continue Monitoring

When the available evidence is not sufficient to make a decision to support the signal, it requires continuous monitoring until sufficient evidence becomes available to either confirm or refute the signal. A subsequent action plan is directed to gather more information to strengthen the signal, as mentioned in Table 2.3.

This can be target follow-up questions, design studies and re-review when sufficient information is available [22].

2.1.11.3   Further Action If a Signal Is Validated

Once the signal is validated, the use or distribution of the medical product is restricted. The risk is communicated to healthcare professionals and PV centers. A warning letter is distributed to the marketing authorization holder and recommended to revise the product information leaflet and label for safety reasons as deemed necessary. The action plan if a signal is validated is presented in Table 2.4.

Classic PV signal detection practices adopt both qualitative and quantitative methods, and there is no single method that can be benchmarked. A combined approach has been valuable in identifying important drug safety signals through SRS databases. Examples include hemolytic anemia

TABLE 2.4

Action Plan If a Signal Is Validated

1. **Risk communication to patients and prescribers**
   - Generate alert, newsletter, article
   - Letter to healthcare professionals
   - Letter to national PV centers
   - Warning to marketing authorization holders, other institutes
2. **Regulatory authorities**
   - Revising product labels (e.g., addition to label or labeling update)
   - Revising patient package inserts
   - Regulatory documents (e.g., annual safety reports, risk management plan, periodic safety update reports)
   - Drug restrictions
   - Drug recalls

TABLE 2.5

List of Signals Identified by a Pharmacovigilance Program in India with Recommendations to CDSCO to Include in Prescribing Information Leaflets

No.	Suspected Drug	ADRs
1	Cefixime	Acute generalised exanthematosus pustulosis (AGEP)
2	Furosemide	Lichenoid dermatitis
3	Itraconazole	AGEP
4	Lithium carbonate	Drug Reaction with Eosinophilia and Systemic Symptoms Syndrome (DRESS)
5	Fluconazole	Hyperpigmentation
6	Oseltamivir	Sinus bradycardia/bradycardia
7	Tinidazole	Fixed drug eruption
8	Mefenamic acid	Fixed drug eruption
9	Doxycycline	Fixed drug eruption
10	Minoxidil	Folliculitis
11	Cephalosporin class	Fixed drug eruption
12	Paracetamol	Fixed drug eruption

Source: https://www.ipc.gov.in/images/PvPI_Newsletter,Vol._13_Issue_1_16.6.2023_compressed.pdf accessed on 19/10/2023.

associated with temafloxacin, ventricular arrhythmias with terfenadine and cisapride, cardiac valvulopathy with fenfluramine and intussusception following administration of the RotaShield rotavirus vaccine. A list of signals identified by PvPI with subsequent actions is shown in Table 2.5.

2.1.12 Recent Trends for Signal Detection in PV

With increasing volume of ICSRs in database and heterogeneous and complex data, it is essential to capture, organize, compile and analyze the data timely and meaningfully to ensure patient safety and identify *'needle from haystack'*. With advancing technology, there is a growing interest to adopt artificial intelligence technology to support and complement decision making during drug safety signal detection and validation. Recently, Baer Ji-H et al. used a novel machine learning algorithm to detect safety signals of two anti-cancer agents from SRS data and compared with traditional DPA methods. The novel method achieved better performance and detected more new ADR signals than traditional DPA methods [23]. On the other hand, there are reports of variable performance of AI tools and not suggestive of complete automation of the PV process [24].

Any novel approach that integrates traditional statistical methods to rigorously identify unusual AE patterns, adds value to drug safety signal detection and validation and enhances the efficiency and effectiveness of PV activities will be a boon to all stakeholders. The new method needs to be validated before implementation in the real world and approved by regulatory authorities. Nevertheless, the drug safety signal detection process requires knowledge and wisdom, and no statistical method or novel tool can replace the importance of medical and scientific judgment of trained PV professionals.

2.1.13 Conclusion

Medicine safety signal detection is both an iterative and dynamic process. PV databases are powerful tool for early detection of safety signals. A signal is a working hypothesis of new possible adverse effects of a drug that needs to be confirmed or refuted. Qualitative assessment along with automated statistical methods offers unique prospects to complement and adds value to qualitative assessment.

With emerging trend and technology, there is a growing interest to adopt artificial intelligence to support and complement decision making during drug safety signal detection and validation. Good quality reports, knowledge along with comprehensive clinical judgment are most essential and will always remain fundamental for final adjudication of causality and signal detection.

REFERENCES

1. Amery W.K. 1999. Why there is a need for pharmacovigilance. *Pharmacoepidemiol Drug Saf* 8 (1): 61–4.
2. Desai M. 2022. Pharmacovigilance and spontaneous adverse drug reaction reporting: Challenges and opportunities. *Perspect Clin Res* 13: 177–9.
3. World Health Organization. 2002. Safety of Medicines: A Guide to Detecting and Reporting Adverse Drug Reactions: Why Health Professionals Need to Take Action. https://apps.who.int/iris/handle/10665/67378
4. Edwards IR, Aronson J.K. 2000. Adverse drug reactions: Definitions, diagnosis, and management. *Lancet* 356: 1255–9.
5. CIOMS. 2010. Practical Aspects of Signal Detection in Pharmacovigilance. Report of CIOMS Working Group VIII, 14.
6. CIOMS. 2010. Practical Aspects of Signal Detection in Pharmacovigilance WG VIII, 116.
7. International Conference of Harmonization, ICH E2B (R3) http://www.ich.org/products/guidelines/efficacy/article/efficacyguidelines.html. Accessed 22 August 2023.
8. World Health organization. 2002. WHO international drug monitoring: Cerivastatin and gemfibrozil. *WHO Drug Info* 16 (2): 8–11.
9. About VigiBase, What is VigiBase? https://who-umc.org/vigibase/. Accessed on 31 August 2023.
10. Lindquist M. 2008. VigiBase, the WHO global ICSR database system: Basic facts. *Drug Inf J* 42 (5): 409–19.
11. Pal SN, Duncombe C, Falzon D, Olsson S. 2013. WHO strategy for collecting safety data in public health programmes: Complementing spontaneous reporting systems. *Drug Saf* 36 (2): 75–81.
12. Bhardwaj P. 2022. National quality registry for India: Need of the hour. *Indian J Community Med* 47: 157–8.
13. Kostoff R. 2001. The extraction of useful information from the biomedical literature. *Acad Med* 76 (12): 1265–70.
14. Caster O, Aoki Y, Gattepaille LM, Grundmark B. 2020. Disproportionality analysis for pharmacovigilance signal detection in small databases or subsets: Recommendations for limiting false-positive associations. *Drug Safety* 43: 479–87.
15. Montastruc JL, Sommet A, Bagheri H, Lapeyre-Mestre M. 2011. Benefits and strengths of the disproportionality analysis for identification of adverse drug reactions in a pharmacovigilance database. *Br J Clin Pharmacol* 72 (6): 905–8.
16. Guideline on Good Pharmacovigilance Practice. 2012. Module IX – Signal Management. European Medicines Agency, 6.
17. Bate A, Evans S.J. 2009. Quantitative signal detection using spontaneous ADR reporting. *Pharmacoepidemiol Drug Saf* 18 (6): 427–36.
18. Bradford-Hill A. 1985. The environment and disease: Association or causation? *Proc R Soc Med* 58: 295–300.
19. CIOMS. 2004. Development and Rational Use of Standardised MedDRA Queries (SMQs).
20. CIOMS. 2010. Practical Aspects of Signal Detection in Pharmacovigilance. WG VIII, 92.
21. Guideline on Good Pharmacovigilance Practice. 2012. Module IX – Signal Management. European Medicines Agency, 7.
22. CIOMS. 2010. Practical Aspects of Signal Detection in Pharmacovigilance WG VIII, 93.
23. Bae B.J-H, Baek Y-H, Lee J-E, Song I, Lee J-H, Shin J-Y. 2021. Machine learning for detection of safety signals from spontaneous reporting system data: Example of nivolumab and docetaxel *front. Pharmacol* 11: 1–13.
24. Aronson J.K. 2022. Artificial intelligence in pharmacovigilance: An introduction to terms, concepts, applications and limitations. *Drug Saf* 45:407–18.

3 Databases and Tools for Signal Detection of Drugs in Post-Marketing Surveillance

Jeesa George, Prizvan Lawrence Dsouza, Yalamanchili Jahnavi, Hemendra Singh, and Pramod Kumar A

3.1 INTRODUCTION

Pharmacovigilance (PV) is defined as the "pharmaceutical science and activities related with the collection, detection, assessment, monitoring and prevention of adverse effects or any other drug-related problems associated with a pharmaceutical medication" (1). Phase IV clinical trials, also known as post-marketing surveillance (PMS), is the process through which the adverse effects of any marketed pharmaceutical medication are monitored and reported (2). These reports contain information on individual patients, medications consumed as over the counter (OTC) or administered by a healthcare professional within the hospital and the adverse drug reaction (ADR) or adverse event (AE). There are various PV databases where these reports are stored, which are extremely useful not only for identifying possible potential drug safety signals but also for exploring specific drug-event associations (3).

One of the fundamental pillars of PMS is signal detection, which involves systematic analysis of data to uncover patterns, anomalies and potential safety signals associated with drugs (4). Signal as defined by the World Health Organization (WHO) is "any reported information on a possible causal relationship between an adverse event and a drug, the relationship being unknown or incompletely documented previously" (5). Signals are generally generated by a review of spontaneous case reports, active surveillance systems and literature search. Since the manual screening of whole PV databases for signal detection becomes no more feasible due to large collections of PV cases, hence the requirement and use of various specialized tools to assess and associate the AE and the drug are essential (6). In this context, we focus on various databases and tools available for signal detection of a drug.

3.2 PHARMACOVIGILANCE DATABASES USED IN SIGNAL DETECTION

3.2.1 U.S. FOOD AND DRUG ADMINISTRATION ADVERSE EVENT REPORTING SYSTEM

The U.S. FDA Adverse Reporting System (FAERS, formerly known as AERS) (https://www.fda.gov/drugs/questions-and-answers-fdas-adverse-event-reporting-system-faers/fda-adverse-event-reporting-system-faers-public-dashboard) is a database that contains information on medication errors and AEs reported to the FDA (7). The main purpose of the database is to support the FDA's post-marketing safety surveillance program pertaining to drug and therapeutic biological products (8). The reports of medication errors and AEs are coded by using the terms from the Medical Dictionary for Regulatory Activities (MedDRA) terminology. Apart from the AEs reported by the manufacturer, the reports can also be submitted by healthcare professionals and the public (9).

The database is centralized and computerized and is widely used by regulatory authorities or agencies, PV experts and various other related organizations for identifying various post-marketing drug safety concerns (10). It consists of the patient's demographics, drug/biologic information administered, AE experienced, patient outcome, source of report, duration of drug therapy (start

DOI: 10.1201/9781032629940-3

date and end date) and indications for the use of the drug/biologic. The drugs are allocated as primary suspect, secondary suspect and concomitant medications, while the AEs are coded using the preferred terms in the MedDRA (7).

Currently, FAERS uses a Multi-item Gamma Poisson Shrinker (MGPS) program replacing the GPS program to systematically screen the huge database for possible signals received through voluntary reporting of ADRs or AEs. MGPS derives its signal scores by adjusted ratios of the observed/expected counts through Empirical Bayes Geometric Mean (EBGM). Considering the lower one-sided 95% confidence limit of the EBGM, the signal is detected when the EB05 score is greater than or equal to the threshold value of 2.0 (3).

Challenges faced while using FAERS data include underreporting, the Weber effect and stimulated reporting (11, 12). Underreporting remains the most significant limitation as it reduces the chances of detecting a signal for a drug (11). Recently, the FDA has made efforts in order to increase the reporting rate of AEs. The Weber effect is seen when there is a peak in the reporting of AEs of a drug for two years post-regulatory approval followed by a continuous decline in reporting (13). A recent FAERS study suggests that there may be fewer concerns with the Webber effect as the regulatory bodies and healthcare professionals emphasized the importance and utility of post-approval AE reporting (12). Unfortunately, there is no readily easy accessibility of organized FAERS data, hence healthcare professionals heavily rely on the safety information from drug label "inserts" that are often based predominantly on pre-approval clinical trial results (12, 13). Lastly, stimulated reporting is based on the sensitization given to the public of an AEs by the FDA resulting in significantly increased AE reporting rates. As alert-driven shifts in reporting may affect the validity of comparative research and related analytical techniques, stimulated reporting has significant implications for the usefulness of post-marketing AE data (12).

Advantages of FAERS

1. *Large and comprehensive dataset*: FAERS contains over 16 million AE reports submitted since 1968. This large dataset provides valuable information on the safety of marketed drugs.
2. *Variety of data sources*: Patients, medical professionals and pharmaceutical companies are just a few of the sources from which FAERS receives reports of AEs. This diversity guarantees that the database has AEs from all around the United States.
3. *Defined data format*: To guarantee that the data is dependable and simple to examine, FAERS employs defined data formats. It is now feasible to compare reports from various sources and spot AE trends thanks to this standardization.
4. *Data accessibility*: The FAERS Public Dashboard allows users to search, filter and analyse the database, which is available to the general public. Users comprehend the data visualizations on this dashboard, which include maps, bar charts and line graphs.

Disadvantages of FAERS

1. *Underreporting of AEs*: Because FAERS depends on voluntary reporting, not all AEs are disclosed. Less serious AEs and AEs that happen after the medication have been withdrawn from the market are probably more likely to be underreported.
2. *Limited knowledge of AE causality*: FAERS does not offer conclusive evidence regarding the causal relationship between a drug and an AE. The database does not account for other possible causes of the AE; it just documents whether the AE happened after the patient took a medication.
3. *Potential bias in reporting*: Depending on the report's source, the quality of AE reports in FAERS may differ. Patient-reported AEs may be less reliable and more subjective, but reports from healthcare professionals may be more thorough and accurate.

The advantages and disadvantages of FAERS are compiled in Table 3.1.

TABLE 3.1

Summarizing the Advantages and Disadvantages of FAERS

Features	Advantages	Disadvantages
Size of dataset	Large and comprehensive	May not be representative of all AEs
Diversity of data sources	Data from healthcare professionals, patients and manufacturers	May be biased towards more severe AEs
Standardized data format	Easy to compare and analyse	May not capture the full complexity of AEs
Data accessibility	Publicly available	Lag between reporting and data entry

3.2.2 EudraVigilance

EudraVigilance (https://www.ema.europa.eu/en/human-regulatory/research-development/pharma-covigilance/eudravigilance/access-eudravigilance-data) is a PV database under the control of the European Medicines Agency, which is used to collect and analyse suspected ADRs caused by medications in European countries. It first operated in 2001 followed by the complete implementation of the PV legislation in the year 2010, enhancing the major revisions in its access policies by considering the use of new individual case safety report standards developed by the International Council for Harmonization of Technical Requirements for Pharmaceuticals for Human Use and the International Organization for Standardization (14). The main aim of wide access was to facilitate effective safety monitoring of authorized medications, improve data availability for research and provide better information on suspected adverse reactions for healthcare professionals and patients. Also, in 2017, this database was easily accessible to academic research institutions in order to carry out extensive research (15). Marketing Authorization Holders (MAHs) have easy access to statistical calculations for Drug-Event Combinations (DECs) from EudraVigilance. Disproportionality analysis is carried out to find the association between the drug and the event (16).

The EU PV safety monitoring activities are supported by the EudraVigilance Data Analysis System (EVDAS) with the main focus on signal detection and evaluation of individual case safety reports (ICSRs) (3). EVDAS includes a frequentist approach to measure disproportionality (proportional reporting ratio [PRR] and the reporting odds ratio [ROR]). As PRR being a very simple calculation, it was previously implemented as the signal detection method in EVDAS. However, ROR being equally simple gives the same performance as PRR, but is the basis of more complex statistical models (3, 17).

Advantages of EudraVigilance

1. *Incorporated information base*: EudraVigilance is an incorporated dataset that gathers ADR reports from all EU states. This centralization guarantees that ADRs are caught from many sources, giving a more exhaustive image of medication security in the EU.
2. *Continuous revealing*: EudraVigilance takes into consideration the ongoing detailing of ADRs. This ongoing announcing ability empowers the EMA and EU states to rapidly distinguish and evaluate potential security concerns, taking into account ideal intercession and moderation measures.
3. *Robotized signal identification*: EudraVigilance utilizes complex mechanized signal identification calculations to recognize potential security worries from the huge measure of ADR information. These calculations dissect ADR designs and distinguish strange or surprising relationships among drugs and ADRs, taking into consideration opportune assessments of potential security signals.
4. *Complete information examination*: EudraVigilance provides a far-reaching set of devices for breaking down ADR information, including factual techniques, information perception and hazard evaluation instruments. These instruments empower the EMA and EU

states to completely assess potential well-being concerns and survey the general security of advertised drugs.

5. *Improved straightforwardness*: EudraVigilance information is openly available, permitting medical care experts, specialists and people in general to get to and dissect ADR information. This straightforwardness advances educated direction and public mindfulness regarding drug well-being issues.

Disadvantages of EudraVigilance

1. *Underreporting of ADRs*: Similar to other passive surveillance systems, EudraVigilance relies on voluntary reporting of ADRs, which may lead to underreporting. This underreporting can be particularly problematic for rare ADRs or those that occur long after medication initiation.
2. *Limited causality information*: EudraVigilance cannot definitively establish the causal relationship between a drug and an ADR. It only records temporal associations between drug use and ADRs, which may be influenced by confounding factors.
3. *Data quality concerns*: The quality of ADR reports in EudraVigilance may vary depending on the reporting source. Healthcare professionals may provide more detailed and accurate reports, while patient-reported ADRs may be more subjective and less reliable.
4. *Data lag*: There may be a lag between ADR occurrence and reporting and between reporting and data entry into EudraVigilance, which can delay the identification of potential safety concerns.

3.2.3 WHO VigiBase

The WHO International Drug Monitoring Programme (https://who-umc.org/pv-products/vigilyze/) started in 1968 with an aim to identify the earliest possible PV signals. The program involves countries from all parts of the world contributing ICSRs to the WHO Global ICSR Database System (18, 19).

VigiBase is a PV database developed and maintained by the Swedish WHO Collaborating Centre (Uppsala Monitoring Centre [UMC]) for monitoring all adverse effects globally. The database system includes an ICH E2B-compatible ICSR database, MedDRA, WHO Drug Dictionaries (WHO-DD and WHO-DDE), International Classification of Diseases (ICD) and medical terminologies WHO Adverse Reaction Terminology (WHO-ART) (19). More than 20 million ICSRs that are forwarded by various countries through spontaneous reporting are collected and curated by the UMC. Even though the data from the database is not completely homogenous regarding the AEs and the pharmaceutical products, it enhances the quantitative screening of Big Data for rapid and effective PV (20).

The VigiBase system is linked to medical and drug classifications including WHO-DD, WHO-ART, WHO ICD and MedDRA making it easier to enter, retrieve and analyse structured data with precision and aggregation at various levels. The advantage of combining reports into one huge international database allows PV researchers to analyse the patterns of harm, which may not be evident in smaller national databases (19–21). With millions of AE reports, qualitatively screening all the AEs through VigiLyze and thorough screening of scientific literature can additionally add insights to potential safety signals. Similarly, the signal detection team also assess the ICSRs to identify any potential signals between the pharmaceutical product and the AE, thus preventing false signals (18, 22, 23).

Advantages of VigiBase

1. *Largest and most comprehensive ADR database in the world*: VigiBase has the largest and most diverse collection of ADR data available, representing over 99% of the world's population. This large data volume allows for more robust and reliable signal detection and analysis.

2. *Global reach and diversity*: VigiBase receives ADR reports from a wide range of sources, including healthcare professionals, patients and drug manufacturers. This global coverage ensures that ADRs from all over the world are included in the database, providing a more representative picture of drug safety.

3. *Structured data and data quality assurance*: VigiBase uses standardized terminology and coding systems to ensure the quality and consistency of its data. This standardization makes it easier to compare and analyse ADR reports from different sources.

4. *Data access and sharing*: VigiBase is a publicly accessible database, with a user-friendly interface that allows for search, filtering and analysis of ADR data. The UMC also offers a variety of data services and tools to support researchers and healthcare professionals.

Disadvantages of VigiBase

1. *Underreporting of ADRs*: VigiBase relies on passive surveillance, which means that ADR reports are submitted voluntarily by healthcare professionals, patients and drug manufacturers. This method may lead to underreporting, as not all ADRs are reported.

2. *Limited information about ADR causality*: VigiBase does not provide definitive information about the causality of ADRs. The database only records whether a suspect ADR is temporally related to the use of a drug, and it does not take into account other potential causes of the ADR.

3. *Possible bias in reporting*: The quality of ADR reports in VigiBase may vary depending on the source of the report. Healthcare professionals may provide more detailed and accurate reports, while patient-reported ADRs may be more subjective and less reliable.

4. *Slowness of data updates*: VigiBase is updated periodically, with new ADR reports being added over time. However, the lag between reporting and data entry can delay the availability of new information.

3.2.4 THE OBSERVATIONAL MEDICAL OUTCOMES PARTNERSHIP (OMOP)

Observational health data can be converted into a standard format using the Observational Medical Outcomes Partnership (OMOP) Common Data Model (CDM). In rare diseases where data are scarce, CDM transformation enables analysis across diverse databases for the development of new, empirical evidence (24). Electronic health records (EHRs) can be used to collect observational health data that can be used for research in order to assess the quality of care and public health in addition to direct healthcare delivery. Validity of the data and lack of reproducibility are two main concerns, which require a framework to improve the secondary use of health data (25). Observational Health Data Sciences and Informatics (OHDSI) (https://ohdsi.github.io/CommonDataModel/) offers the OMOP CDM along with a wide range of open-source tools and techniques to promote significant research work with the available data. OHDSI also offers a terminology browser to browse through the vocabularies integrated into the OMOP CDM (Athena) to assess and display a database's compliance with the OMOP CDM (Achilles) for connecting to the OMOP CDM (26–28). Various tools are for used for extracting and transforming data (OhdsiRTools and FeatureExtraction) along with statistical analyses and machine learning (26). The majority of the information originates from hospital clinical databases or claims databases used for pharmacoepidemiological research. Numerous studies have been conducted in the last ten years, including predictions at the patient level and estimates of the effects at the population level (29, 30).

3.2.5 ORACLE ARGUS

Oracle Argus Safety is a comprehensive PV safety database that provides a centralized platform for collecting, managing and analysing ADR data. It is widely used by pharmaceutical companies, contract research organizations (CROs) and regulatory agencies to monitor the safety of marketed drugs and medical devices.

Key features and benefits of Oracle Argus Safety include:

1. *Global case processing*: Argus supports the processing of ADRs from around the world, including E2B(R3) and E2B(R2) electronic submissions.
2. *Signal detection*: Argus uses advanced algorithms to identify potential safety signals from ADR data.
3. *Detailed analytics*: Argus provides a variety of analytical tools for exploring and visualizing ADR data.
4. *Electronic case intake and expedited reporting*: Argus supports electronic case intake and expedited reporting to regulatory agencies.
5. *Risk management*: Argus helps to identify and manage potential drug safety risks.
6. *Periodic reporting and submissions*: Argus helps to generate periodic safety reports and submissions to regulatory agencies.
7. *Improved safety monitoring*: Argus helps to improve the safety of marketed drugs and medical devices by providing a centralized platform for collecting, managing and analysing ADR data.
8. *Reduced compliance costs*: Argus can help to reduce compliance costs by streamlining the safety reporting process.
9. *Improved decision-making*: Argus can help to improve decision-making by providing insights into drug safety data.
10. *Reduced time to market*: Argus can help reduce the time to market for new drugs and medical devices by providing a centralized platform for managing safety data.

Overall, Oracle Argus Safety is a valuable tool for improving the safety of marketed drugs and medical devices. It is a comprehensive and easy-to-use platform that can help to reduce compliance costs, improve decision-making and reduce the time to market for new products.

3.2.6 ARIS G

Aris G, also known as Aris C, is a PV safety database developed by ArisGlobal, a leading provider of life sciences software solutions. It is a comprehensive and scalable platform that helps pharmaceutical companies, CROs and regulatory agencies to collect, manage and analyse ADR data.

Key features of the Aris G PV database include:

1. *Global case processing*: Aris G supports the processing of ADRs from around the world, including E2B(R3) and E2B(R2) electronic submissions.
2. *Signal detection*: ARISg uses advanced algorithms to identify potential safety signals from ADR data.
3. *Detailed analytics*: ARISg provides a variety of analytical tools for exploring and visualizing ADR data.
4. *Electronic case intake and expedited reporting*: ARISg supports electronic case intake and expedited reporting to regulatory agencies.
5. *Risk management*: ARISg helps to identify and manage potential drug safety risks.
6. *Periodic reporting and submissions*: ARISg helps to generate periodic safety reports and submissions to regulatory agencies.

Benefits of using ARISg PV database include:

1. *Improved safety monitoring*: Aris G helps to improve the safety of marketed drugs and medical devices by providing a centralized platform for collecting, managing and analysing ADR data.

2. *Reduced compliance costs*: Aris G can help to reduce compliance costs by streamlining the safety reporting process.
3. *Improved decision-making*: Aris G can help to improve decision-making by providing insights into drug safety data.
4. *Reduced time to market*: Aris G can help to reduce the time to market for new drugs and medical devices by providing a centralized platform for managing safety data.

Aris G is a cloud-based solution, which means that it is accessible from anywhere in the world. It is also highly scalable, so it can be easily adapted to meet the needs of organizations of all sizes.

Here are some additional benefits of using Aris G PV database:

1. *Compliance with ICH guidelines*: Aris G is compliant with all relevant ICH guidelines for PV.
2. *Ease of use*: Aris G is a user-friendly platform that is easy to learn and use.
3. *Integration with other systems*: Aris G can be integrated with other systems, such as clinical trial management systems (CTMS) and EHRs.
4. *Support*: ArisGlobal provides a comprehensive support package for Aris G users.

Overall, the Aris G PV database is a valuable tool for improving the safety of marketed drugs and medical devices. It is a comprehensive, scalable and user-friendly platform that can help to reduce compliance costs, improve decision-making and reduce the time to market for new products.

3.2.7 SAFIRE

The SAFIRE PV information base is an extensive security dataset that is utilized to gather and break down adverse drug reaction (ADR) information. It is utilized by drug organizations, contract research associations (CROs) and administrative offices to screen the security of promoted medications and clinical gadgets.

SAFIRE is a worldwide information base that gathers ADR information from around the world. An ongoing information base considers the detailing of ADRs as they happen. SAFIRE utilizes progressed calculations to recognize potential well-being signals from the tremendous measure of ADR information. SAFIRE gives a far-reaching set of devices for examining ADR information, including factual strategies, information representation and hazard evaluation instruments.

SAFIRE is consistent with all applicable ICH rules for PV. An easy-to-use stage is not difficult to learn and utilize. SAFIRE can be incorporated with different frameworks, like clinical preliminary administration frameworks (CTMS) and EHRs. SAFIRE is upheld by a group of specialists who can furnish help with utilizing the information base.

Advantages of utilizing the SAFIRE PV dataset:

1. *Further developed security checking*: SAFIRE assists with working on the well-being of showcased medications and clinical gadgets by giving a unified stage to gathering, making do and investigating ADR information.
2. *Diminished consistence costs*: SAFIRE can assist with lessening consistence costs by smoothing out the security announcing process.
3. *Further developed navigation*: SAFIRE can assist with further developing independent direction by providing bits of knowledge into drug well-being information.
4. *Improved straightforwardness*: SAFIRE information is freely available, advancing educated direction and public mindfulness regarding drug well-being issues.

By and large, the SAFIRE PV dataset is an important instrument for working on the security of showcased medications and clinical gadgets. It is an extensive, continuous and freely open information base that can assist with decreasing consistency costs, further developing independent direction

TABLE 3.2

Few Studies Using Databases for Detecting Signals

Sr. No	Name of Database	Method of Signal Detection	Identified Signals	References
1	U.S. Food and Drug Administration Adverse Event Reporting System	Reporting Odds Ratio	Glioblastoma – TNF inhibitors	https://doi.org/10.1002/phar.1731
		Disproportionality Analysis (ROR-1.96SE >1, PRR ≥ 2 and IC-2SD > 0)	Pantoprazole-associated dyspepsia hypocalcemia and hyponatremia	https://doi.org/10.1016/j.ajg.2022.10.012
		Reporting Odds Ratio	Ubrogepant and rimegepant – 10 disproportionality signals for ubrogepant and 25 disproportionality signals for rimegepant were identified. These were mostly related to psychiatric, neurological, gastrointestinal, skin, vascular and infectious type of adverse events	https://doi.org/10.1080/14740338.2023.2223958
2	EudraVigilance		Signal detection activity on EudraVigilance data: Analysis of the procedure and findings from an Italian Regional Centre for pharmacovigilance	https://doi.org/10.1080/14740338.2017.1284200
			Adverse events of acute nephrotoxicity reported to EudraVigilance and VAERS after COVID-19 vaccination	https://doi.org/10.1016/j.vaccine.2023.10.030

and upgrading straightforwardness. A few recent studies regarding the detection of signals using available databases are compiled in Table 3.2.

3.3 STATISTICAL AND DATA MINING TECHNIQUES

The ability to quickly identify potential risks related to a pharmaceutical product is made possible by statistical signal detection (31). Screening enormous databases of spontaneous case reports results in the identification of potential ADRs. There are several quantitative statistical methodologies, mostly relying on two separate approaches of statistical inference (frequentist method and Bayesian method) (7, 32).

3.3.1 METHODS FOR DETECTION OF SAFETY SIGNALS IN HEALTHCARE DATABASES

a. *Disproportionality analysis approach*: This is a widely used traditional approach to detect signals in various databases (7, 32–35). This method presents all spontaneous reports as a sizable contingency table and the values of PRR and ROR are used to predict true or false signals.

b. *Traditional pharmacoepidemiological designs*: These designs are frequently used for ad hoc studies involving a two-step process. Primarily selection of two groups based on exposures (cohort approach) or events (case-based) is done, followed by comparing the rate of the drug-event association in these groups.

c. *Sequence symmetry analysis*: This method is done by comparing the order in which two drug exposures within a specific time window show a potential AE. For example, Drugs A and B are compared, as exposure to drug A is used as exposure of interest, while exposure to drug B is used as a substitute for the potential AE.

    d. *Sequential statistical testing approach*: The purpose of this method is to test the null hypothesis, which states that exposed patients experience more events than unexposed patients using prospective cohort data. Using this method, we can generate potential signals.

    e. *Temporal association rule approach*: This method follows two criteria, which take into account the incident of the AE occurring after the drug exposure and the occurrence must take place during a predetermined time window. A correlation score is calculated for potential AEs that are successively mined for a given drug.

    f. *Supervised machine learning (SML)*: The basic principles of the SML algorithms is that a classifier (random forest model) is trained using a reference set that comprises drug-event associations that are already known. Then, the chosen parameters for each drug event association that were screened from the sample of data used for testing are applied to the trained classifier to forecast those associations that might be new ADRs.

    g. *Tree-based scan statistic method*: This method uses a hierarchical structure of classifications used to code AEs, which is mapped as a tree. The root of a specific event relates to its largest description, the nodes to its sublevel descriptions, the leaves to the codes with the most detailed descriptions and the branches connect the three elements.

3.4 SOFTWARE/TOOLS USED IN PHARMACOVIGILANCE

3.4.1 ORACLE ARGUS SAFETY

Oracle Argus Safety offers a quick and easy approach to checking the international regulatory safety reporting requirements and facilitating the manufacturers to comply. With this tool, signal detection and analysis of the overall safety profile of pharmaceutical products can assessed (36, 37).

3.4.2 EVDAS

EVDAS uses a frequentist approach by measuring the disproportionality. The PRR and ROR have been implemented in EVDAS to detect signals. Calculation of ROR can also be used to adjust confounding factors such as age, gender and other medications (3, 16, 17).

3.4.3 CLINTRAC

ClinTrac is a clinical database system that has been created with the intention of gathering, processing, storing, analysing and transferring clinical data. A practice-unit-focused database forms the system's central component. Every patient treated in a practice is fully chronicled in the practice-unit-centred database. All records are generated and saved by the hardware at the primary practice location. Scheduling, processing clinical records and billing are all managed by the same programme and are fully integrated. This "private" database is easily and discreetly accessible (38, 39).

3.4.4 ENNOV PV ANALYSER

A complete solution for gathering, reporting and analysing data on human and veterinary PVs. In addition to offering sophisticated signal detection and PV data analysis tools, Ennov's PV suite retains the collection, management, assessment and reporting of adverse human or veterinary events in one unified database. Ennov has been designed for the management of regulated processes and content (40).

3.4.5 PHARMAMINER

It is a tool that combines AEs that are semantically related by performing terminological reasoning between drug-event pairs, followed by using the Bayesian and statistical analysis methods on these groups to find possible signals. This tool is intended for automatic signal generation. With large datasets, this tool can be used to accomplish common signal detection techniques (6).

3.5 CONCLUSION

The utilization of worldwide PV datasets, for example, the FAERS and the WHO Worldwide ICSRs datasets are major in recognizing potential security signals related with drugs. These information data bases keep on filling in as fundamental storehouses of certifiable antagonistic occasion information, supporting the early discovery of well-being concerns. This section has given a complete outline of the different datasets and devices accessible for signal recognition in the field of PV and drug well-being. We have investigated a large number of assets, from customary unconstrained detailing frameworks to cutting-edge information mining strategies, exhibiting the development of sign identification techniques.

Notwithstanding these assets, we have dove into the domain of information mining and factual calculations. Techniques like disproportionality examination, Bayesian information mining and AI models have introduced another time of sign location, empowering more effective and exact distinguishing proof of unfriendly occasions. The consolidation of organized information from EHRs and patient-created information from different libraries has improved our capacity to screen and survey drug security continuously.

Furthermore, we have highlighted the importance of data quality, standardization and harmonization in signal detection. The field of signal detection in drug safety is undergoing a transformative journey, driven by innovation and collaboration. The number of databases and tools explored in this chapter reflect the multifaceted nature of PV, where the ultimate goal is to protect the well-being of patients. With ongoing advancements, we are better equipped than ever to detect, evaluate and respond to emerging safety signals, ensuring that the benefits of medications far outweigh their potential risks.

REFERENCES

1. Sardella, Marco., Glyn Belcher, Calin Lungu, Terenzio Ignoni, Manuela Camisa, Doris Irene Stenver, et al. "Monitoring the manufacturing and quality of medicines: A fundamental task of pharmacovigilance." *Therapeutic Advances in Drug Safety*. 12 (2021 Jan): 204209862110384.
2. Alomar, Muaed, Ali M. Tawfiq, Nageeb Hassan, and Subish Palaian. "Post marketing surveillance of suspected adverse drug reactions through spontaneous reporting: Current status, challenges and the future." *Therapeutic Advances in Drug Safety* 11 (2020 Aug): 2042098620938595.
3. Bihan, Kévin, Bénédicte Lebrun-Vignes, Christian Funck-Brentano, and Joe-Elie Salem. "Uses of pharmacovigilance databases: An overview." *Therapies* 75, no. 6 (2020): 591–598.
4. M Coloma, Preciosa, Gianluca Trifirò, Vaishali Patadia, and Miriam Sturkenboom. "Postmarketing safety surveillance: Where does signal detection using electronic healthcare records fit into the big picture?" *Drug Safety [Internet]* 36, no. 3 (2013 Mar): 183–197. Available from: https://doi.org/10.1007/s40264-013-0018-x
5. Chakraborty, Bhaswat S. "Pharmacovigilance: A data mining approach to signal detection." *Indian Journal of Pharmacology* 47, no. 3 (2015): 241.
6. Bousquet, Cédric, Corneliu Henegar, Agnès Lillo-Le Louët, Patrice Degoulet, and Marie-Christine Jaulent. "Implementation of automated signal generation in pharmacovigilance using a knowledge-based approach." *International Journal of Medical Informatics* 74, no. 7–8 (2005): 563–571.
7. Sakaeda, Toshiyuki, Akiko Tamon, Kaori Kadoyama, and Yasushi Okuno. "Data mining of the public version of the FDA adverse event reporting system." *International Journal of Medical Sciences* 10, no. 7 (2013): 796.

8. Rodriguez, Evelyn M., Judy A. Staffa, and David J. Graham. "The role of databases in drug postmarketing surveillance." *Pharmacoepidemiology and Drug Safety* 10, no. 5 (2001): 407–410.

9. Hoffman, Keith B., Andrea R. Demakas, Mo Dimbil, Nicholas P. Tatonetti, and Colin B. Erdman. "Stimulated reporting: The impact of US food and drug administration-issued alerts on the adverse event reporting system (FAERS)." *Drug Safety* 37 (2014): 971–980.

10. Khaleel, Mohammad Ali, Amer Hayat Khan, Siti Maisharah Sheikh Ghadzi, Azreen Syazril Adnan, and Qasem M. Abdallah. "A standardized dataset of a spontaneous adverse event reporting system." *Healthcare* 10, no. 3 (2022 MDPI): 420.

11. Hazell, Lorna, and Saad A. W. Shakir. "Under-reporting of adverse drug reactions." *Drug Safety* 29, no. 5 (2006): 385–396.

12. Hoffman, Keith B., Mo Dimbil, Colin B. Erdman, Nicholas P. Tatonetti, and Brian M. Overstreet. "The Weber effect and the United States food and drug Administration's adverse event reporting system (FAERS): Analysis of sixty-two drugs approved from 2006 to 2010." *Drug Safety* 37 (2014): 283–294.

13. Arora, Ankur, Rajinder K. Jalali, and Divya Vohora. "Relevance of the Weber effect in contemporary pharmacovigilance of oncology drugs." *Therapeutics and Clinical Risk Management* 13 (2017): 1195–1203.

14. Postigo, Rodrigo, Sabine Brosch, Jim Slattery, Anja van Haren, Jean-Michel Dogné, Xavier Kurz, Gianmario Candore, Francois Domergue, and Peter Arlett. "EudraVigilance medicines safety database: Publicly accessible data for research and public health protection." *Drug Safety* 41 (2018): 665–675.

15. Banovac, Marin, Gianmario Candore, Jim Slattery, Francois Houÿez, David Haerry, Georgy Genov, and Peter Arlett. "Patient reporting in the EU: Analysis of EudraVigilance data." *Drug Safety* 40 (2017): 629–645.

16. Sardella, Marco, and Calin Lungu. "Evaluation of quantitative signal detection in EudraVigilance for orphan drugs: Possible risk of false negatives." *Therapeutic Advances in Drug Safety* 10 (2019): 2042098619882819.

17. Vogel, Ulrich, John van Stekelenborg, Brian Dreyfus, Anju Garg, Marian Habib, Romana Hosain, and Antoni Wisniewski. "Investigating overlap in signals from EVDAS, FAERS, and VigiBase®." *Drug Safety* 43 (2020): 351–362.

18. Venulet, Jan, and Margaretha Helling Borda. "WHO's international drug monitoring—the formative years, 1968–1975: Preparatory, pilot and early operational phases." *Drug Safety* 33, no. 7 (2010): e1–e23.

19. Lindquist, Marie. "VigiBase, the WHO global ICSR database system: Basic facts." *Drug Information Journal* 42, no. 5 (2008): 409–419.

20. Khamisy-Farah, R., G. Damiani, J. D. Kong, J-H. Wu, and N. L. Bragazzi. "Safety profile of Dupilumab during pregnancy: A data mining and disproportionality analysis of over 37,000 reports from the WHO individual case safety reporting database (VigiBase™)." *European Review for Medical and Pharmacological Sciences* 25, no. 17 (2021): 5448–5451.

21. Centre UM. About VigiBase [Internet]. who-umc.org. Available from: https://who-umc.org/vigibase/

22. Pal, Shanthi N., Chris Duncombe, Dennis Falzon, and Sten Olsson. "WHO strategy for collecting safety data in public health programmes: Complementing spontaneous reporting systems." *Drug Safety* 36 (2013): 75–81.

23. Vogler, Marcelo, Heloísa Ricci Conesa, Karla de Araújo Ferreira, Flávia Moreira Cruz, Fernanda Simioni Gasparotto, Karen Fleck, Fernanda Maciel Rebelo, Bianca Kollross, and Yannie Silveira Gonçalves. "Electronic reporting systems in pharmacovigilance: The implementation of VigiFlow in Brazil." *Pharmaceutical Medicine* 34 (2020): 327–334.

24. Biedermann, Patricia, Rose Ong, Alexander Davydov, Alexandra Orlova, Philip Solovyev, Hong Sun, Graham Wetherill, Monika Brand, and Eva-Maria Didden. "Standardizing registry data to the OMOP common data model: Experience from three pulmonary hypertension databases." *BMC Medical Research Methodology* 21, no. 1 (2021): 1–16.

25. Jensen, Peter B., Lars J. Jensen, and Søren Brunak. "Mining electronic health records: Towards better research applications and clinical care." *Nature Reviews Genetics* 13, no. 6 (2012): 395–405.

26. Voss, Erica A., Rupa Makadia, Amy Matcho, Qianli Ma, Chris Knoll, Martijn Schuemie, Frank J. DeFalco, Ajit Londhe, Vivienne Zhu, and Patrick B. Ryan. "Feasibility and utility of applications of the common data model to multiple, disparate observational health databases." *Journal of the American Medical Informatics Association* 22, no. 3 (2015): 553–564.

27. Yoon, Dukyong, Eun Kyoung Ahn, Man Young Park, Soo Yeon Cho, Patrick Ryan, Martijn J. Schuemie, Dahye Shin, Hojun Park, and Rae Woong Park. "Conversion and data quality assessment of electronic health record data at a Korean tertiary teaching hospital to a common data model for distributed network research." *Healthcare Informatics Research* 22, no. 1 (2016): 54–58.

28. Hripcsak, George, Jon D. Duke, Nigam H. Shah, Christian G. Reich, Vojtech Huser, Martijn J. Schuemie, Marc A. Suchard et al. "Observational health data sciences and informatics (OHDSI): Opportunities for observational researchers." *Studies in Health Technology and Informatics* 216 (2015): 574.

29. Lamer, Antoine, Osama Abou-Arab, Alexandre Bourgeois, Adrien Parrot, Benjamin Popoff, Jean-Baptiste Beuscart, Benoît Tavernier, and Mouhamed Djahoum Moussa. "Transforming anesthesia data into the observational medical outcomes partnership common data model: Development and usability study." *Journal of Medical Internet Research* 23, no. 10 (2021): e29259.

30. Makadia, Rupa, and Patrick B. Ryan. "Transforming the Premier Perspective® hospital database into the observational medical outcomes partnership (OMOP) common data model." *Egems (Washington, DC)* 2, no. 1 (2014): 1110. doi:10.13063/2327-9214.1110

31. Harpaz, Rave, William DuMouchel, Nigam H. Shah, David Madigan, Patrick Ryan, and Carol Friedman. "Novel data-mining methodologies for adverse drug event discovery and analysis." *Clinical Pharmacology & Therapeutics* 91, no. 6 (2012): 1010–1021.

32. Arnaud, Mickael, Bernard Bégaud, Nicolas Thurin, Nicholas Moore, Antoine Pariente, and Francesco Salvo. "Methods for safety signal detection in healthcare databases: A literature review." *Expert Opinion on Drug Safety* 16, no. 6 (2017): 721–732.

33. Bate, Andrew, Marie Lindquist, I. Ralph Edwards, Sten Olsson, Roland Orre, Anders Lansner, and R. Melhado De Freitas. "A Bayesian neural network method for adverse drug reaction signal generation." *European Journal of Clinical Pharmacology* 54 (1998): 315–321.

34. Maignen, Francois, Manfred Hauben, Eric Hung, Lionel Van Holle, and Jean-Michel Dogne. "Assessing the extent and impact of the masking effect of disproportionality analyses on two spontaneous reporting systems databases." *Pharmacoepidemiology and Drug Safety* 23, no. 2 (2014): 195–207.

35. Evans, Stephen J. W., Patrick C. Waller, and S. Davis. "Use of proportional reporting ratios (PRRs) for signal generation from spontaneous adverse drug reaction reports." *Pharmacoepidemiology and Drug Safety* 10, no. 6 (2001): 483–486.

36. Gill, Supreet Kaur, Ajay Francis Christopher, Vikas Gupta, and Parveen Bansal. "Emerging role of bioinformatics tools and software in evolution of clinical research." *Perspectives in Clinical Research* 7, no. 3 (2016): 115-122. Available from: https://www.ncbi.nlm.nih.gov/pmc/articles/PMC4936069/

37. Argus Safety Quickstart Guide [2016]. [cited 2023 Sep 10]. Available from: https://docs.oracle.com/health-sciences/argus-safety-811/QKSRT/toc.htm#QKSRT102

38. Bryner, U. M. "The practice-unit centered clinical database–the implementation." In *Proceedings of the Annual Symposium on Computer Application in Medical Care*, p. 919. American Medical Informatics Association, 1991.

39. Gregory, Andrew., The Guardian. Omicron Covid variant poses very high global risk, says WHO. The Guardian [Internet]. 2021 Nov 29 [cited 2023 Sep 7]. Available from: https://www.theguardian.com/world/2021/nov/29/omicron-covid-variant-poses-very-high-global-risk-says-who

40. Brown, C. Andrew, Jessica H. Bailey, Margaret E Miller Davis, Paula Garrette, William J. Rudman "Improving patient safety through technology." *Perspectives in Health Information Management* 60, no. 3 (2005 Sep): 11.

4 Role of Electronic Medical Records, Insurance Claims Data, Distributed Data Networks and Social Media in Signal Analysis

Vandana Roy

4.1 INTRODUCTION

"Have any of the readers seen such cases", a call by Australian physician Dr. McBride in 1958, in his letter to the journal where he was reporting cases of phocomelia he had observed in children born to mothers who had taken thalidomide during pregnancy. Keen observation, sharp insight and critical ability to link findings attributing thalidomide as a cause of phocomelia. Was anyone to know that it was the first major signal to be generated about an adverse drug reaction (ADR) that would lead to the International Program of Monitoring Drugs by the World Health Organization in 1968 and eventually to the birth of science of pharmacovigilance (PV) (1).

Many systems for monitoring adverse drug reactions (ADR) are established globally and within countries. In these, reports of adverse drug events (ADE) sent as individual case safety reports (ICSRs) are analyzed for establishing drug causality to the ADR and generating a signal, defined as *a reported information on a possible causal relationship between an adverse event and a drug, the relationship being unknown or incompletely documented previously* (2). ADR signals have led to drug alerts, label changes and drug withdrawals after regulatory approval within countries and globally, playing a major role in ensuring drug safety.

Safety testing of medicinal products for human use is a mandatory requirement both before preapproval and after post-marketing of drugs. Due to inherent limitations in premarketing safety monitoring of drugs (clinical trials with less number of patients with specific inclusion/exclusion criteria, hence not able to detect rare ADRs, non-inclusion of large subsets of populations in clinical trials such as pregnant women, pediatric and geriatric patients, limited duration of clinical trials hence not able to detect delayed ADRs), post-marketing surveillance is essential for detecting rare and delayed ADRs. For many decades, data from preapproval and post-marketing drug approvals, ICSRs in PV programs and published literature formed the core data for generating signals about ADRs. In the last few decades with the advent of information technology and recently artificial intelligence, digital sources of patient data for PV have come forth (3).

This chapter explores the use of data from electronic medical/health records (EMRs/EHRs), medical insurance claims, distributed data networks (DDNs) and social media in signal generation for monitoring drug safety.

DOI: 10.1201/9781032629940-4

4.2 SIGNAL GENERATION

Use of case reports of ADRs for signal generation evolved from paper-based ICSRs in spontaneous reporting systems (SRSs) in the 1960s to primary data collection systems such as the Boston Collaborative Drug Surveillance Program, Prescriptive Event Monitoring Systems in the United Kingdom (UK) and New Zealand. Advances in computer technology led to electronically submitted reports, analyzed as part of signal management systems (4). The 1980s saw increasing use of EMRs, administrative and medical insurance claims in the United States and the UK (5). There are now longitudinal observational databases (LODs) available in North America, Europe and Asia from drug or outcome registries, EMR databases and insurance claims databases (6). Many databases have been linked across regions and continents with the intent to improve efficacy, avoid duplication of effort and save time and costs in the analysis of the data. These are being used in pharmacoepidemiological and PV studies. And among their many uses is the signal generation of ADR.

According to the Council for International Organizations of Medical Sciences (CIOSMS), the criteria for a PV signal is "Information that arises from one or multiple sources (including observations and experiments), which suggests a new potentially causal association, or a new aspect of a known association, between an intervention and an event or set of related events, either adverse or beneficial, that is judged to be of sufficient likelihood to justify verificatory action" (7).

Among the methods to establish drug causality to an ADR, a temporal association between the use of a drug and the ADR has to be there; a dechallenge should result in resolving the problem; a rechallenge should cause the ADR to occur again; and there should be other findings to support the decision and no other likely explanation for the same (8, 9). Causality analysis has to be followed by analyses of similar reports by statistical methods (10). Among the methods, disproportionality analysis is most commonly used. It detects drug-ADE associations that occur at higher-than-expected frequencies. This method does not adjust for small numbers of drug ADE associations, where more advanced statistical methods are used, such as the Multi-Item Gamma Poisson Shrinker (MGPS) used by the U.S. Food and Drug Administration (FDA) or the Bayesian Confidence Propagation Neural Network (BCPNN) used by WHO to analyze SRS databases (11, 12). There are commercially available data mining software programs that can generate proportional reporting ratios (PRR) and or MGPS scores (10). A proprietary software tool developed by the FDA in collaboration, enables mechanism-based safety analysis (MASE) for all drugs currently on the market, along with the identification of potential adverse events for drugs in development (13). Text mining of data present in the text submitted in ADE reports, event descriptions or narratives in EMRs, medical insurance claim bills, social media, internet and medical literature using technology and tools such as artificial intelligence, natural language processing (NLP), change point analysis, visualization tools and geographical information systems technology is aiding the PV process (14).

A basic prerequisite for the generation of signals to ADEs is the availability of adequate and appropriate data. There is increasing interest in using real-world data (RWD) to generate real-world evidence (RWE). RWD has been defined in the 21st Cures FD and C Act as data relating to patient health status and or the delivery of healthcare routinely collected from a variety of sources (EMR), medical claims, billing data, data from product or disease registries, patient-generated data and data gathered from other sources that can inform on health status such as mobile devices. RWE is the clinical evidence about the usage and potential benefits or risks of a medical product derived from analysis of RWD (15).

While the backbone of the PV system is still the collection of ICSRs through SRSs, the use of pharmacoepidemiologic data across large networks of observational databases along with the availability of evolving information technology resources to handle voluminous, high velocity and variable data is enabling the evolution of the use of "Big Data" in PV analytics (16). While sources of data in the health system may not fulfill the classic definition of Big Data, the term is being used for the available digitalized patient data in SRS, EMR, medical insurance claims and DDNs. Another potential source being considered is social media (Figure 4.1).

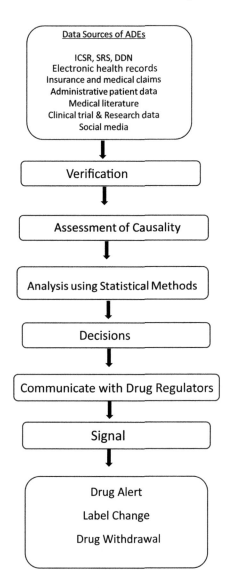

FIGURE 4.1 Process of signal generation. *Abbreviations:* ADE, Adverse Drug Event; DDN, Distributed Data Networks; ICSR, Individual Case Safety Reports; SRS, Spontaneous Reporting System.

4.3   SOURCES OF DIGITALIZED DATA FOR SIGNAL GENERATION

Large repositories of patient data in electronic form that can be used for PV include:

    i. SRS databases
    ii. Healthcare databases containing electronic medical or health records
    iii. Medical insurance claims
    iv. DDNs
    v. Social media content
    vi. Others including, clinical trial data, research studies and published literature

4.3.1 SPONTANEOUS REPORTING SYSTEM DATABASES

This is the oldest, most common way by which reports of ADEs are collected and analyzed for signal generation. They are a rich source of patient information. The ICSRs are now sent electronically to national and global databases of ADRs (Table 4.1). In some low-resource countries or local drug safety systems, paper reports of ICSRs are still used.

4.3.1.1 World Health Organization (VigiBase)

VigiBase is the largest database of reported adverse effects of medicinal products with more than 35 million reports from more than 156 member countries. The aggregate data are mined for safety signals. The database is updated continuously. The VigiBase is dependent on national centers, which use VigiBase data and send reports via VigiLyze for signal detection and signal strengthening. It has data management and quality assurance tools and is linked to multiple medical and drug classifications such as WHO Drug: an international classification of drugs, MedDRA (Medical Dictionary for Regulatory Activities), WHO International Classification of Diseases (ICD), WHO-ART, a tool for classification of adverse reaction terms, which enable structured data entry, retrieval and analysis at different levels of precision (17, 18).

4.3.1.2 FDA Adverse Event Reporting System (FAERS)

This is a computerized database of Spontaneous Reports of ADEs reported to the U.S. FDA by healthcare professionals, pharmaceutical companies, patients and consumers, maintained by the

---

TABLE 4.1

Global and National Adverse Drug Event Reporting Systems

Databases	Organization/Country	Started	Sources
Program for International Drug Monitoring (PIDM) VigiBase	World Health Organization	1968	National Pharmacovigilance Centres
Adverse Event Reporting System* FDA Adverse Event Reporting System (FAERS)	USA	1990* 2012	Healthcare Professionals, Market Authorization Holders, Patients, Consumers
FDAs Vaccine Event Reporting System (VAERS)	USA	1990	
European Medicines Agency (EudraVigilance)	European Union	2001	Market Authorization Holders
Individual country databases			
Database of Adverse Event Notifications (DEAN)	Australia (Therapeutic Goods Administration)	1971	Patients, Consumers, Healthcare Professionals
Japan Adverse Drug Event Report Database (JADER)	Japan	2012	Healthcare Providers, Patients, Consumers, Market Authorization Holders
Pharmacovigilance Program of India	India	2010	Healthcare Professionals, Market Authorization Holders, Patients, Consumers

Note: These are among the largest databases for PV in the world.

* In the United States, originally the SRS established in 1990 was known as Adverse Event Reporting System. Subsequently, the name changed to FDA Adverse Event Reporting System in 2012.

FDA for post-marketing surveillance of all approved drugs and therapeutic biologic products. The system interacts with several related systems including MedWatch and the Vaccine Adverse Event Reporting System (VAERS). It replaced the Adverse Event Reporting System (AERS) in 2012.

To date, its public dashboard is showing a total of 16,470,915 reports, with reports excluding death as 9,257,928 and death reports as 1,646,084 (19).

4.3.1.3 FDA's Vaccine Event Reporting System (VAERS)

VAERS is a national vaccine safety surveillance program in the United States co-monitored by the FDA and the Centers for Disease Control and Prevention (CDC). The VAERS system collects national data from U.S. territories.

The purpose of VAERS is to detect possible signals of adverse events (side effects) that occur after the administration of U.S.-licensed or authorized vaccines. Anyone can report in the system, healthcare providers (HCPs), vaccine manufacturers, vaccine recipients (or their parents/guardians) and state immunization programs. If a safety signal is found then the report is verified by conducting further studies.

The limitations of the system are that it may not be possible from VAERS data to confirm if the vaccine caused the ADE, the tendency for reporting of serious ADRs vs. non-serious ADRs and the possibility that reporting may increase with media reports (20).

4.3.1.4 European Medicines Agency (EudraVigilance)

EudraVigilance is a system for managing and analyzing information on suspected adverse reactions to medicines that have been authorized or are being studied in clinical trials in the European Economic Area. It is operated by the European Medicines Agency (EMA). And it facilitates the electronic exchange of ICSRs between EMA, national competent authorities, marketing authorization holders (MAH) and sponsors of clinical trials (21).

4.3.1.5 Database of Adverse Event Notifications (DEAN)

DEAN is a database of adverse events to medicines, vaccines and biological therapies prescribed or used in Australia. This information is used to detect signals. Pharmaceutical companies must report all serious adverse events suspected of being related to their products (22).

4.3.1.6 Japan Adverse Drug Event Report Database (JADER)

JADER is an SRS of drug adverse events in Japan, managed by the Pharmaceuticals and Medical Devices Agency (23).

4.3.1.7 Pharmacovilance Program of India

This is under the Indian Pharmacopoeia Commission, Ministry of Health and Family Welfare, Government of India. It started in 2010 and receives reports of ADEs from HCPs, patients, ADR monitoring centers and MAH. Until July 2023, it has reported 721559 ICSRs and occupies the eighth position in terms of the number of ADRs being reported to the global database VigiBase (24).

Data mining of SRS databases has helped in identifying drug ADE signals, leading to regulatory action (25). Some examples include the association of ventricular arrhythmias with terfenadine and cisapride, hemolytic anemia with temafloxacin, cardiac valvulopathy with fenfluramine, hepatotoxicity with propylthiouracil, cholestasis with flucloxacillin-induced hepatitis (11, 12, 25, 26). Association between febrile seizures and the Fluzone vaccine in young children, intussusception with RotaShield rotavirus vaccine in 1999 and association between a meningococcal conjugate vaccine and risk of development of Guillain-Barre syndrome are someother examples of signals generated with regulatory action observed through data mining of the VAERS database (12).

Advantages

i. The ADEs are reported spontaneously and voluntarily by individuals.
ii. They contain detailed, relevant information about the case as written by doctors, nurses and pharmacists. Doctors are trained to evaluate medical histories, drug exposures and outcomes. Nurses and pharmacists as HCPs are also in a better informed position for reporting ICSRs.
iii. The ICSRs are peer-reviewed. This forms a quality control.
iv. It may be possible to go back to the case reported to retrieve more information.
v. They are reproducible as a record of the ICSR is available.

Limitations and Challenges

i. Many reports may go unreported as the process is dependent on passive reporting. Studies have shown that up to 98% of serious ADEs may go unreported (27).
ii. Reports may be incomplete. Prescribing decisions are often influenced by factors that affect clinical outcomes, insurance and access to care. Hence, the required information may not be available.
iii. Quality of SRS databanks may vary by region and knowledge and skill of the person entering the data.
iv. There may be duplication or overlap of reports as the same reports may be reported in national and international databases. This is especially true for serious or severe ADRs that are often reported to national or regional authorities, as well as to the global repository VigiBase.
v. Continually reporting of ADRs requires continuous analysis.
vi. Delays may occur before a drug-ADE is detected.
vii. The reports are based on anecdotal data points and may not support conclusions for larger populations.
viii. Usually cannot measure the total number of exposures as there is no denominator to estimate the frequency of adverse events.
ix. Domain-specific knowledge and technical understanding of data mining methodologies are required for interpreting the information in an SRS database.
x. Incorrect hypotheses generated from erroneous or incomplete adverse event report data can be costly, with false positives resulting in resources wasted on unnecessary studies and false negatives leading to harm to patients (3).

4.3.1.8 Current Status

Digitalization of the SRS has helped in the growth of the databases with millions of reports of ADRs being reported. Despite the large databases, SRS may still not be universally considered Big Data, but they are the most important component of the signal-generating system available for ADRs to date.

4.3.2 ELECTRONIC MEDICAL/HEALTH RECORDS

Electronic medical records contain computerized information about a patient that is gathered during routine clinical care. They include details of a patient along with their medical history, treatment history, laboratory investigations and follow-up details in a specific format. This information can be easily shared among different HCPs or be accessed by them at different locations (29). EMR databases contain real-time and real-world patient data, which is extensive concerning large populations of patients unlike SRS databases that are dependent on ICSRs. EMRs are being used as a source of data mining for monitoring drug safety (30, 31).

They provide an opportunity to identify the magnitude of a drug safety problem. A cohort of patients can be identified taking the same drug for the same disease and analyzed to determine the extent of a suspected association between a drug and a particular ADE or health outcome (30). This healthcare data in EMRs can be used to validate signals identified through mining SRS databases.

To facilitate the use of EMRs in PV, EMR databases are now connected through DDNs, for example, Sentinel DDN. This will be reviewed in Section 4.3.4.

Studies have shown that combining safety signals from different sources can improve the accuracy of signal detection (32, 33). A study comparing signal generation of ADE with disease-modifying drugs in rheumatoid arthritis patients by combining FAERS and EMR or each method alone showed that recall greatly increased when FAERS was combined with EMR compared with FAERS alone and EMR alone. In addition, signals detected from EMR considerably overlapped with signals detected from FAERS or ADE knowledge bases (34). Comparison between SRS reports and EMR data for identifying drug–drug interactions (DDIs) signals showed that unstructured narrative data in EMRs can yield better results for some PV procedures than the mostly coded, structured data in SRS reports and data mining a combination of structured and unstructured data for DDIs may provide signals with higher statistical confidence levels (35). Use of EMRs for PV offers the following.

Advantages

i. It is efficient with respect to manpower and time required to complete the analysis.
ii. Quality of data is high because they are created and maintained by HCP.
iii. In EMR, since data collected is longitudinal, ADRs that have a long delay, for example, cancers can be identified, especially in databases that have a low patient turnover and long follow-up.
iv. Since data is real-time and real-world, the data will evolve with new treatments and medications prescribed. Mining EMR may detect new risks due to off-label uses, changes in indication or the ways older drugs are used.
v. Useful for detecting ADRs that are unpredictable drug ADEs, which are underreported in SRS databases because they are not suspected to be drug-induced.

Limitations and Challenges

The primary use of EMRs is for a patient's individual care and administrative billing purposes. Using it for any other purpose is fraught with challenges (28, 30, 35):

i. *Correctness*: It refers to the accuracy of the collected data that is directly linked with its initial documentation. EMR data is collected through routine clinical practice where the clinician's priority is to collect information as per their requirements and administrative needs. Errors can occur, and data accuracy of EMRs ranges between 44% and 100% (36).
ii. *Incompleteness*: EMR may not contain complete patient history as patients may not reveal all to one HCP in one institute.
iii. There is a lack of standardization as EMR data is handled by different people at different locations who are entering, storing and processing the data. This can lead to inconsistency and variability in the clinical notes and terminology used within a healthcare system for individual clinical parameters, making it difficult to extract data and analyze through standardized elements and software.
iv. It is based on unstructured narrative information where the exact meaning may be different from what is interpreted.
v. It is associated with confounding control and biases as a lot more non-standardized background information will be there. The narrative will be dependent upon the writer who may have certain reasons for entering or not writing specific information.
vi. Unpredictability about patient compliance with the treatment limits the use and extension of data.

   vii. Reliability, due to all the above reasons, may be an issue.

  viii. Signal detection may be low for drugs that are used infrequently.

    ix. Sample size may be too small. There may be too few cases in an EMR database to analyze a particular ADE. This may be a problem with orphan drugs and drugs that are used infrequently.

     x. There is a lack of access to a patient's EMR due to patient privacy measures. As a result, data concerning small populations may cause false positives or false negatives.

    xi. At present what methodology is most appropriate for application to a particular database to answer a specific question is still evolving. It is not clear how the inconsistent findings that may arise from the use of different methodologies and databases should be interpreted (14).

   xii. EMR data may be locked in institutional silos on systems that are unique for each provider and institution and suffer from bias related to their primary purpose, i.e., medical and legal records. If a health system is fractionated, it may be difficult to track patients across different healthcare systems.

  xiii. Using patient data may result in confidentiality and privacy issues. A patient's sensitive information about disease and drug use patterns may become public (37, 38).

  xiv. There may be security issues as EMR data are vulnerable to cyber threats (39).

4.3.2.1 Current Status

EMR data as a source for data mining is used to complement the analysis of SRS databases. They can be used for the validation of drug ADE signals that have been initially detected in SRS databases.

4.3.2.2 The Future

EMRs as a data source for signal generation of ADR offer many possibilities. Use of EMRs to conduct post-approval clinical research is being envisaged. EMRs will be used as a primary data source for post-marketing observational and comparative effectiveness studies (31). Data mining will be able to analyze data from millions of EMRs to identify a potential drug-ADE association, unlike a clinical trial that has a limited number of subjects (30). It is thought that this will require fewer resources, decrease costs involved, save time and will have the potential to yield more consistent results based on RWD (40).

But for using EMRs to justify fewer, shorter randomized clinical trials, the quality and validity standards for such data mining studies to be able to identify drug-ADR/ADE associations will have to be very high (30). Regulatory decision-making may not permit delaying decisions until information is available from ancillary sources. Clinical trials may still remain the best source for the needed data (12). Data protection regulations need to be strengthened for using EMRs as a data source for PV.

4.3.3 Medical Insurance Claims Data

Insurance claims data contain patient-specific information in relation to medical history, diagnosis laboratory investigations done, treatment given and outcome. This may be used for data mining for drug safety surveillance (30).

While being a possible data source for monitoring information for drug safety surveillance, it has to be remembered that the insurance claims data will provide an insurance claim auditor's interpretation of healthcare data. The coding and terminology used in these records will be biased by reimbursement policies (12). The information in social security or health maintenance organization databases could be influenced by the lack of administrator or payer incentive to record accurate data (12). A study showed differences in the diagnosis-related group coding and classification assigned by the physician compared with that recorded by hospital administration staff (12). The data provided may also be influenced by evolving clinical practices such as changes in disease management guidelines or shifts in preferential prescribing (12). Medical claims data may be able to follow an individual over time and across sites of care but may be limited if the individuals change insurers (15). The medical insurance claims data offers the following.

Advantages

i. Likely to have information pertaining to serious ADRs/ADEs as these would require management and hence increase costs.
ii. Possibility of rare ADRs getting captured.

Limitations and Challenges

i. Availability of less information to establish causality of ADRs.
ii. The coding and terminology used may not be suitable.

4.3.3.1 Current Status

Presently insurance claims databases provide ancillary support to ADR databases.

4.3.3.2 Future Prospects

Expansion of information provided about medication management of patients in insurance claims may add value to insurance claims as a source of data for generating signals to ADRs.

4.3.4 Distributed Data Networks

DDNs are multidatabases that are linked to each other through IT-distributed network architectures. They offer an opportunity over single databases to accomplish similar tasks and address common questions about medical products. Efficiency is brought into the workflow through well-coordinated networks (41).

The availability of electronic sources of patient data collected routinely in EMR, insurance claims, administrative data, disease or product-specific registries has generated interest in their use for the evaluation of efficacy and safety of health and treatment outcomes. These databases are being used in pharmacoepidemiology and PV. Rather than using individual databases to answer the same question, the databases are being connected. This enables the sample pool of data to increase manifold. As a result, the findings are more generalizable. Rare outcomes and ADEs can be investigated. Since the patient database is large, heterogeneous populations exposed to the same product will be there. Safety evaluations can be done in a shorter period as the sample size will be large (41). This would be of relevance in emergency use authorization of medicinal products, as was observed for vaccine approval during the COVID-19 pandemic. They could be linked to genomic data, biological specimens held in biobanks or social media.

The easiest approach for combining data would be to pool data from individual databases into a central pool, where it can be analyzed. This, however, is not always possible, and the approach is fraught with ethical, legal, logistical and administrative barriers (42, 43). There are issues related to the confidentiality of patient data, data security, unauthorized uses of data and administrative issues.

These issues have led to a distributed approach in which databases comprising any type of data are not combined centrally but stored in different physical locations under the direct control of the participating sites. A common data model (CDM) is followed, which includes a set of standardized data files and variables that may be adopted by all data sources participating in a DDN. A common protocol may be adopted that includes a detailed description of the key design and analytic parameters of the study, which has to be followed by all participating data sources.

A centralized distributed database management system (DDBMS) integrates data logically so it can be managed as if it were all stored in the same location. The DDBMS synchronizes all the data

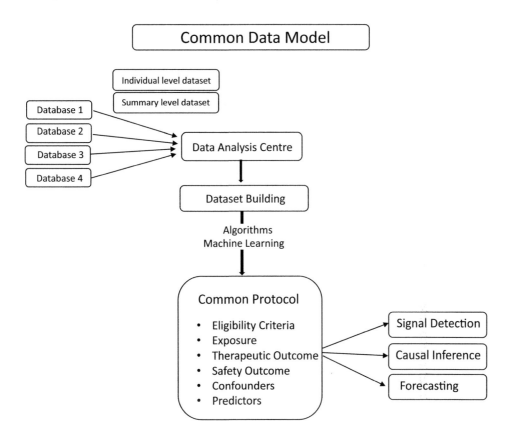

FIGURE 4.2   Distributed data network.

periodically and ensures that data updates and deletes performed at one location will be automatically reflected in the data stored elsewhere (Figure 4.2) (44). The data sent from the individual sites may be an individual-level dataset that contains more information but has issues of privacy and confidentiality, or it could be a summary dataset that has more privacy protection.

The basic features of a DDN are the presence of one or more coordinating center(s), the presence of data partners who maintain physical control of their data and have ability to review and approve each data request as well as review output and freedom to opt out any data request. These features offer data partners autonomy and keep the data close to those who know the data best (43).

Artificial intelligence with ML that can automatically model complex associations in high-dimensional data with minimal human guidance is the technology being applied for data mining from these databases (45). Linkage of patient records allows the integration of several sources of data to give an overall picture of patients across databases. This can be done by preserving patient privacy through the use of anonymized patient identifiers. Globally multidatabase networks have been developed. The number of healthcare databases for post-marketing surveillance is increasing globally (42).

Algorithms in ML are being used to create patient cohorts, generate hypotheses about medicinal product safety and identify high-risk individuals.

i. *Create cohorts*: Create cohorts of individuals with specific biological and clinical features. This enables measuring study eligibility criteria, treatment confounders and outcomes (46).
ii. *Safety signal detection*: Many approaches include approaches used in the analysis of SRS such as disproportionality analysis, pharmacoepidemiologic study designs such as

cohorts of case series and other statistical methods (47). ML is being used in these, for example, BCPNN, to estimate information component, a disproportionality measure (48). Through the estimation of propensity scores, ML can help reduce the effects of potential confounding in signal detection (49). Supervised ML predicts the likelihood of an under-drug event pair being an ADR reaction based on a vector of risk ratios under different cohort designs and deep-learning models to extract mentions of drug-adverse pairs from unstructured clinical text that can be applied also to structured data in EMR and claims data (50–52).

iii. *Causal inference*: It addresses questions of comparative safety and efficacy of products. This is done by framing a hypothesis or question and statistically formulating a causal parameter that answers the question and can be validated from available data. To control for confounding, nuisance functions are estimated. Nuisance functions estimate the probability of the outcome (propensity scores) conditional on a set of observed covariates (53, 54).

iv. *Forecasting*: This can predict the risk of future health outcomes, ADEs and treatment response. The information can be used to put mitigation plans in place to improve patient outcomes (55).

4.3.4.1　Examples of Distributed Data Networks

DDNs have been in existence for more than 20 years. Some examples are as follows.

4.3.4.1.1　*Asian Pharmacoepidemiology Network (AsPEN)*

AsPEN was established in 2008 as a multinational research network of Asia-Pacific countries. It is a collaboration of eight countries involving 12 databases. It works as a Special Interest Group of the International Society for Pharmacoepidemiology and covers a population of 220 million.

Its main functions are to conduct research in pharmacoepidemiology and rapid identification and validation of safety issues (56, 57).

Studies conducted by AsPEN include the use of antipsychotics and the risk of hyperglycemia, heart failure with thiazolidinediones across ethnic groups and cardiac safety of methylphenidate among pediatric patients (58–60).

4.3.4.1.2　*Canadian Network for Observational Drug Effects Studies (CNODES)*

CNODES was created in 2011 and is a distributed network of Canadian research teams whose main purpose is to provide evidence to Canadian stakeholders on the safety and effectiveness of drugs. It has nine data partners and has obtained access to the administrative healthcare data of millions of medication users across Canada and internationally. It covers a population of 35 million currently. It is one of the four collaborating centers supported by the Drug Safety and Effectiveness Network (DSEN) of the Canadian Institutes of Health Research (42, 61). The queries are sent by people to DSEN, which then sends them to the CNODES coordinating center. The CNODES teams analyze DDN and report back (62).

Some examples of studies conducted by CNODES are statin use and risk of heart failure, incretin-based drug use and risk of heart failure (63, 64).

4.3.4.1.3　*Healthcare Systems Research Network (HCSRN)*

It was established in 1994 and was formerly known as the Health Maintenance Organization Research Network (HMORN). It is a consortium of 18 integrated delivery systems and health plans and covers a population of 16 million. It is multipurpose and has many collaborative projects built and maintained into it, for example, Vaccine Safety Datalink, Cancer Research Network, Mental Health Research Network and Cardiovascular Research Network (65–68). Examples of studies conducted by HCSRN are lipid-lowering drug use and the risk of rhabdomyolysis, ADHD medication and the risk of serious congenital malformations (69, 70).

4.3.4.1.4 National Patient-Centered Clinical Research Network (PCORnet)

This was started in 2013. It is a network of networks and includes more than 80 partners. Its networks are Clinical Data Research Networks (13), comprising healthcare delivery systems, People-Powered Research Networks (20), led by patient and caregiver organizations, and Health Plan Research Networks (2) of national insurers and a coordinating center in the USA. It has the data of more than 100 million individuals. It is establishing data networks and procedures for evaluating and ensuring the relevance and reliability of data in clinical research in the US (71).

Examples of studies conducted are aspirin dosing and secondary prevention of atherosclerotic cardiovascular disease (72, 73).

4.3.4.1.5 Pharmacoepidemiologic Research on Outcomes of Therapeutics by a European Consortium (PROTECT)

This was initiated in 2009 and was there till 2015. It was a joint undertaking by the European Union and the pharmaceutical industry as part of the Innovative Medicines Initiative (IMI). Its 35 partners were coordinated by the European Medicines Agency (EMA). The partners include academics, regulators and small and medium enterprises, and the European Federation of Pharmaceuticals Industries and Associations contributed to PROTECT. A network of electronic healthcare databases was established to conduct multicountry and multidatabase drug safety studies. Several partners are in collaboration with other public partners in the European Research Network for Pharmacoepidemiology and PV (42, 74). An example of a study conducted by PROTECT is antiepileptic drug use and risk of suicidal behavior (75).

4.3.4.1.6 Sentinel System

It is a post-marketing surveillance system launched in 2008 in the United States. It was a result of a mandate issued by the FDA for the prospective monitoring of drug safety in a system that relied on spontaneous reporting. It is a network of connected data networks of 18 data partners and collaborating institutions that rely primarily on the use of electronic health data (EMR and US medical claims) (76). It has electronic health claims information that is continuously quality-checked by the FDA of more than 290 million people. Under the Sentinel initiative, the Center for Drug Evaluation and Research (CDER) and the Center for Biologics Evaluation and Research (CBER) perform safety monitoring studies (77). CDER and CBER perform studies in collaboration with other federal partners including the Centers for Medicare and Medicaid Services (CMS) and the Veterans Health Administration. CDER uses the Clinical Practice Research Datalink that captures UK longitudinal patient-level EHR data (REF), CDER uses RWD from the Centers for Disease Control and Prevention National Electronic Surveillance System – Cooperative Adverse Drug Event Surveillance Project. This system is operating in 60 hospital emergency departments across the USA to specifically evaluate drug abuse, misuse and the potential for self-harm. CBER is connected to vaccine safety using the Post Licensure Market Rapid Immunization Safety Monitoring (PRISM) system and for blood-derived products with the blood surveillance continuous Active Surveillance Network. CBER has expanded the Sentinel Biologics Effectiveness and Safety (BEST) system.

It is an example of a complex, longitudinal, observational database network for safety assessment (78). It conducts rapid descriptive and inferential analysis using preprogrammed and customizable analytic tools (79). With data systems encompassing more than half and up to two-thirds of the US population, the distributed database, which is updated quarterly, has information on over 7 billion medical encounters and 6 billion outpatient pharmacy dispensations and is growing at nearly 1 billion encounters per year (71, 80) It conducts assessments of hundreds of products, conditions and product outcome pairs each year.

The Sentinel initiative has been used by the FDA to assess the risk of seizures with ranolazine. FAERS identified a potential signal among ranolazine users. It was evaluated under the Sentinel initiative where a population of older patients, renal disease condition and use of antiepilepsy drugs were identified for further evaluation (81). Evaluation of the risk of venous thromboembolism after oral contraceptives under the Sentinel initiative indicated no substantial increase in risk (82). Analysis of risk of strokes with antipsychotic medications concluded that increased strokes previously observed in the elderly/demented may not exist in the non-elderly/non-demented (83).

4.3.4.1.7   Innovation in Medical Evidence Development and Surveillance (IMEDS)

This program modeled after the Sentinel initiative is available to industry and other researchers as a platform that allows access to a curated Big Data resource. This data resource can be used to evaluate the benefit-risk balance and risk management actions for drugs and biologics. All information available in Sentinel is available in this database for regulated industries through IMEDS.

It provides a collaborative platform to coordinate public-private partnerships, The foundation partners with the Harvard Pilgrim Health Care Institute. The Harvard Pilgrim Analytic Center collaborates with the independent data partners (integrated delivery systems and/or insurers) to provide scientific input (84).

4.3.4.1.8   Vaccine Safety Datalink (VSD)

This was established in 1990 under the funding sourced from the US Centers for Disease Control and Prevention. It monitors the safety of vaccines. The databases it uses are EHRs forming a network of nine delivery systems and health plans and has a population of 9 million under it in the United States. The program is able to provide real-time surveillance of vaccine safety for both routine vaccinations and emergency vaccination campaigns. The program has demonstrated non-association of childhood vaccines with autism or other developmental disabilities. It also monitors vaccine safety in pregnant women (85). Examples of vaccines evaluated are the safety of H1N1 and seasonal influenza vaccines, quadrivalent human papillomavirus vaccination and the risk of Guillain-Barre syndrome (86, 87).

4.3.4.1.9   Observational Medical Outcomes Partnership (OMOP)

This was a partnership between the U.S. FDA and the Pharmaceutical Research and Manufacturers of America. It was established to inform the appropriate use of electronic healthcare databases for studying the effects of medical products. The initiative created a common data model (CDM) and a suite of CDM-compatible analytic tools. The model enables the capture of patient information in a standardized way across different institutions and organizations. The work of OMOP continues within other research organizations (88, 89).

4.3.4.1.10   Observational Health Data Sciences and Informatics Network (OHDSI)

This is an international network of researchers and observational health databases with a central coordinating center housed at Columbia University. It is a multistake, interdisciplinary collaborative program to bring the value of healthcare data through using large-scale analytics. It has adopted the CDM model developed by OMOP. All the solutions are open source.

Its vision is to create reliable clinical evidence that is freely accessible by utilizing the clinical experience of millions of patients globally (90).

4.3.4.1.11   Others

There are many other data partnership networks and consortia that have emerged and generated safety and efficacy data.

4.3.4.1.11.1   Innovative Medicines Initiative   Initiative Medicines (IMI) was started in 2008 as a public–private partnership (PPP) between the European Union and the European pharmaceutical industry. One of the objectives is the development of tools, standards and approaches to assess the efficacy, safety and quality of regulated health products (91).

4.3.4.1.11.2   Joint International Society for Pharmacoepidemiology – International Society of Pharmacoeconomics and Outcomes Research (JISP-ISPOR)   This is a task force on RWE in healthcare decision-making. It provides guidance on the design and reporting of pharmacoepidemiological analysis of longitudinal healthcare databases (92–94).

4.3.4.1.11.3   Exploring and Understanding Adverse Drug Reactions (EU-ADR)   The EU ADR project was launched in 2008 with the purpose of understanding ADRs by data mining of clinical records and biomedical knowledge. The aim was to produce a computerized integrated system for the early detection of drug safety signals by utilizing the information from various EHR databases (95).

4.3.4.1.12   Study-Specific Databases

The DDN approach has been used in many study-specific networks (Table 4.2). These databases are being established by using standardized protocols for data collection among different partners. The terminologies to be used while filling the data are defined. The specified data has to be anonymized and shared. Emphasis on International Good Practice Guidelines with the maintenance of transparency in conduct and reporting of studies is encouraged. EMA has made it a requirement for companies to register and disclose protocols and study reports of Post Authorization Safety Studies (PASS).

These systems are being viewed as a national resource, and public-private partnerships have been created to permit use by stakeholders other than drug regulatory authorities. The outcomes are available as an open source on many networks for use.

The network's databases are being used in different areas of PV, management of ICSRs, aggregate reporting and signal management.

DDNs can support analyses where a lot of data is needed, which may not be possible with a single database. Broadly these are (i) assessments of rare exposures, for example, use of drugs for orphan diseases; (ii) assessment of rare outcomes, for example, the occurrence of Guillain-Barre syndrome following human papillomavirus vaccination; (iii) assessment of treatment heterogeneity, for example, to study drug effects and safety in different subsets of populations such as children, elderly and people with specific diseases; and (iv) post-marketing surveillance of newly approved medical products (42, 87).

Health and Administrative registries in Denmark and Sweden were used to study the prevalence of congenital malformations among infants exposed and not exposed to varenicline in utero. The data on varenicline use and congenital malformations in offspring were obtained by linkage to nationwide registries of dispensed prescriptions and hospital admissions over a period from 2007 to 2012. The analysis showed that the use of varenicline during pregnancy does not appear to increase the risk of major congenital malformations or other adverse birth outcomes (29).

Claims data and EMR were used to study pregnancy and infant outcomes including major congenital malformations among women with chronic inflammatory arthritis (cIA) or psoriasis (PsO) with and without etanercept use. In this claims-based data, delineated pregnancy exposures and outcomes of live or nonlive births among women with cIA and PsO (etanercept exposed, unexposed) with general population (GP) as comparators were analyzed. Infant outcomes were determined for live-born infants covered by the mother's insurer. Medical records were obtained from all accessible mother–infant pairs. Overall the study did not identify any new safety concerns associated with the use of etanercept during pregnancy (103).

During the COVID pandemic, multiple emergency use authorization of medicines and vaccines was granted by drug regulatory bodies in different countries. Subsequently, the evaluation of safety and efficacy was done by analysis of data through DDNs.

Comparison of Longitudinal Observation Databases (LODs) to spontaneous reports for signal detection has shown that at present, LODs can complement the existing SRS for the generation of signals.

TABLE 4.2

Study-Specific Distributed Database Networks

Organization Network	Features	Reference
Arrhythmogenic Potential of Drugs ARITMO Project	A project to assess the arrhythmogenic potential of drugs. The analyses of healthcare data, epidemiological data and pharmacovigilance data were performed in a distributed fashion by using a common remote research environment (RRE) that is located at Erasmus MC, the Netherlands. The RRE allows for loading, retrieving, extracting and transforming of the data.	(96)
Accelerated Development of Vaccine Benefit–Risk Collaboration in Europe(ADVANCE)	The ADVANCE project is a partnership between European pharmaceutical companies, the medicine regulators and the public health bodies that collect data on infectious diseases. The ADVANCE project was initiated after the 2009 H1N1 pandemic to gather evidence about the safety and effectiveness of the approved new H1N1 flu vaccines. ADVANCE developed and tested new methods and data analysis tools to create a wide system that can generate solid evidence on how vaccines are faring once on the market. ADVANCE has led to Vaccine Monitoring Collaboration for Europe (VAC4EU), an international non-profit collaboration platform that acts as a vaccine safety and monitoring system.	(97, 98)
The Safety of Non-Steroidal Anti-Inflammatory Drugs (SOS)	This study assessed the relative cardiovascular and gastrointestinal safety of non-steroidal anti-inflammatory drugs (NSAIDs) from clinical trials and published observational studies, followed by the design and conduct of a multicountry study in existing healthcare databases in the UK, the Netherlands, Germany and France, comprising medical information on at least 35 million persons.	(99)
European Concerted Action on Congenital Anomalies and Twins (EUROCAT)	It is a high-quality network of population-based congenital anomaly registries across Europe for the monitoring, surveillance and research of congenital anomalies.	(100)
Safety Evaluation of Adverse Reactions in Diabetes (SAFEGUARD)	A multidisciplinary consortium of experts has been established to collaborate on evaluating the cardiovascular, cerebrovascular and pancreatic safety risk of the type 2 diabetes mellitus drugs (T2DM). This will be done by investigating (1) published clinical trials and observational studies; (2) spontaneously reported adverse event reports in national and international pharmacovigilance databases; (3) data from nine population-based healthcare databases in six countries capturing longitudinal drug exposure and event data on more than 1.7 million T2DM patients.	(101)
Global Research in Pediatrics (GRIP)	This is a network of experts in pediatric medicine research. It comprises institutions and major networks representing hundreds of clinical sites across Europe, the United States and Asia. Integration of the WHO, EMA and the National Institutes of Health – National Institute of Child Health and Human Development (NIH-NICHD)- associated networks, including the FDA, is an asset for an effective implementation of the network activities without duplication and rapid conversion into activities that will provide children with safe and effective medicines.	(102)

In an EU ADR project, a comparison was made between the use of EHR databases and SRS for drug safety signal detection. The observations showed that SRS could explore as potential signals a larger number of drugs for the six events in comparison to EU ADR (630–3393 vs. 87–856), particularly for those events commonly thought to be potentially drug-induced (i.e., 3393 vs. 228) (104). Use of DDNs in signal detection offers the following advantages and challenges.

Advantages

i. Provides a large source of data
ii. Efficiency due to nonduplication of data analysis and saving of time
iii. Global analysis of data for generating signals

The overall availability of large networked data systems coupled with the ability to gain insights into a study question within days, rather than months or years, is the promise of Big Data most likely to be fulfilled (80).

Limitations and Challenges

i. *Nature of data*: Broadly two types of data are entered: structured and unstructured. There may be an intermediate type. Structured data is well-defined, objective and easier to analyze. Unstructured data as is there in case histories, discharge summaries and investigation reports is difficult to process and analyze. Most DDNs use structured data from their source systems to analyze. But in DDNs utilizing EMR databases, structured and unstructured data often coexist, and it is difficult to extract information from these and analyze. The unstructured data at the moment is underutilized due to difficulties in extracting objective information from it (105). The use of ML and NLP in developing methods to promote the use of text in analyses is increasingly being done (106).

ii. *Standardization of data*: Variations in data quality, data completeness, coding terminologies used, content of data entered and patient population can affect the quality of analysis and time required for the same. This problem can be there in any system requiring electronic entry and use of data. The problem however gets compounded in multidatabase studies. The use of CDM which specifies the way data must be entered by different databases facilitates reproducible implementation of analysis across different data sites. It can reduce the time required for the correction of coding errors and programming (107). Data sites are gradually shifting to use standardized CDM.

iii. *Granularity of shared data*: The ways in which individual datasets of different sites are shared with the central coordinating center can vary. Data can be shared in different ways. (i) The site shares its individual-level dataset with the analysis center that combines all individual datasets to make a single centralized dataset. This is the most granular level of data sharing. It leads to maximum analytic flexibility but the least protection of patient privacy (42). The other way is for the site to share only the summary-level data. Here the analysis will be restricted as it will be difficult to combine and process the individual summary-level data from different sites (108).

iv. *Incomplete linkage of data* from primary sources.

v. *Privacy and confidentiality concerns* of patient data shared: measures to deidentify patient data are required (42).

vi. *Data governance and management*: To ensure data confidentiality, data use agreements and established collaboration between sites and trust have to be there. A multicenter ethics review will be required. More communication between the coordinating and other sites may be required.

vii. *Quality of data:* Heterogeneity between databases due to differences in data encoding, content and variations in patient characteristics can exist. Quality assessments are needed continuously in a systematic way (109–111).

viii. *The lack of learned reporters* who suspect a medicine's adverse event relationship and are able to provide detailed medical information and a rationale for that suspicion.

ix. *Delays* in update of data.

x. *Regulatory considerations* over tools that measure objective medical data.

xi. *Technology use:* Siloed storage and the diverse nature of the databases in DDNs create unique challenges for using ML. Different models will be required based on the heterogeneity of datasets (108).

xii. *Statistical analysis*: The conduct of statistical analysis may be more complex due to non-access to some site data and the presence of confounders.

xiii. *Confounding*: They may still be confounding as they are controlled at individual sites. Methods to control confounding situations will have to be incorporated into the analysis.

xiv. *Financial resources:* A large amount of financial resources are required for the creation of DDNs and maintaining them.

4.3.4.2 Possible Ways of Overcoming Problems

Having a common data model with or without a common protocol: By using a CDM as in Sentinel and OMOP DDNs. Some information can be lost during standardization, especially when data elements are available in different coding systems (42). This may require input from researchers from different sites. A common protocol will aid the process of analysis.

Having a common protocol without a common data model: This can lead to variations in the interpretation of the protocol, which may increase the heterogeneity across sites, affecting the quality of analysis. Each site will require programming resources to conduct its own analysis and intensive coordination across sites will be required to ensure consistency in analysis. For DDNs that have different languages or coding systems, a common protocol can increase the work (42). It has a detailed statistical analysis plan and a phased analysis, where analysis reviewed at specific stages may help.

Developing a CDM: Developing a CDM requires resources in terms of time, technical know-how, human resources and finances. Maintenance of the CDM requiring upgrading when new analytics are required will require more effort. The design of the CDM will depend upon what the researchers at the site want, i.e., the extent of preprocessing of the information upfront and how much to be done when conducting the study. Preparing a CDM for databases that contain the same type of information and coding system is easier than developing a CDM for disparate data sources. While more resources are required in the beginning, the CDM achieves an economy of scale when the number of studies supported by the CDM is sufficiently large (42, 112).

4.3.4.3 Current Status

DDN complements traditional SRS for signal detection, especially for those adverse events that are frequent in the general population and are not commonly thought to be drug-induced. These datasets have more use for exploratory assessment of drug safety rather than for the usual signal evaluation.

Besides medicinal product safety surveillance, the DDNs are being used for pharmacoepidemiological research, comparative effectiveness research, patient-centered outcomes research, public health surveillance and quality improvement (42).

4.3.4.4 Future Prospects

Standardization of protocols and terminology used, with advancement in analytics, will advance the potential use of DDN in generating signals in drug safety. Considerations of performance, price and privacy of data will determine the generalizability of the DDNs model framework.

Patient data will be accessed through various mobile wearable devices. The Innovative Medicines Initiative WEB RADR developed a mobile application that allows patients to directly report potential medicine side effects and also receive reliable information on their drugs (113). In the future,

DDNs will include data of the same individual collected through different modalities, such as EMRs, claims bases and mobile devices. Ways to ensure transparency and reproducibility of methods used and observations will increase confidence in the process. A national or international infrastructure to support multiple DDNs will facilitate evidence-based delivery of healthcare (42, 92).

4.3.5 SOCIAL MEDIA

Another data source for PV that is under consideration is social media. Social media refers to a variety of technologies that facilitate the sharing of ideas, thoughts and information through virtual networks and communities. It comprises heterogeneous data sources ranging from postings on social media applications such as Facebook and Twitter to chatrooms and blogs to compilation of search engine logs. Millions of people use and interact with each other. The participants are from diverse backgrounds. The data on social media is large and continuously growing. In early 2023, more than 4.7 billion people used social media, equal to roughly 60% of the world's population, 94.8% of users accessed chat and messaging apps and websites, followed closely by social platforms, with 94.6% of users (114).

Social media is used by the public for sharing information. In relation to health, patients share experiences concerning medication use and problems, ADEs and reports of the use of medicine prescribed or self-medicated. Patients and their caregivers consult the internet to seek information and interact with others through online discussion sites for health information. Social media, including social networks, chatrooms, health blogs and patient community websites provide a more patient-centered model of ADE reporting than SRS or EMR databases. A large volume of diverse data accumulates with speed. The concept of "proto AE" or a post that resembles an AE has arisen (115).

These can enable in identifying patterns in behavior in relation to drug use, DDIs and ADRs. A recent survey showed that about 3%–4% of internet users had publicly shared their concerns about ADR on social media sites (116).

Data mining using NLP of this unstructured data is being used to extract and classify information from social media sources to enable further processing of ADEs (115, 117). NLP algorithms are being continuously developed, and one output from the WEB RADR project has been the development of a "gold standard" reference that can be used to enhance the analytics of social media (118).

Social media reports are in real time, and they provide an opportunity for earlier detection of drug safety issues that may occur by analyzing SRS databases. A series of shared similar experiences in the community may trigger a spike in reports, serving as an early warning system for ADEs. ADEs not reported to SRS may be captured. ADEs for drugs that treat conditions that have a social stigma, which may be underreported in other sources, may be picked up. Even DDI not yet documented, such as between dietary supplements and drugs, may be observed (35).

Mentions of drugs and drug combinations have been detected on Twitter and Instagram. Co-mentions of drugs and potential ADEs were identified for daily, weekly and monthly periods (119–121). Websites that host large patient communities such as MedHelp (122), Daily Strength (123) or PatientsLikeMe (124) may be of use for data mining for drug-ADE signals and other health information (35). Search engine logs can be analyzed to detect potential drug safety signals by performing text mining and frequency analysis of search queries that concern drug ADE associations (14, 35). One study that analyzed the search logs of 80 million users was able to detect a disproportionate number of searches regarding a specific drug in comparison with a previous period (125). Search logs may detect increased public activity in response to a potential ADE, with people searching for information about a drug (125).

Drug regulators have shown interest in mining this data source for ADEs (126, 127). The Association of British Pharmaceutical Industry published guidance on the management of adverse events and product complaints sourced from digital media (128). The US FDA has not issued such guidelines but has regulations for publishing promotional material and risk/benefit information on social media (129).

In Europe, the guideline on good PV practices module VI states that if an MAH becomes aware of a report of a suspected adverse reaction described in any noncompany-sponsored digital medium, the report should be assessed to determine whether it qualifies for submission as an ICSR. Unsolicited cases of suspected adverse reactions from the internet or digital media should be handled as spontaneous reports and the same time frames as for spontaneous reports should be applied (130).

The use of social media in PV has been analyzed in many studies. A two-year retrospective feasibility study evaluated the contribution of social media data (Twitter, patient and other online forums) to safety signal detection. The social media proto AEs were on a per-product basis, 13.58% serious and in FAERS were at least 28.2% serious for nine of ten products in the analysis dataset. Five of six social media signals of disproportionate reporting were identified sooner in traditional PV data sources. The study did not demonstrate the utility of performing routine signal detection on social media data (131). It has been observed that the volume of events from social media is substantially smaller than the reporting volume from the company safety database or FAERS. The IMI WEB RADR project included 38 products across multiple manufacturers and was able to assess a dataset of 40,000 tweets reporting 56,000 product event combinations in the WHO VigiBase. Other studies have concluded that analysis of social media data did not identify new or previously identified signals (117, 131). Another study showed the performance of established statistical signal detection algorithms on Twitter/Facebook for a range of drugs and adverse events, using currently available methods for adverse event recognition was poor (132).

Other studies have demonstrated that combining safety signals from SRS with internet search logs may improve the accuracy of signal detection (125, 133). A study showed that combining SRS with signals from social media Twitter improved the accuracy of signal detection. The combined system did not achieve better performance data than FAERS alone, indicating that Twitter data are not ready yet for integration into a purely data-driven combination system (134). More studies are needed to better inform the exact role of social media data for signal generation.

The WEB RADR has given a recommendation that data mining activities be classed as a secondary use of data with no planned interactions with patients (135).

Social media as a data source for signal detection in PV offers the following.

Advantages

i. Contain lifestyle information like exercise patterns, eating habits, socioeconomic issues and/or drug abuse behavior that may not be there in EHR.
ii. Systems may capture information pertaining to at-risk populations in whom clinical drug trials do not take place routinely and in whom drug safety is not established, for example, pregnant women. A study using data from Twitter accounts of pregnant women observed a higher medication intake in women who reported birth defects (136). Another study developed a NLP method to identify tweets by users whose children had a birth defect (137).

Challenges

The nature of social media presents several challenges for the extraction of signals related to PV. These include the following:

Specificity: The primary purpose for which most social media posts or communication are there is other than to report a suspected link between a specific health intervention and an adverse event. Most reports that mention the use of a drug and potential adverse events do not meet the basic regulatory definitions of an ICSR. An ICSR is considered valid if it contains an identifiable patient, an adverse event, an identifiable suspect drug ADE and has an identifiable reporter. The information given in narratives may be ambiguous, and it may be difficult to know what it means. For example, a statement regarding how a patient is feeling "I am not feeling well" could mean the patient is unwell or not feeling well as a result of a medicine intake.

Reliability of information: Very few social media posts are relevant to PV, approximately 0.2% of tweets mention a medicine. Information is represented in unstructured text, medicine names are often misspelled, abbreviated or discussed using slang (3); mentions of medical events may not be firsthand accounts.

The *informal* nature of social media reports can make it difficult to distinguish between AE's potential benefits and underlying disease symptoms. Information may not be reliable and can result in false positives.

Reproducibility: Social media posts experience high deletion rates with more than 40% of posts from one study being deleted from the platforms after the study was published (138). To ensure reproducibility, researchers will have to preserve their own copies of data used for a particular study.

Follow-up information: Data from social media are obtained through third-party providers in aggregate without any identifiers. This limits the ability to follow up for further information.

Validity: Patients and caregivers are generally not qualified to diagnose a medical condition; therefore, they are more likely to make posts on social media that are inaccurate, misinformed or incomplete. A suspected medical diagnosis by someone who is not an HCP cannot be taken as proof of an ADE (14).

Difficulty in extracting an ADE from unstructured text.

Less detectability: Data mining for terms used in social media text searches may infrequently (less than 2%) identify drug ADE associations occurring at a rate five to ten times lower than that detected by searching structured data sources (10,139). Information from social media that has not been screened by an HCP may have a lot of background noise, which will hinder signal detection. The overall value proposition of social media has been questioned due to the low prevalence of posts relevant to PV and low coverage for many drugs (140).

Ethics: Social media posts may contain sensitive information about users such as illicit drug use or mental health issues. Balance has to be maintained between ethical concerns, privacy of data and making research reproducible and available.

Legal: Many platforms limit sharing of data collected from their users and require that content be deleted upon user request.

Statistical challenges include lack of specificity, verification difficulties, low validity and bias. The use of a proportional reporting ratio as a method of signal detection in social media has not yet been confirmed. Other methods have to be specified (141).

Geographical: It may be impossible to determine if a social media report is from a geographic region that is relevant to a particular regulatory agency's PV efforts.

Duplication of reports: Data mining social media may identify a duplicate patient-reported record of a suspected ADE that has already been reported through conventional PV channels.

Bias: Social media introduces an inherent bias toward patients younger in age, as well as cultural and socioeconomic groups who have access to electronic devices and the technical ability to use them. Mining social media may exclude older or sicker patients who experience a greater risk for ADEs associated with their medications.

Low value: The low-quality data contained within this source will not provide value for signal detection for drugs that are commonly used. Such reports may lack sufficient details to conduct a medical assessment or describe events already included in the approved product label. These reports are unlikely to provide new medical data relevant to benefit: risk assessment.

Confounders: Mining internet search logs for drug safety signals is the assumption that a search for drug information was made because the patient experienced that event. The assumption that all searches involving terms for a drug and ADE serve as a proxy for drug exposure will confound data mining results. The search may be for information-seeking queries rather than a report of an actual event. This was realized when the web service Google Flu Trends signaled a flu outbreak based on a surge of flu-related internet searches. This was false as most people were researching for information reported in media, rather than those who were ill (142).

It is **difficult to investigate every potential symptom** reported as an ADE. It may be considered too intrusive, with privacy issues and lack of contact information or time required.

4.3.5.1 Current Status

Despite the hype and research activity around social media data streams being able to enhance signal detection capability, evidence of value and practical impact is limited to date. At present social media sites are being explored for potential as sources for data mining for ADEs. More studies are needed to establish their role.

4.3.5.2 Future Prospects

More advanced report identification methods will increase the value of social media data. Social media generated in the context of a drug or health-oriented platform, e.g., drugs.com vs. a general platform, e.g., Twitter may hold some value. FDA has collaborated on the development of an exploratory data mining tool called MedWatcher Social that monitors several social media platforms for potential drug ADE associations. The system is rooted in the well-established FDA MedWatch system. MedWatcher Social uses an indicator score approach to identify adverse events (143). Ongoing initiatives such as IMI's WEB RADR and the EMA's goal to measure the impact of PV practices will likely identify the best uses of these data for PV.

4.4 BIG DATA AND ROLE OF TECHNOLOGY IN USE OF ELECTRONIC MEDICAL RECORDS, DISTRIBUTED DATA NETWORKS, MEDICAL INSURANCE CLAIMS AND SOCIAL MEDIA IN PHARMACOVIGILANCE

Technology has been a catalyst in the PV process. The evolution of information technology, artificial intelligence specifically ML learning and deep neural networks is assisting the use of Big Data, data with a large volume, high velocity, variety and veracity. Improvements in computing power and speed have allowed the automation of drug safety surveillance signal detection in large complex databases. Large networks of SRS databases and digital sources connected electronically, which permits the identification of possibly related discrete data points, identify new associations among drugs, ADEs and risk factors (10). The Big Data approach to PV is exhibiting speed, cost-effectiveness and capacity to identify unsuspected relationships regarding drug-ADE associations (11, 30).

For the industry currently the available data may not be able to substitute the need for clinical trials and post-marketing surveillance for drug safety monitoring. But the development of automation, cognitive technologies and advanced analytics for PV may help in strengthening the PV system and decrease costs. Data mining may assist the pharmaceutical industry in detecting drug safety signal results earlier, conducting risk assessments and gathering RWD for the interpretation of clinical results (144). Proactively collecting RWD from SRS reports, EMR, literature and social media may contribute to a pharmaceutical company's risk management plan (145).

Data mining methods may help understanding the biological basis for signals by incorporating other databases on drug chemistry, physiology, pharmacogenomics of drugs and how drug-receptor interactions can cause ADE. This can lead to predictive PV (14). Newer technology platforms being developed involving the use of mobiles to capture data about ADEs and patient-related outcomes outside of the hospital are a resource that can be expected to add value as a data source for monitoring ADEs (146).

4.4.1 OVERALL CHALLENGES IN USE

Even if a signal is identified through the process of data mining, it will have to be verified to establish its causality. Other types of data will have to be looked at. This was seen when the FDA found an association between the use of statins and amyotrophic lateral sclerosis. Data mining identified 91 cases, while clinical trial data involving 200,000 patient years did not validate this observation, showing 9 cases with the drug and 10 with placebo (147).

Standardization: Absence of established validated statistical methods, variable terminology and few validation studies. Standardized vocabulary is essential for analyzing data from variable sources (14, 30).

Quality of data: The quality of the data source is dependent on the way the data has been structured and entered. This is also dependent on the expertise of the person entering the data. Since most data entry is done manually, it can result in inconsistent quality (11). The lack of established standards and validation methods, confounding variables, false signals, data inconsistencies, bias or too much or too little data are other issues related to quality. This happens because the data used has been repurposed from its original function (14). For example, MeSH indexes in MEDLINE are presumably of higher quality than social media because they are structured and created by linguistic and cognitive experts (11).

Privacy: Data privacy is an important issue. The user is responsible for maintaining the anonymity of information about the patient's identity. Currently, work is needed to ensure confidentiality of patient data.

Legal: Laws on personal data protection have to be developed and complied with. Regulations should be made to see how sharing of data across borders can be done while safeguarding the rights of patients as well as coordinators of databases.

Technical: Technical resources and software to handle data related to PV can produce objective analysis and conclusions across different databases. Deficiencies in existing programs have to be overcome.

Costs: Developing IT resources and implementing them will require financial resources. This may be difficult in low-to-middle-income countries.

Ethics: Using patient data across EMR and medical insurance claims databases may involve the use of personal details that may not be acceptable to the individual person. Ways to see how this can be handled need to be developed.

4.5 CONCLUSION

The use of all data sources for signal generation of ADEs must lead to outcomes that impact individual patient's well-being and public health. A transparent and scientific framework for evaluating new data sources and technologies and measuring their impact on PV processes relative to sources and approaches currently used is needed. The exact role of these newer sources as to whether they will complement or replace existing approaches in PV requires assessment. Reliability, reproducibility, validity, cost-effectiveness, usability, sustainability of all new data sources and processes in detecting emerging safety issues have to be established. There is a need to develop a standard medical terminology as well as design an information technology system that enables data submission, retrieval, processing and evaluation in a consistent manner.

The use of consumer-wearable technologies such as fitness devices and smartphones and smart digital technology such as thermometers, glucose and heart monitors of connected devices may provide more detailed health and behavioral information. Data captured through apps will be both objective and subjective. And it is potentially more representative and systematic than social media data streams. It is anticipated that patients whose healthcare is complex, chronically ill and part of organized communities will be early adopters of sharing their information for research purposes despite loss of privacy.

Real-time data proximal to the patient's experience following the use of a medication offers a promise for PV. Notwithstanding the future possibilities, at present ICSRs in SRS are the bedrock on which signal generation of ADEs depends.

REFERENCES

1. World Health Organization. Pharmacovigilance: Ensuring the Safe Use of Medicines. Geneva: World Health Organization; 2004.
2. World Health Organization. WHO Guidelines on Safety Monitoring of Herbal Medicines in Pharmacovigilance Systems. Geneva: World Health Organization; 2004.
3. Lavertu A, Vora B, Giacomini KM, Altman R, Rensi S. A new era in pharmacovigilance: toward real world data & digital monitoring. Clin Pharmacol Ther. 2021;109:1197–1202.
4. Hauben M, Noren GN. A decade of data mining and still counting. Drug Saf. 2010;33:527–534.
5. Jones JK, Van de Carr SW, Rosa F, et al. Medicaid drug event data: an emerging tool for evaluation of drug risk. Acta Med Scand Suppl. 1984;683:127–134.
6. Hennessy S. Use of healthcare databases in pharmacoepidemiology. Basic Clin Pharmacol Toxicol. 2006;98:311–313.
7. Report of CIOMS Working Group VIII. Practical Aspects of Signal Detection in Pharmacovigilance. Geneva, 2010. Available from https://cioms.ch/wp-content/uploads/2018/03/WG8-Signal-Detection.pdf. [Accessed on January 24, 2024]
8. World Health Organization. Uppsala Monitoring Centre. The use of the WHO-UMC system for standardised case causality assessment. Uppsala. Available from https://www.who.int/docs/default-source/medicines/pharmacovigilance/whocausality-assessment.pdf. [Accessed on January 25, 2024]
9. Adverse Drug Reaction Probability Scale (Naranjo) in Drug Induced Liver Injury Available https://www.ncbi.nlm.nih.gov/books/NBK548069/. [Accessed on January 25, 2024]
10. Ventola CL. Big data & PV: data mining for adverse drug events & interactions. P & T. 2018;43:340–351.
11. Duggirala HJ, Tonning JM, Smith E et al. Use of data mining at the food & drug administration. J Am Med Inform Assoc. 2016;23(20):428–434.
12. Coloma PM, Trifiro G, Patadia V, Sturkenboom M. Post marketing safety surveillance: where does signal detection using electronic health records fit into the big picture? Drug Saf. 2013;36:183–197.
13. Molecular Health Launches Molecular Analysis of Side Effects™ (MASE), a Next-Generation Drug Safety Assessment and Prediction Technology. Available from https://www.prnewswire.com/news-releases/molecularhealth-launches-molecular-analysis-of-side-effects-mase-a-next-generation-drug-safety-assessment-and-prediction-technology-137556013.html. [Accessed on January 25, 2024]
14. Harpaz R, Dumochel W, Shah NH. Big data and adverse drug reaction detection. Cin Pharmacol Ther. 2016;99(3):268–270.
15. Food & Drugs Administration USA. Frameworks for FDAs Real World Evidence Program. 2018. Available from www.fda.gov. [Accessed on January 19, 2024]
16. Triforo G, Coloma PM, Rijnbeek PR et al. Combining multiple healthcare databases for post marketing drug and vaccine safety surveillance: why and how? J Intern Med. 2014;275:551–561.
17. World Health Organization. WHO Guidance for pharmacovigilance. Available from https://www.who.int/teams/regulation-prequalification/regulation-and-safety/pharmacovigilance/networks/pidm. [Accessed on January 14, 2024]
18. World Health Organization. VigiBase/UMC. Available from https://who-umc.org/vigibase/. [Accessed on January 14, 2024]
19. Food & Drug Administration Adverse Event Reporting System (FAERS). Available from https://www.fda.gov/drugs/questions-and-answers-fdas-adverse-event-reporting-system-faers/fda-adverse-event-reporting-system-faers-public-dashboard. [Accessed on January 25, 2024]
20. Vaccine Adverse Event System (VAERS). Available from https://vaers.hhs.gov/faq.html. [Accessed on January 14, 2024]
21. EudraVigilance. Available from https://www.ema.europa.eu/en/human-regulatory-overview/research-and-development/pharmacovigilance-research-and-development/eudravigilance. [Accessed on January 26, 2024]
22. Database of Adverse Event notification (DEANS). Available from https://www.tga.gov.au/safety/safety/safety-monitoring-daen-database-adverse-event-notifications/database-adverse-event-notifications-daen-medicines. [Accessed on January 10, 2024]
23. Japanese Adverse Event reporting Database (JADER). Available from https://www.pmda.go.jp/files/000226212.pdf. [Accessed on January 10, 2024]

24. Pharmacovigilance Program of India (PVPI). Available from https://ipc.gov.in/PvPI/pv_home.html. [Accessed on January 10, 2024]

25. Wysowski DK. Adverse drug event surveillance and drug withdrawals in the United States, 1969-2002. The importance of reporting suspected reactions. Arch Intern Med. 2005;165:1363–1369.

26. Pinnow E, Amr S, Bentzen SM, Brajovic S, Hungerford L, St. George DM, Pan GD. Postmarket safety outcomes for new molecular entity (NME) drugs approved by the food and drug administration between 2002 and 2014. Clin Pharmacol Ther. 2018;104:390–400.

27. Lester J, Neyerapally GA, Lipowski E, Graham CF, Hall M, Pan GD. Evaluation of FDA safety related drug label changes in 2010. Pharmacoepidemiol Drug Saf. 2013;22:302–305.

28. Hazell L, Shakir SA. Underreporting of adverse drug reactions: a systematic review. Drug Saf. 2006;29:385–396.

29. Shah SM, Khan RA. Secondary use of electronic health record: opportunities & challenges. IEEE Access. 2016;4. doi: 10.1109/ACCESS 2020.3011099

30. More TJ, Furberg CD. Electronic health data for post market surveillance: a vision not realized. Drug Saf. 2015;38(7):601–610.

31. Cowle MR, Blomster JI, Curtis LH et al. Electronic health records to facilitate clinical research. Clin Res Cardio. 2017;106:1–9.

32. Li Y, Ryan PB, Wei Y, Friedman C. A method to combine signals from spontaneous reporting systems and observational healthcare data to detect adverse drug reactions. Drug Saf. 2015;38:895–908.

33. Harpaz R, DuMouchel W, Schuemie M, Bodenreider O, Friedman C, Horvitz E et al. Toward multi-modal signal detection of adverse drug reactions. J Biomed Inform. 2017;76:41–49.

34. Wang L, Rastegar-Mojarad M, JI Z, Liu S, Liu K, Moon S, Shen F, Wang Y, Yao L, Davis III JM, Liu H. Detecting pharmacovilance signals combining electronic medical records with spontaneous reports: a case study of conventional disease modifying antirheumatic drugs for rhreumatoid arthritis. Front Pharmacol. 2018. doi: 103389/fphar2018.00875

35. Vilar S, Friedman C, Hripcsak G et al. Detection of drug–drug interactions through data mining studies clinical sources, scientific literature and social media. Brief Bioinform. 2018 Doi. doi: 10.1093/bib/bbx010

36. Hogan WR, Wagner MM. Accuracy of data in computer based patient records. J Am Med Inform Assoc. 1997;4:342–355.

37. Richter G, Borzikowsky C, Lieb W, Schreiber S, Krawczak M, Buyx A. Patient views on research use of clinical data without consent: legal, but also acceptable? Eur J Human Genet. 2019;27:841–847.

38. Shalowitz DI, Miller FG. Disclosing individuals results of clinical research. Implications of respect for participants. JAMA. 2005;294:737–740.

39. Fuentes MR. Cybercrime & other threats faced by the healthcare industry. Trend Micro. 2017;106:2–29.

40. Observational Health Data Sciences & Informatics. Who we are. 2018. Available from www.obdsi.org/who-we-are.accessed. [Accessed on January 26, 2024]

41. Wong J, Alhambra DP, Rijnbeek PR, Desai RJ, reps JM, Toh S. Applying machine learning in distributed data networks for pharmacoepidemiologic and pharmacovigilance studies: opportunities, challenges and considerations. Drug Saf. 2022;45:493–451.

42. Toh S, Pratt N, Klungel O, Gagne JJ, Platt RW. Distributed networks of databases analyzed using common protocols and/or common data models. Pharmacoepidemiology. John Wiley & Sons Ltd; 2019. pp. 617–638.

43. Brown JS, Holmes JH, Shah K, Hall K, Lazarus R, Platt R. Distributed health data networks: a practical and preferred approach to multi institutional evaluations of comparative effectiveness, safety and quality of care. Med Care. 2010;48:S45–S51.

44. Moore L. Distributed database. Available from https://www.techtarget.com/searchoracle/definition/distributed-database [Accessed on January 10, 2024]

45. Beam AL, Kohane Is. Big data and machine learning in healthcare. JAMA. 2018;319:1317–1318.

46. Ricchesson RL, Sun J, Pathak J, Kho AN, Denny JC. Clinical phenotyping in selected national networks: demonstrating the need for high throughput, portable and computational methods. Artif Intell Med. 2016;71:57–61.

47. Arnaud M, Begaud B, Thurin N, Moore N, Pariiente A, Salvo F. Methods for safety signal detection in healthcare databases: a literature review. Expert Opin Drug Saf. 2017;16:721–732.

48. Zorych I, Madigan D, Ryan P, Bate A. Disproportionality methods for pharmacovigilance in longitudinal observational databases. Stat Methods Med Res. 2013;22:39–56.

49. Wang SV, Maro JC, Gagne JJ, Patorno E, Kattinakere S, Stojanovic D et al. A general propensity score for signal identification using tree based scan statistics. Am J Epidemiol. 2021;56:356––368.

50. Reps JM, Garibaldi JM, Aickelin U, Gibson JE, Hubbard RB. Supervised adverse drug reaction signalling framework imitating Bradford Hill's causality considerations. J Biomed Inform. 2015;42:95–97.

51. Liu F, Jagannatha A, Yu H. Towards drug safety surveillance and pharmacovigilance: current progress in detecting medication and adverse drug events from electronic health records. Drug Saf. 2019;42:95–97.

52. Henry S, Buchan K, Filnnino M, Stubbs A, Uzuner O. 2018 n2c2 shared task on adverse events and medication extraction in electronic health records. J Am Med Inform Assoc. 2020;27:3–12.

53. Schneeweiss S, Suissa S. Advanced approaches to controlling confounding in pharmacoepidemiologic studies. Pharmacoepidemiology. John Wiley & Sons Ltd.; 2019. pp. 1078–1107. doi: 10.1002/9781119413431.ch43

54. Brown E, You SC, Sena A, Kostka K, Abedtash H, Abraho MTF et al. Deep phenotyping of 34,128 patients hospitalized with COVID-19 and a comparison with 81,596 influenza patients in America, Europe and Asia: an international network study. MedRxiv Prepr Serv Health Sci. 2020. doi: 10.1101/2020.04.22.20074336.

55. Soyiri In, Reidpath DD. An overview of health forecasting. Environ Health Prev Med. 2013;18:1–9.

56. Lai EC, Man KK, Chaiyakunapruk N et al. Brief report: databases in the Asia Pacific region: the potential for a distributed network approach. Epidemiology 2015;26:815––820.

57. Asian Pharmacoepidemiology Network. Available from https://www.aspensig.asia [Accessed on January 14, 2024]

58. Pratt N, Anderson M, Bergman U et al. Multicountry rapid adverse drug event assessment: the Asian Pharmacoepidemiology Network (AsPEN) antipsychotic and acute hyperglycemia study. Pharmacoepidemiol Drug Saf. 2013;22:915–924.

59. Roughead EE, Chan EW, Chii NK et al. Variation in association between thiazolidinedione and heart failure across ethnic groups: retrospective analysis of large healthcare claims databases in six countries. Drug Saf. 2015;38:823–831.

60. Shin JY, Roughead EE, Parks BJ, Pratt NL. Cardiovascular safety of methylphenidate among children and young people with attention deficit hyperactivity disorder (ADHD): nationwide self-controlled case series study. BMJ. 2016;353:i2550.

61. Canadian Network for Observational drug effect studies. Available from https://www.cnodes.ca/. [Accessed on January 14, 2024].

62. Suissa S, Henry D, Caetano P, Dormuth CR, Ernst P, Hemmelgarn B, Lelorier J, Levy A, Martens PJ, Paterson JM, Platt RW, Sketris I, Teare G; Canadian Network for Observational Drug Effects. CNODES: The Canadian network for observational drug effect studies. Open Med. 2012;6:e134–e140.

63. Dormuth CR, Hemmelgarn BR, Paterson JM et al., Canadian Network for Observational Drug Effects. Use of high potency statins and rates of admission for acute kidney injury: multicenter, retrospective observational analysis of administrative databases. BMJ. 2013;346:f880.

64. Filion KB, Azoulay L, Platt RW et al. A multicenter observational study of incretin based drugs and heart failure. N Engl J Med. 2016;374:1145–1154.

65. Steiner JF, Paolino AR, Thompson EE, Larson EB. Sustaining research networks: the twenty year experience of the HMO research network. EGEMS (Wash DC). 2014;2:1067.

66. Chen RT, Glasser JW, Rhodes PH et al., Vaccine Safety Datalink Team. Vaccine Safety Datalink project: a new tool for improving vaccine safety monitoring in the United States. Pediatrics. 1997; 99: 765–773

67. Chubak J, Zieball R, Grreenlee RT, Honda S, Hornbrook MC, Epstein M, Nekhlyudov L, Pawloski PA, Ritzwoller DP, Ghai NR, Feigelson HS, Clancy HA, Doria Rose VP, Kushi LH. The cancer research network: a platform for epidemiologic and health services research on cancer prevention, care and outcomes in large, stable populations. Cancer Causes Control. 2016;27:1315–1323.

68. Lu CY, Zhang F, Lakoma MD et al. Changes in antidepressants use by young people and suicidal behaviour after FDA warnings and media coverage: quasi experimental study. BMJ. 2014;348:g3596.

69. Graham DJ, Staffa JA, Shatin D et al. Incidence of hospitalized rhabdomyolysis in patients treated with lipid lowering drugs. JAMA. 2004; 292:2585–2590.

70. Habel LA, Cooper WO, Sox CM et al. ADHD medications and risk of serious cardiovascular events in young and middle age adults. JAMA. 2011;306:2673–2683.

71. Fleurence RL, Curtis LH, Califf RM, Platt R, Selby JV, Brown JS. Launching PCORnet, a national patient centered clinical research network. J. Am Med Inform Assoc. 2014;21:578–582.

72. Hernandez AF, Fleurance RL, Rotham RL. The ADAPTABLE trial and PCORnet: shining light on a new research paradigm. Ann Intern Med. 2015;163:635–636.

73. Patient Centered Clinical Research Network Available from https://www.pcori.org/funding-opportunities/announcement/optimizing-infrastructure-conducting-patient-centered-outcomes-research-cycle-1-2021. [Accessed on January 14, 2024]

74. Reynolds RF, Kurz X, de Groot MC, Schlienger RG, Bensouda LG, Lessenot ST, Klungel OH. The IMI PROTECT project: purpose, organizational structure and procedures. Pharmacoepidemiol Drug Saf. 2016;25(suppl 1):5–10.

75. Schuerch M, Gasse C, Robinson NJ et al. Impact of varying outcomes and definitions of suicidality: comparisons from UK Clinical Practice Research Datalink (CPRD) and Danish National Registries (DNR). Pharmacoepidemiol Drug Saf. 2016;25(Suppl 1):142–155.

76. Food & Drug Administration. FDAs Sentinel initiative-back-ground. Available from https://www.fda.gov/safety/FDAsSentinelInitiative/ucm149340htm. [Accessed on September 25, 2023]

77. Ball R, Robb M, Anderson Sa, Dal Pan G. The FDAs sentinel initiative: a comparative approach to medical product surveillance. Clin Pharmacol Ther. 2016;99:265–268.

78. Califf RM, Robb MA, Bindman AB et al. Transforming evidence generation to support health & healthcare decisions. N Engl J Med. 2016;375:2395–2400.

79. Gagne JJ, Han X, Hennessy S, Leonard CE, Chrischilles EA, Carnahan RM, Wang S V, Fuller C, Iyer A, Katcoff H, Woodworth TS, Archdeacon P, Meyer TE, Sebbastian S, Toh S. Successful comparison of US Food & Drug Administration Sentinel analysis tools to traditional approaches in quantifying a known drug-adverse event association. Clin Pharmacol Ther. 2016;100:558–564.

80. Bates A, Reynolds FR, Caubel P. The hope, hype and reality of big data for pharmacovigilance. Ther Adv Drug Saf. 2018;9:5–11.

81. Eworuke E. Integrating Sentinel into Routine Regulatory Drug Review: A Snapshot of the First Year Risk of Seizures Associated with Ranolazine, 2017. Retrieved from https://www.sentinelinitiative.org/sites/default/files/communications/publications-presentations/Sentinel-ICPE-2017-Symposium-Snapshot-of-the-First-Year_Ranexa-Seizures.pdf.

82. Moeny D. Integrating Sentinel into Routine Regulatory Drug Review: A Snapshot of the First Year Venous Thromboembolism after Continuous or Extended Cycle Oral Contraceptive Use [Power Point Presentation], 2017. Retrieved from https://www.sentinelinitiative.org/sites/default/files/communications/publications-presentations/Sentinel-ICPE-2017-Symposium-Snapshot-of-the-First-Year_Lybrel.pdf.

83. Taylor L.G. Integrating Sentinel into Routine Regulatory Drug Review: A Snapshot of the First Year Antipsychotics and Stroke Risk [Power Point Presentation], 2017. Retrieved from https://www.sentinelinitiative.org/sites/default/files/communications/publications-presentations/Sentinel-ICPE-2017-Symposium-Snapshot-of-the-First-Year_Antipsychotic_stroke.pdf.

84. June S. Wasser Executive Director Reagan-Udall Foundation for the FDA) Available from https://reaganudall.org/sites/default/files/2019-12/IMEDS%20DIAglobal%20Article%208-2017.pdf. [Accessed on November 25, 2023].

85. McNeil MM, Gee J, Weintraub ES, Belongia EA, Lee GM, Glanz JM, Nordium JD, Klein NF, Baxter R, Naleway AC, Jackson LA, Omer SB, Jacobsen SJ, DeStefanio F. The vaccine safety datalink: success and challenges monitoring vaccine safety. Vaccine. 2014;32:5390–5398.

86. Lee GM, Greene SK, Weintraub ES et al. Vaccine safety datalink project. H1N1 and seasonal influenza vaccine safety in the vaccine safety datalink project. Am J Prev Med. 2011;41:121–128.

87. Gee J, Sukumaran L, Weintraub E. Vaccine safety datalink project. Risk of Guillain Barre syndrome following quadrivalent human papillomavirus vaccine in the vaccine safety datalink. Vaccine. 2017;35:5756–5758.

88. Hripcsak G, Duke JD, Shah NH, Reich CG, Husen V, Schuemie MJ, Suchard MA, Park RW, Wong ICK, Rijnbeek PR, Lei JV, Pratt N, Noren GN, Li YC, Stang PE, Madigan D, Ryan PB. Observational Health Data Sciences and Informatics (OHDSI): Opportunities for observational researchers. Stud Health Technol Inform. 2015;216:574–578.

89. Observational Medical Outcomes Partnerships (OMOP). Available from https://chime.ucsf.edu/observational-medical-outcomes-partnership-omop. [Accessed on November 23, 2023]

90. Observational Health Data Sciences & Informatics Available from https://www.ohdsi.org/. [Accessed on November 21, 2023]

91. Innovative Medicines Initiative. Available from https://www.ihi.europa.eu/about-ihi/imi-ihi. [Accessed on November 21, 2023]

92. Wang SV, Schneeweiss S. On behalf of the joint ISPE-ISPOR special task force on real world evidence in healthcare decision making: reporting to improve reproducibility and facilitate validity assessment for healthcare database studies VI.0. Pharmacoepidemiol Drug Saf. 2017;26:1018–1032.

93. Berger ML, Sox H, Willke R. Good practices for real world data studies of treatment and/or comparative effectiveness recommendations from the joint ISPOR ISPE special task force on real world evidence in healthcare decision making. Pharmacoepidemiol Drug Saf. 2017;26:1033–1039.

94. Bate A. Guidance to reinforce the credibility of healthcare database studies and ensure their appropriate impact. Pharmacoepidemiol Drug Saf. 2017;26:1013–1017.

95. Coloma PM, Schuemie MJ, Trifiro G, Gini R, Herings R, Hippisley-Cox J, Mazzaglia G, Giaquinto C, Corrao G, Pedersen L, van den Lei J, Sturkenbroom M; EU-ADR Consortium. Combining electronic healthcare databases in Europe to allow for large scale drug safety monitoring: the EU ADR project. Pharmacoepidemiol Drug Saf. 2011;20:1–11.

96. Arrhythmogenic Potential of DrugsARITMO. Project Available from https://cordis.europa.eu/project/id/241679/reporting. [Accessed on November 29, 2023]

97. Development of Vaccine Benefit-Risk Collaboration in Europe. Available from ADVANCE)http://www.advance-vaccines.eu/. [Accessed on November 19, 2023]

98. Vaccine Monitoring Collaborating for Europe (VAC4EU). Available from https://vac4eu.org/overview/. [Accessed on December 12, 2023]

99. The Safety of Non-Steroidal Anti-inflammatory drugs (SOS). Available from https://www.sos-nsaids-project.org/. [Accessed on November 21st 2023]

100. European Concerted Action on Congenital Anomalies and Twins. EUROCAT. Available from http://www.euromedicat.eu/. [Accessed on November 21st 2023]

101. CORDIS. Safety Evaluation of Adverse Reactions in Diabetes (SAFEGUARD).: project information. Available from https://cordis.europe.eu/project.eu/project/rcn/100121/factsheet/en. [Accessed on November 21, 2023]

102. CORDIS. Global research in pediatrics (GRIP): project information. Available from https://cordis.europpa.eu/project/rcn/97619/factsheet/en. [Accessed on November 21, 2023]

103. Carman WJ, Accortt NA, Anthony MS, Iles J, Enger C. Pregnancy & infant outcomes including major congenital malformation among women with chronic arthritis or psoriasis with and without etanercept use. Pharmacoepidemiol Drug Saf. 2017;26(9):1109–1118.

104. Trifiro G, Patadia V, Schuemie MJ, Coloma PM, Gini R, Herings R, Hippislay Cox J, Mazzaglia G, Giaquinto C, Scotti L, Pedersen L, Avillach P, Sturkenboom MCJM, van der Lei J, EuADR Group. EU-ADR healthcare spontaneous reporting system database preliminary comparison of signal detection. Stud Health Technol Inform. 2011;166:25–30.

105. Tayefi M, Ngo P, Chomutare T, Dalianis H, Salvi E, Budrionic A et al. Challenges and opportunities beyond structured data in analysis of electronic health records. WIREs Comput Stat. 2022;2:e1549.

106. OHDSI Natural Language Processing Working Group. Available from https://www.ohdsi.org/web/wiki/doku.php?id=projects:workgroups:nip-wg#objective. [Accessed on November 25, 2023]

107. Kent S, Burn E, Dawood D, Jonsson P, Ostby JT, Hughes N et al. Common problems, common data model solutions: evidence generation for health technology assessment. Pharmacoeconomics. 2021;39:275–285.

108. Toh S, Gagne JJ, Rassen JA, Fireman BH, Kulldorff M, Brown JS. Confounding adjustment in comparative effectiveness research conducted within distributed research networks. Med Care. 2013;51:S4–S10.

109. Platt RW, Brown JS, Henry DA, Klungel OH, Suissa S. How pharmacoepidemiology networks can manage distributed analyses to improve replicability and transparency and minimize bias. Pharmacoepidemiol Drug Saf. 2019;2:2.

110. Brown JS, Kahn M, Toh S. Data quality assessment for comparative effectiveness research in distributed data networks. Med Care. 2013;51:S22–S29.

111. Adimadhyam S, Barreto EF, Cocoros NM, Toh S, Brown JS, Maro JC et al. Leveraging the capabilities of the FDAs sentinel system to improve kidney care. J Am Soc Nephrol. 2020;31:2506–2516.

112. Gagne JJ. Common models, different approaches. Drug Saf. 2015;38:683–686.

113. WEB RADR; recognizing ADRs. Available from https://web-radr.eu/. [Accessed on January 10, 2024]

114. Social media. Definition, importance, top websites Apps. Available from investopedia.com/terms/s/social-media.asp. [Accessed on January 10, 2024]

115. Pierce CE, Bouri K, Pamer C et al. Evaluation of Facebook and Twitter monitoring to detect safety signals for medical products: an analysis of recent FDA safety alerts. Drug Safety. 2017;40:317–331.

116. Fox S, Duggan M. Health online. 2013;1:5–55.

117. Bhattacharya M, Snyder S, Malin M et al. Using social media data in routine pharmacovigilance: a pilot study to identify safety signals and patient perspectives. Pharmaceut Med. 2017;31:167–174.

118. Tim A, Casperson JLP, Juergen D. Strategies for distributed curation of social media data for safety and pharmacovigilance, In: International Conference on Data Mining. CSREA Press; 2016. pp. 118–124.

119. Correia RB, Li L, Rocha LM. Monitoring potential drug interactions and reactions via network analysis of Instagram user timeliness. Pac Symp Biocomput. 2016;21:492–503.

120. Carbonell P, Mayer MA, Bravo A. Exploring brand name drug mentions on Twitter for pharmacovigilance. Stud Health Technol Inform. 2015;210:55–59.

121. Hamad AA, Wu X, Erickson R et al. Twitter K-H networks in action: advancing biomedical literature for drugs search. J. Biomed Inform. 2015;56:157–168.

122. Liu X, Chen H. A research framework for pharmacovigilance in health social media: identification and evaluation of patient adverse drug event reports. J Biomed Inform. 2015;58:268–279.

123. Nikfarjam A, Sarkar A, O'Connor K, Ginn R., Gonzalez G. Pharmacovigilance from social media: mining adverse drug reaction mentions using sequence labeling with word embedding cluster features. J Am Med Inform Assoc. 2015;22:671–681.

124. Frost J, Okun S, Vaughan T, Heywood J, Wicks P. Patient reported outcomes as a source of evidence in off label prescribing: analysis of data from PatientsLikeMe. J Med Internet. 2011;13(1)e6. doi: 10.2196/jmir.1643

125. White RW, Harpaz R, Shah NH, et al. Toward enhanced pharmacovigilance using patient generated data on the internet. Clin Pharmacol Ther. 2014;96:239–246.

126. Wittich CM, Blurke CM, Lanier WL. Ten common questions (and their answers) about off label drug use. Mayo Clin Proc. 2012; 87(10):982–990. doi:10.1016/j.mayocp.2012.04.017.

127. Kuehen BM. Scientists mine web search data to identify epidemics and adverse events. JAMA. 2013;309:1883–1884.

128. Naik P, Umrath T, van Stekelenborg J, Ruben R, Abdul-Karim N, Boland R et al. Regulatory definitions and good pharmacovigilance practices in social media: challenges and recommendations. Ther Innov Regul Sci. 2015;49:840–851.

129. Food, Administration D. Guidance for industry: fulfilling regulatory requirements for post marketing submissions of interactive promotional media for prescription human and animal drugs and biologics. 2014.

130. European Medicines Agency. Guidelines on good pharmacovigilance practices (GVP) Module VI-Collection, management and submission of reports of suspected adverse reactions to medicinal products (Rev 2). In European Medicines Agency HoMA (ed). Rev 2nd ed. 2017;106.

131. Rees S, Mian S, Grabowski N. Using social media in safety signal management: is it reliable? Ther Adv Drug Saf. 2018;9:591–599.

132. Caster O, Dietrich J, Kurzinger ML, Lerch M, Maskell S, Noren GN, Lessenot ST, Vroman B, Wisniewski A, van Stekelenborg J. Assessment of the utility of social media for broad ranging statistical signal detection in pharmacovigilance: results from the WEB-RADR project. Drug Safety. 2018;41:1355–1369.

133. Xiao C, Li Y, Baytas IM, Zhou J, Wang F. An MCEM framework for drug safety signal detection and combination from heterogeneous real world evidence. Sci Rep. 2018;8:1806.

134. Ying Li, Yeper AJ, Xiao C. Combining social media and FDA adverse event reporting system to detect adverse drug reaction. Drug Safety. 2020;43:893–903.

135. Brosch S, de Farren AM, Newbould V, Farkes D, Lengsavath M, Treguno P. Establishing a framework for the use of social media in pharmavovigilance in Europe. Drug Saf. 2019;42(8):921–930.

136. Golder S et al. Pharmacoepidemiologic evaluation of birth defects from health related postings in social media during pregnancy. Drug Saf. 2019;42:389–400.

137. Klein A.Z., Sarker A, Weissenbacher D, Gonzalez-Hernandez G. Towards scaling Twitter for digital epidemiology of birth defects. NPJ Digit Med. 2019;2:96.

138. Magge A, Sarker A, Nikfarjam A, Gonzalez Hernandez G. Comment on: Deep learning for pharmacovigilance: recurrent neural network architectures for labelling adverse drug reactions in Twitter posts. J Am Med Inform Assoc. 2019;26:577–579.

139. Lee JY, Lee YS, Kim DH, Lee HS, Yang BR, Kim MG. The use of social media in detecting drug safety related new black box warnings, labelling changes or withdrawals: Scoping Review. JMIR Public Health Surveill. 2021;7(6):e30137. doi: 10.2196/30137

140. van Stekelenberg J et al. Recommendations for the use of social media in pharmacovigilance: lessons from the IMI WEB RADR. Drug Saf. 2019;42:1393–1407.

141. Caster OLM, Vroman B, Van Stekelenborg J. Performance of disproportionality analysis for statistical signal detection in social media data (abstract). Pharmacoepidemiol Drug Saf. 2016;25:411.

142. Butler D. When Google got flu wrong. U.S outbreak foxes a leading web based method for tracking seasonal flu. Nature. 2013;494:155–156.

143. Nguyen AT., Raff E, Lien J, Mekaru SR. Improved automatic pharmacovigilance: An enhancement to the MedWatcher social system for monitoring adverse events. Med Watch. doi: 10.1101/717421

144. Lu Z. Information technology in pharmacovigilance: benefits, challenges and future directions from industry perspectives. Drug Health Patient Saf. 2009;1:35–45.
145. Andrews E, Dombeck M. The role of scientific evidence of risks and benefits in determining risk management policies for medications. Pharmacoepidemiol Drug Safe. 2004;13:599–608.
146. Dhruva SS et al Aggregating multiple real world data sources using a patient centered health data sharing platform. NPJ Digit Med. 2020;3:60.
147. Colman E, Szarfman A, Wyeth J et al. An evaluation of a data mining signal for amyotrophic lateral sclerosis and statins detected in FDAs spontaneous adverse drug event reporting system. Pharmacoepidemiol Drug Saf. 2008; 17:1068–1076.

5 Uncovering False Alarms

Examining Biases and Confounders in Disproportionality Analysis

Lipin Lukose, Mansi Pawar, Aina M Shaju,
Gouri Nair, and Subeesh K Viswam

5.1 INTRODUCTION

Spontaneous reporting systems (SRSs) are essential in detecting and assessing adverse drug reactions (ADRs) that may not have been identified during pre-market clinical trials. These systems rely on healthcare professionals (HCPs), pharmaceutical manufacturers, and consumers to voluntarily report adverse events (AEs) associated with newly marketed medications and identify delayed effects and effects that may arise with concomitant usage of other medicines [1]. Various methods are utilised to quantify the risk, of which case-non-case studies are most widely used as disproportionality measures with pharmacovigilance (PV) databases. Different databases are used for monitoring by different countries to identify AEs that exist at the national and international levels. Some national databases include the Food and Drug Administration (FDA) Adverse Event Reporting System (FAERS) database in the United States and the Japanese Adverse Drug Event Report (JADER) database, which are publicly available. The international ones include EudraVigilance or the World Health Organisation (WHO) database, VigiBase, and the European Medicines Agency (EMA) database. These databases help to generate a PV signal of unknown or underestimated AEs of marketed medications. In this context, a signal refers to a higher observed reporting of an AE with a particular medication than expected, thus leading to a "disproportionate" reporting compared to all the other reactions recorded in the database. Once a signal is generated, different methods are employed to quantify the potential risk and establish an association between the two.

Two methods are mainly used to study the disproportionality of ADR reports, namely, frequentist methods such as reporting odds ratio (ROR) or the proportional reporting ratios (PRRs) [2, 3] and Bayesian methods [4, 5], such as information component (IC), Multi-Item Gamma Poisson Shrinker (MGPS) method, or the Bayesian Confidence Propagation Neural Network (BCPNN) method. Frequentist methods are most commonly used for signal detection studies and to quantify the strength of the association with ROR, the most widely used [6]. ROR is defined as the ratio of the odds of an ADR being reported for a drug of interest to the odds of the same ADR being reported for all other drugs, which is stipulated with the help of a 2×2 contingency table describing the number of adverse reactions of interest reports in the database and the exposure to the drug of interest.

The interpretation is similar to the traditional odds ratio, with the requirement that the lower bound of the 95% CI be greater than 1 to be deemed statistically significant. ROR is often preferred over other methods due to better performance in terms of signal precocity [6] and the flexibility in the usage of multivariable logistic regression, which allows consideration of confounding and other interaction effects [3]. Similarly, PRR is the ratio of the proportion of reports for a specific event associated with a particular drug to the proportion of reports for the same event associated with all other drugs [2]. However, using these metrics might be limited due to the inherent nature of SRS.

One of the major advantages of conducting disproportionality analysis (DA) using SRS is that it allows the study of rare AEs since only a few cases of exposure-event status are required to conduct the analysis and are cost-effective. Also, it helps to identify newer AEs post-marketing of medications not recorded in the clinical trials, which justifies the conduct of confirmatory studies. For example, a study identified the occurrence of bladder cancer with the use of pioglitazone from the FAERS database [7], which led to its confirmation by advanced study methods and the implementation of regulatory measures [8]. This might not be the scenario always, as many times it would not be feasible and would not lead to firm conclusions. Compared to clinical trials, the SRS contains a cohort of real users (patients with multiple comorbidities and concomitant medications, population who are often excluded in the clinical trials, etc.) representative of the medication use in the population. These methods also serve as a platform for initial evidence synthesis and help the regulators identify and inform consumers and HCPs about medications if serious AEs appear during the post-marketing phase [2]. With methodological advancements in recent years, various designs are being utilised to stratify the signals [9, 10] and address drug-host factor interactions [11]. Also, disproportionality studies using SRS in combination with bioinformatics and in vitro studies are being used for drug repurposing [12] and the provision of biological plausibility for the AE [13]. Even though the study method is helpful in hypothesis generation, the major limitations associated deem it lower on the evidence synthesis scale.

A significant limitation of SRS is the underreporting of an AE. It is unlikely that all AEs experienced by patients are reported due to the voluntary nature of SRS. A systematic review by Hazell and Shakir [14] revealed that the underreporting rate was as high as 94% in the included studies. Furthermore, the estimates provided by the FDA suggested that merely 1%–10% of ADRs are reported [15]. This may hinder the actual measurement of adverse reactions, subsequently limiting the sensitivity and specificity of disproportionality measures [16]. In addition, there is no certainty regarding the strength of the association (especially in cases where clinical details are missing) and the exposure population's denominator (such as the number of prescriptions written in the same geographic area for the unknown) within the same time frame. As a result, it is impossible to compute the incidence and reporting rates of AEs using SRS. ADRs may also go unreported due to limited time for HCPs, insufficient PV training, and patient unfamiliarity with reporting procedures [17].

Addressing confounding variables is crucial to mitigating these limitations and enhancing the validity of disproportionality measures. In disproportionality analyses, confounders can be defined as *"variables involved in the association between exposure and adverse reaction reporting and those involved in the association between exposure and the effect itself, such as the indication of the drug"* [18]. The most frequently encountered confounding factors include age, gender, concomitant drug therapy, and indication. Various studies have highlighted these issues and proposed methods to address them [9, 19]. For example, a study conducted in the French national PV database found a positive association of hypoglycemia with usage of angiotensin receptor blockers (ARBs). However, the indication itself, i.e., diabetes, was a confounder in this study as these inhibitors are more frequently prescribed to diabetic patients who are also treated with anti-diabetic agents (ADAs). This likely led to an increase in the reports of hypoglycemia associated with ARBs, and thus, a false-positive signal was obtained. To address this factor, the researchers stratified the reports based on the presence or absence of anti-diabetic drugs, which led to the disappearance of the signal [20].

Various factors, such as media coverage of an adverse reaction (notoriety effect), time since authorisation (Weber effect), severity of the AE, type of reporter, and reporting patterns, can also cause an overestimation or underestimation of risk. These factors lead to biases, thus distorting the true association of the drug-ADR complex. A more accurate risk–benefit profile can be established by recognising and adjusting for confounders and biases. For example, a study might find a correlation between a drug and a rare AE. However, without considering factors, such as age, sex, underlying health conditions, and those listed above, the true strength of the association would not be determined [21].

This chapter aims to address some of the biases and confounding factors that need to be considered while conducting DA using spontaneous reporting databases (Figures 5.1 and 5.2). However,

BIASES IN DISPORPORTIONALITY ANALYSIS

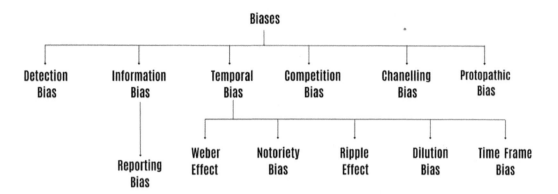

FIGURE 5.1 List of biases to consider while conducting disproportionality analysis in a spontaneous reporting database.

this is not an exhaustive list of all factors that could impact DA. Further biases and confounders may exist beyond what has been described here.

5.1.1 BIASES

There are a number of biases, such as detection bias, information bias, reporting bias, temporal biases, notoriety bias, dilution bias, time frame bias, and competition bias, which are explained in detail below.

CONFOUNDERS IN DISPORPORTIONALITY ANALYSIS

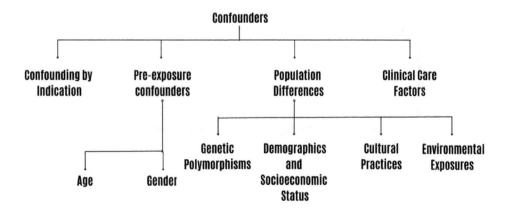

FIGURE 5.2 List of confounders to consider while conducting disproportionality analysis in a spontaneous reporting database.

5.1.1.1 Detection Bias

Detection bias is defined as an increase in the frequency of AEs due to tests performed to confirm or rule out a suspected AE. In light of reported literature evidence, the performance of laboratory tests can lead to increased surveillance for specific AEs with the agent of interest. An example of detection bias was demonstrated in patients receiving incretin-based therapy, such as GLP-1 inhibitors. A piece of published information [22] reported 30 cases of acute pancreatitis with exenatide. This GLP-1 receptor agonist subsequently led to an increase in the frequency of measurement of lipase and amylase in patients receiving GLP-1 agonists. The targeted testing further led to identifying and reporting of more suspected pancreatitis cases, inflating the number of reported AE cases and thus the higher association of the AE with GLP-1 agonists [23].

Another example of detection bias in SRSs would be increased reports of endometrial cancer with hormone replacement therapy (HRT) [24]. This may be due to increased screening for endometrial cancer among HRT users, leading to more reporting of the cases (Figure 5.3). Further, since HRT is also associated with intrauterine bleeding, the symptom could increase endometrial cancer screening and report a surge in the safety signal.

The effects of detection bias can lead to a sudden surge in the reported safety signal, thus depicting a stronger association than actuality. The reporting time frames and implementation of screening programmes should be considered while conducting safety signal analysis to counter this bias.

5.1.1.2 Information Bias

Missing data is a considerable drawback of PV databases, and even though it does not affect the crude analysis, it can prevent stratification and confounding assessment. This limits the potential to measure the true effects of the association. Missing information, such as age, sex, indication, concomitant medications, country, type of reporter, and specific doses, can severely limit the potential for evidence generation. For example, an analysis of the reports in the FAERS database revealed that nearly 42%, 11%, and 46% of the reports were missing age, sex, and outcomes of the reported events [25].

The practices and differences in the reporting conventions varying from institute to country may lead to differences in the recorded information. For example, the US FDA only requires other countries to report an AE if it is severe and is unlabeled in the FDA label [26], thus leading to fewer

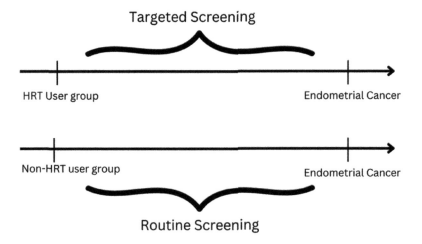

FIGURE 5.3 Impact of detection bias on spontaneous reporting databases. The figure depicts the impact of targeted screening in the HRT user group which leads to a sudden rise in the identified cases in this group as compared to the non-HRT group who undergo routine screening, if undertaken. This targeted increase only in one group leads to a sudden rise in the signal.

reports being recorded. Also, differences in reporting AEs by various stakeholders and the potential of old cases being reported as new leads to duplication of reports if not adequately tracked [27]. Therefore, "de-duplication" is one of the first steps to be carried out while analysing reporting databases. Several tools, such as OpenVigil [28] and AERSmine [29], and algorithms such as VigiMatch help to identify and remove the duplicates while keeping the complete and informative reports [30]. Similarly, multiple events for the same person, i.e., recurrently, would count as several independent observations due to the lack of unique patient identifiers.

Various types of information bias, such as misclassification and reporting bias, exist, with the latter most prominent in the reporting databases. In reporting databases, misclassification may occur due to reporting an AE with a specific drug. In contrast, the drug may not be the causative agent because of a lack of information on medical history or concomitant medications. For example, a lack of medical history can lead to ARB-induced coughing being reported, not comprehending that it may be caused by an unlisted previously administered angiotensin-converting enzyme (ACE) inhibitor [27].

5.1.1.3 Reporting Bias

Ideally, all ADRs should be reported to PV databases, but the reality falls far short, with less than 10% reported [15]. For example, when comparing the overall ADRs between active surveillance and spontaneous reports for two ophthalmic drugs, a difference of nearly 55% was observed between the two [31], which emphasises the underreporting culture present. Even though one would expect all ADRs to be reported equally, unexpected and unlabeled serious reactions often have higher reporting rates than anticipated ones [26]. One of the primary reasons for this is the belief of HCPs and patients to refrain from reporting an already known or well-established ADR. Multiple studies have also reported that HCPs and patients tend to notify ADRs when they are unexpected, severe, or serious, leaving aside those which are not [32, 33]. A specific example can be seen in a study by Mastuda et al., who observed the reporting of a serious:non-serious ratio and found that the ratio constantly increased with the increasing time of the marketing period, hinting at the biased reporting [34]. Also, a mandatory reporting requirement for the industry on serious AEs leads to a rise in the reported ones. Although it is presumed that underreporting will affect various drugs consistently, this is only sometimes the case. This can be observed by nearly the twofold difference in the reporting of ADRs between artificial tears and anti-glaucoma medications, due to the severity of occurring ADRs being one of the factors for the difference [31]. However, in a study by Hoffman et al. [35] to determine the impact of stimulated reporting on the FAERS database, it was found that issuing FDA alerts does not impact the majority of modern-day FAERS reporting. At the same time, a few subsets of drugs may still show this.

Reporting bias is also attributable to the type of reporter reporting the AE. The reporters' occupation can also impact the completeness of the reports, with consumers, pharmacists, and HCPs having a difference in the reporting of AEs [36]. Thus, the errors in reporting exposure, diagnosis, concomitant drugs, and patient characteristics often depend on the reporter type.

In order to limit this bias, categorising the reports based on the type of reporter and reporting form used would help reduce the impact of heterogeneities in ADR reports [36]. Another way of mitigating this bias might be by limiting the analysis to reports with a lower risk of error, such as those by HCPs, or the reports considering the drug as a "primary suspect" or biologically plausible. Focus should also be placed on the definition of exposure using medication categories to prevent exposure errors. For instance, information bias could emerge from omitting the drug combination anatomical therapeutic chemical (ATC) codes. Another potential source for bias may be using proxies based on medication exposures in search of clinical data such as medical history and indications.

Thus, a disproportionality signal may be the consequence of either an actual increase in the frequency of the adverse reaction or an artefactual increase in reports not associated with an increase in incidence, compromising signal detection specificity. The reporting pattern is extensively per-suaded by confounders, which might jeopardise the accuracy of the signal being detected. The product age (i.e., the time the drug has been on the market) and stimulated reporting should be acknowledged as two of the most significant confounders.

5.1.1.4 Temporal Biases

5.1.1.4.1 Weber Effect

When a drug receives market approval, a substantial increase in spontaneously reported ADRs dur-ing the first two years of marketing is observed due to the incomplete safety profile and increased exposure. This eventually declines due to decreased enthusiasm for reporting ADRs already known. This phenomenon is known as the "Weber effect" [37]. Thus, the Weber effect can be defined as a variation in reporting AEs once a new drug is introduced on the market or a new indication/safety label that results in an increased amount of clinically mild or trivial reaction reports that subside with time. The Weber effect was first observed with nonsteroidal anti-inflammatory drugs (NSAIDs) [38], and since then, it has been replicated multiple times through various studies [39, 40]. On analysis of nine NSAIDs, Dr JCP Weber observed that the number of AEs usually peaked near the second year of marketing and saw a subsequent decline. This was in line with two temporal changes: an increase in patient exposure to the drug leading to a rise in reports and subsequent fall as the HCPs became acquainted with the medication and lost interest in reporting events. Another example of the Weber effect was demonstrated by Hartnell et al. [41], who observed the reporting pattern of ADRs with serotonin reuptake inhibitors (SSRIs). Out of the five SSRIs analysed, the Weber effect was observed with only fluvoxamine and not others, hinting at the non-consistent phenomenon.

The Weber effect can also be observed for older drugs approved for a new indication or a change in the dosage form Graff et al. [42], who examined the reporting pattern of ADRs associated with two formulations of omeprazole (tablet and capsule), illustrated an example of the Weber effect due to a change in the formulation. In 1999, the tablet formulation of omeprazole (Losec MUPS) was granted marketing authorisation, followed by the company's withdrawal of the capsule formulation (Losec capsules). This forced switch in the formulation led to a sharp rise in the number of ADR reports (early Weber effect) associated with Losec MUPS, with gastrointestinal complications and decreased efficacy being the reported ADRs. A pharmacokinetic study conducted to determine any differences in the formulations demonstrated bioequivalence, making the reporting of gastrointes-tinal complications and decreased efficacy due to psychological effects in patients, thereby giving a case example of the impact of the Weber effect. Another study which evaluated the effect of a change in indication reported that three of the five drugs included in the analysis showed a surge in the ADR reports during the following year [43].

Though the Weber effect was ideally observed for older drugs, the modern drugs approved exhibit significant variability and do not show the traditional Weber curve as shown in various stud-ies [44, 45]. Also, approval of a new indication for an old drug leading to an increase in AE report-ing, forming a pattern which does not resemble the Weber curve, has been shown in recent studies [46]. This may be due to the change in the reporting system of the FDA in 2014–15 [47], as well as an effort by regulatory agencies to increase awareness regarding the benefits of post-marketing AE reporting. Publicity, regulatory actions, a new indication, changes to the formulation, and changes in marketing tactics can all impact AE reporting, which can explain the variety of reporting patterns [43]. A study also pointed out that the Weber effect is inconsistent with oncology drugs [48]. Thus, the class of drugs and changes in the regulatory system must be considered while determining if the Weber effect is present.

DA, in general, is ineffective at identifying long-term consequences of medications, as reporting can get remote from the market with time. Thus, the Weber effect can lead to over- and underes-timating ROR values. In evaluating serious AEs, the inclusion of a high proportion of non-serious events due to the Weber effect can lead to a masking effect of the signal, which can cause an underestimation of the ROR. This could be addressed through a temporal analysis by restricting the case/non-case analysis to severe events only [18].

5.1.1.5 Notoriety Bias

Media attention and publicity arising from advertising or regulatory measures (e.g., dear doctor letters or warnings about drug-related safety risks) may also cause an increase in reporting and a higher-than-expected ROR. This phenomenon is known as "notoriety bias". The phenomenon was depicted by Moore et al. [49] by providing an example of sertindole. Sertindole, an atypical anti-psychotic, was introduced to the UK market in early 1996. When the drug was initially licensed, regulators did not impose strict warnings or electrocardiographic (ECG) monitoring. This was due to preclinical evidence and premarketing clinical studies demonstrating the drug's potential to prolong the QT interval without increasing the risk of cardiac mortality. However, when the drug's license was expanded to encompass other European markets in 1997, the authorities requested a change in the summary of product characteristics (SPC) to incorporate ECG monitoring before and during sertindole therapy. Following this, a "*Dear Doctor Letter*" was circulated, emphasising the need for ECG monitoring during the use of sertindole. After the "*Dear Doctor Letter*" circulation, there was a surge in the reporting proportion for QT-interval prolongation and asymptomatic arrhythmia for sertindole compared to the other neuroleptics. Subsequently, sertindole was suspended due to differential reporting of sudden death and fatal arrhythmia compared to the other atypical neuroleptics in the class, as it seemed to have a tenfold greater PRR than the other drugs.

The sertindole report rates in the UK reflected several events, including the initial "Weber effect", a significant increase in reports following the "Dear Doctor letter", and minor rises in report rates in response to adverse publicity around suspending sertindole. However, after considering additional sources of data such as prescription monitoring studies (PEM) and European sertindole exposure studies (ESES), spontaneous reports, and drug utilisation, the associated bias for sertindole was recognised, and the suspension was lifted after almost three years.

The notification of exenatide-induced pancreatitis demonstrates another example of notoriety bias. Exenatide was approved by the FDA in 2005, and the first case report of pancreatitis associated with exenatide appeared in February 2006. By 2007, the FDA received 30 post-marketing reports of acute pancreatitis, which led them to issue an alert and instruct the manufacturers to add the AE on the label [50]. Following this, an upsurge in the reports of exenatide-induced pancreatitis was observed, which led to positive signals through DA being reported. Multiple observational studies proved no association was present, and it was concluded that the ROR values were significantly overestimated due to notoriety bias caused by the FDA alerts [51].

5.1.1.6 Ripple Effect

The study by Raschi et al. [51] also described an increase in reported cases of pancreatitis associated with sitagliptin after the alert on exenatide. After the issuance of the alert on exenatide by the FDA, an increase in the reporting by almost 40% was observed with sitagliptin, even though no notification was made from the FDA. Thus, a pancreatic DA first surfaced in 2008, before the FDA issued an alert in 2009. This escalation in case reports and ROR among other drugs of the same pharmacological or therapeutic class following the warning of the initial drug suspect is termed a "ripple effect". An alert highlighting an ADR can set off a *ripple effect*, causing a surge in reporting the same reaction for other drugs from the pharmacological or therapeutic class [52].

Another instance which illustrates the ripple effect was demonstrated by Pariente et al. [52] pertaining to statins and rhabdomyolysis. On 8 August 2001, the pharmaceutical firm Bayer decided to withdraw cerivastatin from the market after cases of rhabdomyolysis were reported in patients, some of which were fatal. Between 1999 and 2001, 25 cases of rhabdomyolysis associated with statins (excluding reports associated with cerivastatin) were also reported. The ROR for rhabdomyolysis associated with statins, excluding cerivastatin, was 5.8 (95% CI 3.8, 9.0). Two years following cerivastatin withdrawal, between 2001 and 2003, 63 cases of rhabdomyolysis were reported in the database. The ROR for rhabdomyolysis with statins other than cerivastatin rose to 9.4 (95% CI 7.0, 13.0) after cerivastatin withdrawal, highlighting the ripple effect associated with drug withdrawal.

Notoriety bias and the ripple effect can often lead to an overestimation of ROR, leading to false-positive signals, as depicted in the examples above. Though many studies have reported the existence of these biases, a recent study by Neha et al. concluded the non-existence of notoriety bias in the FAERS database and the overall signal estimates to be unaltered by safety alerts [19]. However, few drugs exhibit increased reporting, and caution should be taken to mitigate the effect.

5.1.1.7 Dilution Bias

When all the drugs in the same class are similarly affected by media attention, the drugs with longer marketing lives (more pre-existing reports) will be less likely to generate disproportionality after notoriety than newer drugs within the same class, leading to false-positive signals. The distortion in disproportionality caused due to this effect is known as *"dilution bias"*.

Dilution bias can be depicted through a case of paroxetine and the risk of suicide [53]. In 2002, escitalopram was first introduced in the UK to treat major mood disorder (MMD). In 2003, a television show was broadcast about the effects of one of the SSRIs, paroxetine, and the risk of suicide. This resulted in an increase in the reports of suicide associated with paroxetine as a result of regular monitoring of suicidal effects with the drug and the drug class. On conducting DA to quantify the effect, it was found that escitalopram, not paroxetine, appeared to be associated with an approximately 2.5-fold higher risk of suicide compared with other drugs of the same class. Even though post-broadcast, the highest number of suicide cases were reported with paroxetine, the PRR for escitalopram was reported to be the highest (2.06 [95% CI 0.28, 14.87]). During the period, escitalopram accounted for nearly 68% of all reports and 75% of suicides, whereas for older drugs, it was 3.6% of all reports and 13% of reports of suicides. This increase in the risk of suicide with escitalopram, even though a higher number of suicide reports were associated with paroxetine, can be attributed to the dilution effect.

Generally, while computing the disproportionality measures for a drug, all reports are considered irrespective of the marketing duration, which can lead to a dilution of ROR due to the high reporting period for older drugs. For example, considering the contingency table, all ADRs associated with a drug (b) would be much higher than those of a newer drug in the market. As ROR is calculated by [ad/bc], the high number of "b" reports for older drugs compared to the newer ones can lead to a *"dilution effect"*, despite the ADR being more prominent in the older drug. Even though the broadcast had little absolute impact on the number of reports with escitalopram, a more significant relative effect was observed with the drug due to its short duration in the market, which led to falsified signal strength.

Several strategies can be employed to mitigate the notoriety bias, dilution bias, and ripple effect. One method is to perform a temporal study of the ROR, concentrating on pre- and post-media coverage, to see whether there is a significant shift in the reporting patterns. Another approach is to restrict the study to the period preceding the media coverage, allowing for a more precise estimate of the true ROR. Furthermore, examining the reporting rate pattern can provide insights into potential alerts or spikes in reports. If an unanticipated pattern appears, it warrants closer investigation into the role of the media [16, 18].

5.1.1.8 Time Frame Bias

Time frame bias can be defined as a difference in the outcome or association of the AE with the drug concerning the different time frames utilised to quantify the effect. This may occur when one tries to assess the safety profile of a class of drugs using DA. An example of time frame bias was illustrated by the reporting of proportions of embolic and thrombotic events associated with anticoagulants over a period of time [54]. Choosing 2011 as the time point makes it possible to infer that dabigatran has the lowest disproportionate reporting rate for embolic and thrombotic events compared to all other anticoagulants. However, if 2012 and 2013 are considered, warfarin is, without a doubt, the lowest.

Further, on analysing data from 2014, apixaban has the lowest metric of disproportionate reporting. To mitigate this issue, the period chosen for analysis should minimise the disparities in reporting patterns caused by duration on the market or other known variables. The same fixed-length post-approval analysis time frame for each medicine can be used to standardise the time a product spends on the market. However, the variation in the time frame between the medications or other factors regulating reporting would not be accounted for.

The temporal analyses can also be affected by the interval between the occurrence of the reaction and its entry into the PV databases, especially in the case of multinational databases such as FAERS and VigiBase.

5.1.1.9 Competition Bias

Traditionally, the detected signal depends on the denominator, i.e., those involving the AE associated with other drugs, and the numerator, which involves the AE with the drug of interest. In some cases, the rate of unexposed cases (c) may be higher as the AE under interest may be reported with other drugs than the drug of interest, leading to an underestimation of the ROR. This may lead to the masking of the true signal, and the masking phenomenon that occurs due to a bias in associating the true exposure effect is called competition bias. For instance, if the AE had a strong correlation with other drugs, this would increase the background reporting rate and the unexposed cases, lowering the signal-generating process' sensitivity [55]. Competition bias may be further classified based on the drug and the event.

Drug competition bias is defined as the phenomenon when there is a "reduced likelihood to find a signal of disproportionate reporting" for an AE associated with other classes of drugs, even though it is plausible when a drug or a drug class is already known to cause that adverse effect and is notified more in the SRS [16]. A study conducted in the France spontaneous reporting database by Pariente et al. [55] demonstrated the effect of competition bias on signal strength. In the study, the authors extracted all the spontaneous reports submitted to the database during 1986–2001. Signals of disproportionate reporting were considered potential signals if not reported in the literature before 1 January 2002. The event of interest was selected as gastric and oesophageal haemorrhages, as these are known AEs of anti-thrombotic agents and NSAIDs.

Further, all reports of gastric and oesophageal haemorrhage associated with antithrombotic agents and NSAIDs were removed to find potential agents with the AE. After removing these cases, positive signals associated with prednisone, isotretinoin, and rivastigmine were identified. Hence, the study showed the effect of competition bias on signal detection and helped to establish three potential signals that would only have been identified with accounting for the bias.

Event competition bias is defined as when there is a reduced likelihood of reporting other AEs associated with a drug or a drug class when a well-known AE is documented to be associated with the drug or drug class [16]. For example, Salvo et al. [56] illustrated the event competition bias. As statins are known to cause rhabdomyolysis/myopathy, the study hypothesised that the association between statins and rhabdomyolysis/myopathy would prevent the emergence of reporting signals for additional statin-related occurrences that had not yet been investigated. In the study, all disproportionate reporting signals were considered potential signals if they had yet to be reported in the literature before 1 January 2002 but were subsequently confirmed. After removing the cases associated with rhabdomyolysis/myopathy, 11 new disproportionality signals were identified, further strengthening the presence of event competition bias.

Another potential pathway that leads to competition bias is the submission of precautionary reports by Market Authorization Holders (MAHs) due to regulatory requirements. These reports usually do not establish a causal relationship, are poorly documented, and are sent out from organised data collection systems, such as patient support programmes (PSPs), non-interventional studies, or compassionate use programmes. Precautionary reports for a product of interest raise the background-reporting rate and, hence, the threshold for detecting the event of interest. In parallel, for events that are strongly correlated with precautionary reporting, the drug-event combination reports (a) are increased to the point that false signals are produced. In addition to the masking effect, precautionary reports may delay the identification of safety signals by raising the time-to-detection, necessitating more reports with the event of interest over time to establish a potential positive signal.

An example of this phenomenon was depicted by evaluating the ADRs associated with erythropoietin, bisphosphonate and the endothelial receptor antagonist using the Dutch ADR database [57]. The study found 89%, 79%, and 59% of precautionary reports present for erythropoietin, bisphosphonate, and endothelin receptor antagonists, respectively. On removing the precautionary reports from the analyses, the study identified 10, 7, and 4 new safety signals for the 3 agents. Furthermore, 27, 101, and 26 spontaneous drug reports were not detected after the removal of the precautionary reports from the analysis. Thus, the example shows how the bias can lead to over- and underreporting signals.

In order to mitigate competition bias, sensitivity analysis should be performed, in which all reports related to potential drug competitors are excluded from the dataset. This would help to reduce the number of exposed cases needed to show a signal. Another way of mitigating the bias is by conducting the analysis restrictive to reports under the same system order classification than the AE of interest.

5.1.1.10   Channelling Bias

Channelling bias may be considered a type of allocation bias when therapies are given to individuals with significant prognosis differences. Drugs with similar actions that are introduced at various points in the market may be directed to various patient populations. For instance, patients who have not responded satisfactorily to therapy with an established, early-entry medicine are more likely to receive a late-entry drug. These newer drugs often emphasise distinct benefits such as fewer side effects, encouraging physicians to prescribe them to patients who had adverse effects with the previous drugs. Similarly, a new drug claiming greater efficacy may be intended for patients who did not respond to earlier therapies. An example of channelling bias was demonstrated after introducing "Osmosin", a controlled-release indomethacin marketed to have fewer gastrointestinal side effects. A post-introduction study using the PEM method revealed a high prevalence of gastrointestinal side effects after using the drug. Later, the drug was withdrawn from the market due to a high incidence of gastrointestinal ulcerations, bleeding, or perforation. Despite the claim, this might have been due to more at-risk patients, i.e., at risk of developing gastrointestinal effects, being prescribed the drug, potentially hinting at the bias present [58].

Comorbid conditions, such as diabetes, cardiovascular conditions, or kidney diseases, can also lead to preferential therapy in patients and the reporting of associated AEs. For example, in patients with chronic kidney disease (CKD), dosage adjustments must be made for several dipeptidyl peptidase-4 (DPP-4) inhibitors like sitagliptin, saxagliptin, and vildagliptin, whereas linagliptin can be administered without dosage adjustment. This might lead to clinicians opting to use linagliptin, which can lead to channelling bias [59].

Another example of channelling bias was demonstrated by Gatti et al., who characterised AEs associated with sacubitril/valsartan. They showed that most cardiovascular AEs with significant disproportionality were associated with heart failure or comorbidities and co-administered medications rather than with the inherent toxicity of sacubitril/valsartan. Considering the intended therapeutic purpose and recommendations provided by the European Society of Cardiology (ESC) guidelines, it is probable that the individuals who were given sacubitril/valsartan treatment were those at advanced stages of cardiac insufficiency. As a result, these patients innately are at a heightened risk of experiencing major cardiovascular events, leading to preferential prescribing and channelling bias [60].

Even though channelling bias is unlikely to be entirely avoided even after multiple adjustment models, one way to minimise this bias is by performing active comparator restrictive DA (ACR-DA). In a study performed by Alkabbani et al., the impact of ACR-DA on the signal strength of five ADRs associated with SGLT2 inhibitors was demonstrated, wherein DPP-4 inhibitors were used as the clinically active comparator. Reports from the FAERS database were used to assess if SGLT2 inhibitors were associated with five potential ADRs, i.e., acute kidney injury (AKI), genito-urinary tract infections (GUTIs), diabetic ketoacidosis, fractures, and amputation. For each ADR, the PRR was calculated and adjusted using the three types of reference groups: no SGLT2 inhibitor (background risk reference), other diabetes drugs (therapeutic class reference), and DPP-4 inhibitors (active comparator reference). Based on ACR-DA, safety signals were not detected for AKI or fractures associated with SGLT2 inhibitors compared to DPP-4 inhibitors. However, safety signals for GUTIs, diabetic ketoacidosis and amputations were associated with the agent. As evident, PRR and ROR estimates were attenuated using the ACR-DA with DPP-4 inhibitor as an active comparator. Thus, to mitigate the effect of channelling bias, this method may reduce the detection of false-positive safety signals when an appropriate active comparator is chosen [61].

5.1.1.11 Protopathic Bias

Several epidemiological studies have demonstrated the increased risk of CKD with acetaminophen and aspirin. However, most of the patients enrolled in these studies had CKD history or reached end-stage renal disease [62]. Some studies have also suggested the role of acetaminophen in causing end-stage renal disease [63]. This might have been due to the preferential switching of NSAIDS to acetaminophen due to the relatively safe profile on renal function. Another study found similar results even on enrolling patients with early-stage renal failure. Despite enrolling them in the earlier stages, these patients had pre-existing renal or systemic disease and were likely to be symptomatic in the years following their enrolment in the study [64]. An increase in analgesic use among these individuals due to underlying conditions, predisposing to renal insufficiency, may indicate the observed association's strength. This variant of confounding by indication, where the drugs used to treat a manifestation that occurs along the prognosis of the disease get reported as the causative agent itself, is called protopathic bias [65]. The term "protopathic bias" is used when the initial symptoms of the outcome being studied are the very reasons for the administration of a treatment.

Salas et al. [66] depicted an example of protopathic bias through the reported potential association between aspirin use and Reye's syndrome. Aspirin might be preferentially given to children displaying specific prodromal symptoms that precede Reye's syndrome. Suppose aspirin is selectively prescribed to children displaying a specific set of early symptoms preceding Reye's syndrome. In that case, it might falsely appear that aspirin usage is associated with the syndrome. This scenario exemplifies protopathic bias. In some situations, distinguishing between protopathic bias and confounding by indication can be challenging. Consider the association between cimetidine use and subsequent development of gastric carcinoma. In this case, confounding by indication and protopathic bias could influence the association.

Another example that can be used to explore the impact of protopathic bias is the reports of NSAIDs and spontaneous abortions (Figure 5.4). A retrospective cohort study was conducted further to evaluate this association, and various timelines of NSAID exposure before the abortion were evaluated. The study found the disappearance of the association, excluding the exposure to non-selective NSAID the day before the spontaneous abortion [67]. Notably, a consistent pattern of reducing the hazard ratio was observed with non-selective agents on evaluating the day, two days, and three days before the spontaneous abortion. Thus, the study proved the effect of "protopathic bias", as referenced by Faillie, on the association of NSAIDs with spontaneous abortion [8]. The

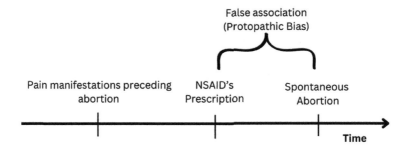

FIGURE 5.4 Depiction of protopathic bias using an example of NSAIDs and spontaneous abortion. The figure depicts how across a timeline, it might seem as if NSAIDs are causing the abortion, but as NSAIDs are also prescribed for the pain that may manifest preceding abortion, this creates a false association, leading to a false-positive signal.

bias may have resulted from using NSAIDs to alleviate the associated pain with the miscarriage preceding an abortion.

5.1.2 CONFOUNDERS

5.1.2.1 Confounding by Indication

Confounding by indication occurs when drugs are prescribed to treat manifestations of a particular condition or given preferentially with a certain condition, and the manifestation that occurred due to or with the condition is reported as the ADR itself. For example, during the 1990s, several case reports were published showing an increased risk of hypoglycaemia with the use of ACE inhibitors (ACEi), but this innately was a false signal as ACE inhibitors were usually prescribed in patients with diabetes having concomitant ADAs, and this could have led to the hypoglycaemic effects (Figure 5.5). For example, a study investigated the effect of ARB-associated hypoglycaemic reports while stratifying based on ADA agents. All reports of hypoglycaemia were categorised as case and others as non-case and further mined for the presence of ADAs, ACEis, and ARBs. Diazepam was used as a negative control, and cibenzoline and disopyramide as

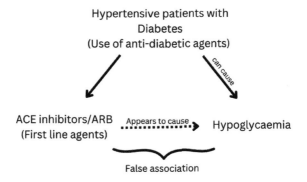

FIGURE 5.5 Depiction of confounding of indication using an example of false association of ACE inhibitors/ARB with hypoglycaemia. The figure depicts how hypertensive patients with diabetes taking anti-diabetic medications also take up ACE inhibitors/ARB for their hypertension. The manifestation of hypoglycaemia in these patients may be wrongly classified as due to anti-hypertensive medications, but in reality might be just due to the disease condition or use of anti-diabetic medications.

positive controls. Crude signal strength analysis revealed positive signal strength with the agents. However, the signal disappeared on stratification based on the presence of ADAs, thus suggesting the role of confounding by indication [20].

Depiction of confounding of indication using an example of false association of ACEi/ARB with hypoglycaemia. The figure depicts how hypertensive patients with diabetes taking anti-diabetic medications also take up ACEi/ARB for their hypertension. The manifestation of hypoglycaemia in these patients may be wrongly classified as due to the anti-hypertensive medications, but in reality might be just due to the disease condition or use of anti-diabetic medications.

Another example of confounding by indication and its impact on signal detection was shown by Ying Li et al. In the study, the researchers deemed to find the association of various drugs associated with two major ADRs, rhabdomyolysis and pancreatitis. The researchers reported false signals for pancreatitis with clonidine, fentanyl, meperidine, metoclopramide, norepineph-rine, nystatin, simethicone, vancomycin, sodium acetate, and calcium acetate, which occurred due to confounding by indication [68]. The reference group's choice is essential to limit this bias' effect. Suppose all other drugs are chosen as the reference group in the presence of this bias. In that case, it may be problematic, as the signal might reflect the risk associated with the drug indi-cation rather than the drug exposure. Furthermore, subgrouping and stratification can also help to determine the presence of indication bias, as was depicted in the ARB-associated hypoglycaemia example.

The study by Peng et al. retrospectively analysed AEs associated with the drug baricitinib using a self-reporting database. However, confounding by indication may have influenced the results, as baricitinib is approved for treating rheumatoid arthritis (RA). Patients with RA likely have different underlying health conditions compared to the general population. Therefore, the AEs observed in the study, such as infections, thrombotic events, and opportunistic infections, may be related to the RA disease itself rather than caused by baricitinib. To determine if baricitinib certainly increases the risk of these AEs, a randomised controlled trial comparing baricitinib to placebo would be needed to control for confounding by indication present in this retrospective study of a drug primar-ily prescribed for RA patients [69].

5.1.2.2   Pre-Exposure Confounders

5.1.2.2.1   Age

Age can act as a confounder in these analyses, given that AE reporting rates tend to be higher in the elderly compared to younger populations. For instance, a retrospective study by Ooba et al. (2010) analyzed over 300,000 case reports in a Japanese AE database and demonstrated signifi-cantly higher RORs for multiple AEs in patients over 65 years old. The authors controlled for these age-related differences by calculating adjusted ORs [70]. Another example is the study by Sakai et al. [45], which examined the association between prostaglandin eye drops and pregnancy-related AEs using data from the JADER database and the FAERS. The focus was on latanoprost usage during pregnancy and its correlation with pregnancy loss. The study identified a drug safety signal linked to latanoprost and pregnancy loss, supported by significant RORs. However, the analysis also revealed that the age of individuals experiencing pregnancy loss due to latanoprost was higher than the average age for all pregnancy-related reports, suggesting that age could potentially confound the observed association [71].

Similarly, red man syndrome is known to occur more frequently in older adults compared to younger populations. Since AE reporting rates also tend to be higher among the elderly, this could lead to disproportionately more red man syndrome cases being reported with certain drugs purely due to the older age distribution of the treated population [72]. The failure to adjust for age-related differences in reporting could result in distorted associations between drugs and red man syndrome. This highlights the need to consider age as a potential factor influencing the apparent link between a drug and AEs in such studies.

5.1.2.2.2 Gender

If not adequately controlled for, differences in AE occurrence and reporting between genders can confound apparent treatment-event associations in PV analyses. For instance, a study by Xu et al. [47] analysed adverse drug events in the FAERS database and found sex differences in AEs for 307 out of 668 frequently used drugs in the United States. Their analysis used Chi-squared tests, RORs, and logistic regression to detect and quantify sex-specific AE signals. This highlights the need to identify sex-specific risks to improve outcomes through personalised medicine approaches. However, the study did not explicitly control for potential confounding by sex in the analysis through techniques like stratification or multivariate adjustment [73].

Another study by Gouverneur et al. [48] found more neurocognitive AEs associated with PCSK9 inhibitor use reported in women (58% of cases) compared to men. Nevertheless, 58% of the overall study population were women. This underlying gender distribution of the sample needs to be considered, as it could confound the apparent association between the drug and AEs. The higher female proportion of neurocognitive events may simply reflect the higher percentage of women in the study rather than indicating a true causal effect of the drug [74].

5.1.2.2.3 Population Differences

Genetic polymorphisms, cultural practices, environmental exposures, and socioeconomic factors can vary significantly among populations, impact susceptibility to ADRs and reporting rates, and confound DA if not adequately controlled.

5.1.2.2.4 Genetic Polymorphisms

The influence of genetic polymorphisms on susceptibility to ADRs is a crucial aspect of PV research. For example, carbamazepine-induced Stevens-Johnson syndrome and toxic epidermal necrolysis have been strongly associated with HLA-B*1502 allele, which occurs in about 10% of Han Chinese and Thai populations but is extremely rare among Caucasians [75]. Failing to adjust for this allele frequency when comparing carbamazepine AEs between Asian and Western countries could lead to confounded, higher disproportionality signals in Asian populations [76].

5.1.2.2.5 Demographics and Socioeconomic Status

Population differences in demographics, socioeconomic status, behaviours, healthcare access, and disease prevalence can confound by creating systematic differences between comparison groups that distort the true association between exposure and outcome. The study by Cohen et al. demonstrates how population differences can confound apparent associations between vaccines and AEs in spontaneous reporting databases like the Vaccine Adverse Event Reporting System (VAERS). They found that multiple sclerosis (MS) cases reported after hepatitis B (HB) vaccination in VAERS disproportionately came from France versus other regions. This skewed geographic distribution differed from MS cases after other vaccines, indicating a potential population difference influencing reporting. The authors suggest that litigation cases and intense media coverage in France created a perception of risk that impacted HB-MS reporting rates. Thus, population factors like publicity and legal situations confounded the HB-MS association in VAERS [77].

5.1.2.2.6 Cultural Practices

Cultural and lifestyle factors that vary among populations, such as using traditional or herbal medicines, can also confound AE data if not accounted for. Chinese populations have a higher prevalence of traditional Chinese medicine (TCM) use than Western countries [78]. However, traditional or herbal medicines have the potential to confound AE data in DA. These medicines can lead to AEs that may not be captured or reported accurately, resulting in a skewed analysis of safety signals. For example, the study by Walji et al. highlights the underreporting of AEs related to herbal products in the Canadian reporting system [79]. Consequently, it is crucial to consider the influence of traditional or herbal medicines when conducting DA of AE data. However, PV databases often do not systematically capture or adjust this.

5.1.2.2.7 Environmental Exposures

Environmental exposures can act as confounding variables in epidemiological studies investigating the relationship between drug exposure and adverse health outcomes. Confounding occurs when the effect of the exposure on the outcome is mixed up with the effect of a third factor associated with both the exposure and the outcome. Using a global safety reporting database, Sipos et al. [50] analysed adverse hepatic events associated with second-generation cephalosporin antibiotics. However, confounding environmental exposures could have influenced the apparent association between antibiotics and liver injury [80]. Research has unveiled that exposure to aflatoxin in some areas of Africa increases susceptibility to liver damage from drugs metabolised by the liver [73, 81]. Therefore, the increased rates of hepatic AEs for antibiotics use seen in the Sipos et al. [50] study may reflect this environmental exposure rather than a true causal effect of the antibiotics. A randomised controlled trial comparing cephalosporin antibiotics to placebo would be needed across geographic regions with differing aflatoxin exposures to control for this potential confounding. Without controlling for this environmental factor, the higher PRRs for cholestatic hepatitis and other hepatic reactions associated with certain cephalosporins observed in the study, particularly in the elderly, may be confounded by the underlying aflatoxin exposure levels of the treatment populations rather than caused directly by the antibiotics.

5.1.2.2.8 Clinical Care Factors

Concomitant medications patients take can confound the apparent association between a particular drug and an AE. Patients, especially those with multiple chronic conditions, are often on complex medication regimens involving several interacting drugs. This polypharmacy makes it challenging to isolate the effect of an individual medication on AEs. As the study by Papay et al. demonstrates, differences in patterns of concomitant medication use across populations and healthcare settings can impact AE profiles identified in safety studies. When comparing a European and global study, they found discrepancies in medications associated with Stevens-Johnson syndrome/toxic epidermal necrolysis, partly attributable to geographic variations in prescription drug utilisation [82]. Factors like newer drug approvals, the exclusion of in-hospital cases, and the lower use of some medications in Europe contributed to the discrepancies. Failing to account for these factors appropriately can lead to misleading interpretations, including missed genuine safety concerns or falsely identifying risks. This highlights the vital need to account for concomitant medications and contextual prescription patterns when evaluating AE associations through continued safety monitoring across diverse real-world populations. Therefore, rigorous statistical adjustments, signal refinement, and clinical expertise are essential to ensure accurate and reliable results in DA, enabling the accurate assessment of a medication's safety.

5.1.3 Strategies to Overcome

In most scenarios, the biases present can be hard to identify and control. One of the methods to control for the presence of undetected biases would be through the use of "control groups", the association or lack of association with the effect of interest, of which is present in the literature. This is done by measuring the signal strength for those agents known to cause the effect of interest (positive controls) and those not associated with the effect of interest (negative controls) and assessing the presence and absence of a signal for positive and negative controls, respectively. The consistency obtained between expected and observed signals would suggest the absence of significant biases in the study. The study conducted by Gregoire et al. [20] to mitigate confounding by indication explains the same (refer to Section 1.21). A summary of the effects and the various methods that can be used to control for confounding factors and biases is given in Table 5.1.

TABLE 5.1

Strategies to Overcome Confounders and Biases in Case-Non-Case Analysis

Confounding Factor	Effect(s)	Method to Overcome	Reference
Detection bias	Overestimated ROR	Consideration of the timeline of the screening programme while conducting the analysis and limiting the analysis to pre-screening	[24, 83]
Reporting bias	Reduced signal detection sensitivity leading to incorrect ROR values	Categorisation of reports based on the reporter and restriction to medically confirmed ones Appropriate classification and restriction to reports where the drug is considered a "suspect"	[84]
Weber effect	Variation in ROR overtime Over- and underestimation of ROR	Restricting the analysis to serious cases, performing a time related analysis	[24]
Notoriety bias and ripple effect	Overestimation of ROR	Performing time-related analysis, restricting the analysis to time before the media coverage	[24, 85]
Dilution bias	Overestimation of ROR for newer drugs	Performing time-related analysis, restricting the analysis to time prior to the media coverage	[85]
Competition bias	Reduction or loss of identification of association of a certain drug-adverse event due to the presence of other known drugs causing that adverse event	Removal of potential drug competitors while analysing the association (sensitivity analyses)	[24]
Channelling bias	Differential prescription due to disease severity leading to falsified or higher ADR reporting because of its use in certain populations	Active comparator restrictive disproportionality analysis	[69]
Protopathic bias	Misclassification of symptoms as the adverse event caused by the drug, when the drug is used to treat specific disease condition	Knowledge of biological plausibility of the association and interpretation while accounting the prognosis of the condition	[86]
Confounding by indication	Distort the true relationship between a treatment/exposure and an outcome in observational studies by introducing bias due to underlying differences in patient characteristics	Propensity score-based approach and sensitivity analysis	[87, 88]
Concomitant medications	It can obscure the accurate assessment of drug-event associations, as concurrent medication use may independently influence adverse event reporting and create erroneous safety signals	Include terms for the most common co-medications in the logistic regression model to adjust for their confounding influence on adverse event reports	[89]
Genetic polymorphisms	Impact susceptibility to adverse drug reactions and reporting rates	Adjust for allele frequencies in different populations when comparing adverse event rates	[90]

(Continued)

TABLE 5.1 *(Continued)*

Strategies to Overcome Confounders and Biases in Case-Non-Case Analysis

Confounding Factor	Effect(s)	Method to Overcome	Reference
Population differences	Can distort the identification of genuine drug-event associations, as variations in reporting behaviours across populations may lead to misleading signals	Control for demographic, socioeconomic, and healthcare access factors in analysis	[24]
Cultural practices	Vary between populations, influencing adverse event reporting	Consider the use of traditional or herbal medicines when interpreting adverse event data	[91]
Environmental exposures	Can lead to false associations between drugs and adverse events, as environmental factors might independently influence both drug usage and the occurrence of adverse outcomes	Subgroup analyses excluding litigation cases and stratifying by location and time period are needed to control for these confounding factors	[92]
Age	Reporting rates might be higher in the one-age category, leading to over or under estimation of outcomes	Calculate adjusted odds ratios to account for age-related differences	[93]
Gender	This could lead to incorrectly attributing outcome discrepancies between groups to the intervention rather than recognising them as arising from gender-related inequalities	Employ stratification or multivariate adjustment to control for gender-related confounding	[94]

5.1.4 CONCLUSION

Spontaneous reporting databases are an important tool to identify rare events associated with commonly or uncommonly used drugs and give insights into real-life PV data. Several methods are utilised to analyse the disproportionality measures, case-non-case studies being one of the most widely used. Despite the limitations and biases, these studies allow for the early detection of PV signals. As described in the chapter, it is vital to be aware of various biases that can occur using PV data and strategies to mitigate them. Various statistical advances are currently being undertaken to address these issues, but it is essential to know the existence of these biases to address them. In all, PV databases are essential for detecting and quantifying the burden of AEs associated with marketed drugs, and coupled with pharmacological and clinical analysis, they remain one of the primary evidence bases for drug safety.

REFERENCES

1. Klein, E., and D. Bourdette. 2013. "Postmarketing adverse drug reactions: A duty to report?" *Neurol Clin Pract* 3 (4):288–294. doi: 10.1212/CPJ.0b013e3182a1b9f0
2. Evans, S. J., P. C. Waller, and S. Davis. 2001. "Use of proportional reporting ratios (PRRs) for signal generation from spontaneous adverse drug reaction reports." *Pharmacoepidemiol Drug Saf* 10 (6):483–486. doi: 10.1002/pds.677

3. van Puijenbroek, E. P., A. Bate, H. G. Leufkens, M. Lindquist, R. Orre, and A. C. Egberts. 2002. "A comparison of measures of disproportionality for signal detection in spontaneous reporting systems for adverse drug reactions." *Pharmacoepidemiol Drug Saf* 11 (1):3–10. doi: 10.1002/pds.668

4. Bate, A., M. Lindquist, I. R. Edwards, S. Olsson, R. Orre, A. Lansner, and R. M. De Freitas. 1998. "A Bayesian neural network method for adverse drug reaction signal generation." *Eur J Clin Pharmacol* 54 (4):315–321. doi: 10.1007/s002280050466

5. DuMouchel, W. 1999. "Bayesian data mining in large frequency tables, with an application to the FDA spontaneous reporting system." *Am Stat* 53 (3):177–190.

6. Chen, Y., J. J Guo, M. Steinbuch, X. Lin, C. R. Buncher, and N. C. Patel. 2008. "Comparison of sensitivity and timing of early signal detection of four frequently used signal detection methods: An empirical study based on the US FDA Adverse Event Reporting System database." *Pharm Med* 22:359–365.

7. Piccinni, C., D. Motola, G. Marchesini, and E. Poluzzi. 2011. "Assessing the association of pioglitazone use and bladder cancer through drug adverse event reporting." *Diabetes Care* 34 (6):1369–1371. doi: 10.2337/dc10-2412

8. Faillie, J. L., P. Petit, J. L. Montastruc, and D. Hillaire-Buys. 2013. "Scientific evidence and controversies about pioglitazone and bladder cancer: Which lessons can be drawn?" *Drug Saf* 36 (9):693–707. doi: 10.1007/s40264-013-0086-y

9. Lukose, L., P. M. Shantaram, A. Raj, G. Nair, A. M. Shaju, and K. Subeesh V. 2023. "Purine antimetabolites associated *Pneumocystis jirovecii* pneumonia." *Pharmacoepidemiol Drug Saf.* doi: 10.1002/pds.5647

10. Lv, B., Y. Li, A. Shi, and J. Pan. 2023. "Model Driven Method for exploring Individual and Confounding Effects in Spontaneous Adverse Event Reporting Databases."

11. Lu, Z., A. Suzuki, and D. Wang. 2023. "Statistical methods for exploring spontaneous adverse event reporting databases for drug-host factor interactions." *BMC Med Res Methodol* 23 (1):71.

12. Böhm, R., C. Bulin, V. Waetzig, I. Cascorbi, H. J. Klein, and T. Herdegen. 2021. "Pharmacovigilance-based drug repurposing: The search for inverse signals via OpenVigil identifies putative drugs against viral respiratory infections." *Br J Clin Pharmacol* 87 (11):4421–4431. doi: 10.1111/bcp.14868

13. Jacob, A. T., A. H. Kumar, G. Halivana, L. Lukose, G. Nair, and V. Subeesh. "Bioinformatics-Guided Disproportionality Analysis of Sevoflurane-Induced Nephrogenic Diabetes Insipidus Using the FDA Adverse Event Reporting System (FAERS) Database."

14. Hazell, L., and S. A. Shakir. 2006. "Under-reporting of adverse drug reactions: A systematic review." *Drug Saf* 29 (5):385–396. doi: 10.2165/00002018-200629050-00003

15. CDER, FDA. 2020. "World of Drug Safety Unit List: Overview of Drug Safety." https://www.accessdata.fda.gov/scripts/cderworld/index.cfm?action=drugsafety:main&unit=1&lesson=1&topic=8&page=4#:~:text=Underreporting%3B%20FDA%20does%20not%20get.

16. Raschi, E., E. Poluzzi, F. Salvo, A. Pariente, F. De Ponti, G. Marchesini, and U. Moretti. 2018. "Pharmacovigilance of sodium-glucose co-transporter-2 inhibitors: What a clinician should know on disproportionality analysis of spontaneous reporting systems." *Nutr Metab Cardiovasc Dis* 28 (6):533–542. doi: 10.1016/j.numecd.2018.02.014

17. Varallo, F. R., O. Guimarães Sde, S. A. Abjaude, and C. Mastroianni Pde. 2014. "[Causes for the under-reporting of adverse drug events by health professionals: A systematic review]." *Rev Esc Enferm USP* 48 (4):739–747. doi: 10.1590/s0080-623420140000400023

18. Faillie, J.-L. 2019. "Case–non-case studies: Principle, methods, bias and interpretation." *Therapies* 74 (2):225–232. doi: 10.1016/j.therap.2019.01.006

19. Neha, R., E. Beulah, B. Anusha, S. Vasista, C. Stephy, and V. Subeesh. 2020. "Aromatase inhibitors associated osteonecrosis of jaw: Signal refining to identify pseudo safety signals." *Int J Clin Pharm* 42 (2):721–727. doi: 10.1007/s11096-020-01018-z

20. Grégoire, F., A. Pariente, A. Fourrier-Reglat, F. Haramburu, B. Bégaud, and N. Moore. 2008. "A signal of increased risk of hypoglycaemia with angiotensin receptor blockers caused by confounding." *Br J Clin Pharmacol* 66 (1):142–145. doi: 10.1111/j.1365-2125.2008.03176.x

21. Skelly, A. C., J. R. Dettori, and E. D. Brodt. 2012. "Assessing bias: The importance of considering confounding." *Evid Based Spine Care J* 3 (1):9–12. doi: 10.1055/s-0031-1298595

22. Ahmad, S. R., and J. Swann. 2008. "Exenatide and rare adverse events." *N Engl J Med* 358 (18):1970–1971; discussion 1971-2.

23. Nauck, M. A., and N. Friedrich. 2013. "Do GLP-1-based therapies increase cancer risk?" *Diabetes Care* 36 (Suppl 2):S245–S252. doi: 10.2337/dcS13-2004

24. Faillie, J. L. 2019. "Case-non-case studies: Principle, methods, bias and interpretation." *Therapie* 74 (2):215–222. doi: 10.1016/j.therap.2019.01.006

25. Baah, E. M. 2020. "Analysis of data on adverse drug events reported to the Food and Drugs Administration of the United States of America." *Open J Stat* 10 (02):203.

26. Stephenson, W. P., and M. Hauben. 2007. "Data mining for signals in spontaneous reporting databases: Proceed with caution." *Pharmacoepidemiol Drug Saf* 16 (4):359–365.

27. Bate, A., M. Lindquist, I. R. Edwards, S. Olsson, R. Orre, A. Lansner, and R. M. De Freitas. 1998. "A Bayesian neural network method for adverse drug reaction signal generation." *Eur J Clin Pharmacol* 54 (4):315–321. doi: 10.1007/s002280050466

28. Böhm, R., L. von Hehn, T. Herdegen, H. J. Klein, O. Bruhn, H. Petri, and J. Höcker. 2016. "OpenVigil FDA – Inspection of U.S. American adverse drug events pharmacovigilance data and novel clinical applications." *PLoS One* 11 (6):e0157753. doi: 10.1371/journal.pone.0157753

29. Sarangdhar, M., S. Tabar, C. Schmidt, A. Kushwaha, K. Shah, J. E. Dahlquist, A. G. Jegga, and B. J. Aronow. 2016. "Data mining differential clinical outcomes associated with drug regimens using adverse event reporting data." *Nat Biotechnol* 34 (7):697–700. doi: 10.1038/nbt.3623

30. Norén, G N., R. Orre, A. Bate, and I. R. Edwards. 2007. "Duplicate detection in adverse drug reaction surveillance." *Data Min Knowl Discov* 14:305–328.

31. Contreras-Salinas, H., L. M. Baiza-Durán, M. A. Bautista-Castro, D. R. Alonso-Rodríguez, and L. Y. Rodríguez-Herrera. 2022. "Underreporting and Triggering Factors for Reporting ADRs of Two Ophthalmic Drugs: A Comparison between Spontaneous Reports and Active Pharmacovigilance Databases." *Healthcare*.

32. Hughes, M. L., and M. Weiss. 2019. "Adverse drug reaction reporting by community pharmacists – The barriers and facilitators." *Pharmacoepidemiol Drug Saf* 28 (12):1552–1559. doi: 10.1002/pds.4800

33. Jacob, D., B. Marrón, J. Ehrlich, and P. A. Rutherford. 2013. "Pharmacovigilance as a tool for safety and monitoring: A review of general issues and the specific challenges with end-stage renal failure patients." *Drug Healthc Patient Saf* 5:105–112 doi: 10.2147/dhps.S43104.

34. Matsuda, S., K. Aoki, T. Kawamata, T. Kimotsuki, T. Kobayashi, H. Kuriki, T. Nakayama, S. Okugawa, Y. Sugimura, M. Tomita, and Y. Takahashi. 2015. "Bias in spontaneous reporting of adverse drug reactions in Japan." *PLoS One* 10 (5):e0126413. doi: 10.1371/journal.pone.0126413

35. Hoffman, K. B., A. R. Demakas, M. Dimbil, N. P. Tatonetti, and C. B. Erdman. 2014. "Stimulated reporting: The impact of US food and drug administration-issued alerts on the adverse event reporting system (FAERS)." *Drug Saf* 37 (11):971–980. doi: 10.1007/s40264-014-0225-0

36. Toki, T., and S. Ono. 2020. "Assessment of factors associated with completeness of spontaneous adverse event reporting in the United States: A comparison between consumer reports and healthcare professional reports." *J Clin Pharm Ther* 45 (3):462–469. doi: 10.1111/jcpt.13086

37. Arora, A., R. K. Jalali, and D. Vohora. 2017. "Relevance of the Weber effect in contemporary pharmacovigilance of oncology drugs." *Ther Clin Risk Manag* 13:1195–1203. doi: 10.2147/tcrm.S137144

38. Weber, J. C. P. 1984. "Epidemiology of adverse reactions to nonsteroidal anti-inflammatory drugs." *Adv Inflamm Res* 6:1.

39. Hartnell, N. R., and J. P. Wilson. 2004. "Replication of the Weber effect using postmarketing adverse event reports voluntarily submitted to the United States Food and Drug Administration." *Pharmacotherapy* 24 (6):743–749. doi: 10.1592/phco.24.8.743.36068

40. Wallenstein, E. J., and D. Fife. 2001. "Temporal patterns of NSAID spontaneous adverse event reports: The Weber effect revisited." *Drug Saf* 24 (3):233–237. doi: 10.2165/00002018-200124030-00006

41. Hartnell, N. R., J. P. Wilson, N. C. Patel, and M. L. Crismon. 2003. "Adverse event reporting with selective serotonin-reuptake inhibitors." *Ann Pharmacother* 37 (10):1387–1391. doi: 10.1345/aph.1C522

42. de Graaf, L., M. A. Fabius, W. L. Diemont, and E. P. van Puijenbroek. 2003. "The Weber-curve pitfall: Effects of a forced introduction on reporting rates and reported adverse reaction profiles." *Pharm World Sci* 25 (6):260–263. doi: 10.1023/b:phar.0000006518.22231.ea

43. Chhabra, P., X. Chen, and S. R. Weiss. 2013. "Adverse event reporting patterns of newly approved drugs in the USA in 2006: An analysis of FDA adverse event reporting system data." *Drug Saf* 36 (11):1117–1123. doi: 10.1007/s40264-013-0115-x

44. Hoffman, K. B., M. Dimbil, C. B. Erdman, N. P. Tatonetti, and B. M. Overstreet. 2014. "The Weber effect and the United States Food and Drug Administration's Adverse Event Reporting System (FAERS): Analysis of sixty-two drugs approved from 2006 to 2010." *Drug Saf* 37 (4):283–294. doi: 10.1007/s40264-014-0150-2

45. Modgill, V., L. Dormegny, and D. J. Lewis. 2020. "Reporting rates of adverse reactions to specialty care medicines exhibit a direct positive correlation with patient exposure: A lack of evidence for the Weber effect." *Br J Clin Pharmacol* 86 (12):2393–2403. doi: 10.1111/bcp.14342

46. McAdams, M. A., L. A. Governale, L. Swartz, T. A. Hammad, and G. J. Dal Pan. 2008. "Identifying patterns of adverse event reporting for four members of the angiotensin II receptor blockers class of drugs: Revisiting the Weber effect." *Pharmacoepidemiol Drug Saf* 17 (9):882–889. doi: 10.1002/pds.1633

47. Center for Drug, Evaluation, and Research. 2019. *Questions and Answers on FDA's Adverse Event Reporting System (FAERS)*. U.S. Food and Drug Administration.

48. Arora, A., R. K. Jalali, and D. Vohora. 2017. "Relevance of the Weber effect in contemporary pharmacovigilance of oncology drugs." *Ther Clin Risk Manag* 13:1195–1203. doi: 10.2147/tcrm.S137144

49. Moore, N., G. Hall, M. Sturkenboom, R. Mann, R. Lagnaoui, and B. Begaud. 2003. "Biases affecting the proportional reporting ratio (PPR) in spontaneous reports pharmacovigilance databases: The example of sertindole." *Pharmacoepidemiol Drug Saf* 12 (4):271–281. doi: 10.1002/pds.848

50. Gale, E. A. 2009. "Collateral damage: The conundrum of drug safety." *Diabetologia* 52 (10):1975–1982. doi: 10.1007/s00125-009-1491-8

51. Raschi, E., C. Piccinni, E. Poluzzi, G. Marchesini, and F. De Ponti. 2013. "The association of pancreatitis with antidiabetic drug use: Gaining insight through the FDA pharmacovigilance database." *Acta Diabetol* 50 (4):569–577. doi: 10.1007/s00592-011-0340-7

52. Pariente, A., F. Gregoire, A. Fourrier-Reglat, F. Haramburu, and N. Moore. 2007. "Impact of safety alerts on measures of disproportionality in spontaneous reporting databases: The notoriety bias." *Drug Saf* 30 (10):891–898. doi: 10.2165/00002018-200730100-00007

53. Pariente, A., A. Daveluy, A. Laribière-Bénard, G. Miremont-Salame, B. Begaud, and N. Moore. 2009. "Effect of date of drug marketing on disproportionality measures in pharmacovigilance: The example of suicide with SSRIs using data from the UK MHRA." *Drug Saf* 32 (5):441–447. doi: 10.2165/00002018-200932050-00007

54. Michel, C., E. Scosyrev, M. Petrin, and R. Schmouder. 2017. "Can disproportionality analysis of post-marketing case reports be used for comparison of drug safety profiles?" *Clin Drug Investig* 37 (5):415–422. doi: 10.1007/s40261-017-0503-6

55. Pariente, A., P. Avillach, F. Salvo, F. Thiessard, G. Miremont-Salamé, A. Fourrier-Reglat, F. Haramburu, B. Bégaud, and N. Moore. 2012. "Effect of competition bias in safety signal generation: Analysis of a research database of spontaneous reports in France." *Drug Saf* 35 (10):855–864. doi: 10.1007/bf03261981

56. Salvo, F., F. Leborgne, F. Thiessard, N. Moore, B. Bégaud, and A. Pariente. 2013. "A potential event-competition bias in safety signal detection: Results from a spontaneous reporting research database in France." *Drug Saf* 36 (7):565–572. doi: 10.1007/s40264-013-0063-5

57. Scholl, J. 2022. "Signal Detection in Pharmacovigilance: Time for a New Era?".

58. Petri, H., and J. Urquhart. 1991. "Channeling bias in the interpretation of drug effects." *Stat Med* 10 (4):577–581. doi: https://doi.org/10.1002/sim.4780100409

59. Tseng, C. H., K. Y. Lee, and F. H Tseng. 2015. "An updated review on cancer risk associated with incretin mimetics and enhancers." *J Environ Sci Health C Environ Carcinog Ecotoxicol Rev* 33 (1):67–124. doi: 10.1080/10590501.2015.1003496

60. Gatti, M., I. C. Antonazzo, I. Diemberger, F. De Ponti, and E. Raschi. 2021. "Adverse events with sacubitril/valsartan in the real world: Emerging signals to target preventive strategies from the FDA adverse event reporting system." *Eur J Prev Cardiol* 28 (9):983–989. doi: 10.1177/2047487320915663

61. Alkabbani, W., and J. M. Gamble. 2023. "Active-comparator restricted disproportionality analysis for pharmacovigilance signal detection studies of chronic disease medications: An example using sodium/glucose cotransporter 2 inhibitors." *Br J Clin Pharmacol* 89 (2):431–439. doi: 10.1111/bcp.15178

62. McLaughlin, J. K., L. Lipworth, W. H. Chow, and W. J. Blot. 1998. "Analgesic use and chronic renal failure: A critical review of the epidemiologic literature." *Kidney Int* 54 (3):679–686. doi: 10.1046/j.1523-1755.1998.00043.x

63. Perneger, T. V., P. K. Whelton, and M. J. Klag. 1994. "Risk of kidney failure associated with the use of acetaminophen, aspirin, and nonsteroidal antiinflammatory drugs." *N Engl J Med* 331 (25):1675–1679.

64. Fored, C. M., E. Ejerblad, P. Lindblad, J. P. Fryzek, P. W. Dickman, L. B. Signorello, L. Lipworth, C.-G. Elinder, and W. J. Blot, and J. K. McLaughlin. 2001. "Acetaminophen, aspirin, and chronic renal failure." *N Engl J Med* 345 (25):1801–1808.

65. Signorello, L. B., J. K. McLaughlin, L. Lipworth, S. Friis, H. T. Sørensen, and W. J. Blot. 2002. "Confounding by indication in epidemiologic studies of commonly used analgesics." *Am J Ther* 9 (3):199–205. doi: 10.1097/00045391-200205000-00005

66. Salas, M., A. Hofman, and B. H. Stricker. 1999. "Confounding by indication: An example of variation in the use of epidemiologic terminology." *Am J Epidemiol* 149 (11):981–983. doi: 10.1093/oxfordjournals.aje.a009758

67. Daniel, S., G. Koren, E. Lunenfeld, and A. Levy. 2015. "NSAIDs and spontaneous abortions – True effect or an indication bias?" *Br J Clin Pharmacol* 80 (4):750–754. doi: 10.1111/bcp.12653

68. Li, Y., H. Salmasian, S. Vilar, H. Chase, C. Friedman, and Y. Wei. 2014. "A method for controlling complex confounding effects in the detection of adverse drug reactions using electronic health records." *J Am Med Inform Assoc* 21 (2):308–314. doi: 10.1136/amiajnl-2013-001718

69. Peng, L., K. Xiao, S. Ottaviani, J. Stebbing, and Y.-J. Wang. 2020. "A real-world disproportionality analysis of FDA Adverse Event Reporting System (FAERS) events for baricitinib." *Expert Opin Drug Saf* 19 (11):1505–1511. doi: 10.1080/14740338.2020.1799975

70. Ooba, N., and K. Kubota. 2010. "Selected control events and reporting odds ratio in signal detection methodology." *Pharmacoepidemiol Drug Saf* 19 (11):1159–1165. doi: 10.1002/pds.2014

71. Sakai, T., C. Mori, H. Koshiba, R. Yuminaga, K. Tanabe, and F. Ohtsu. 2021. "Pregnancy loss signal from prostaglandin eye drop use in pregnancy: A disproportionality analysis using Japanese and US spontaneous reporting databases." *Drugs – Real World Outcomes* 9 (1):43–51. doi: 10.1007/s40801-021-00287-y

72. M. Shaju, A., N. Panicker, V. Chandni, V. M. L. Prasanna, G. Nair, and V. Subeesh. 2022. "Drugs-associated with red man syndrome: An integrative approach using disproportionality analysis and Pharmip." *J Clin Pharm Ther* 47 (10):1650–1658.

73. Xu, Y., Y. Y. Gong, and M. N. Routledge. 2018. "Aflatoxin exposure assessed by aflatoxin albumin adduct biomarker in populations from six African countries." *World Mycotoxin J* 11 (3):411–419. doi: 10.3920/wmj2017.2284

74. Gouverneur, A., P. Sanchez-Pena, G. Veyrac, J.-E. Salem, B. Bégaud, and J. Bezin. 2021. "Neurocognitive disorders associated with PCSK9 inhibitors: A pharmacovigilance disproportionality analysis." *Cardiovasc Drugs Ther.* doi: 10.1007/s10557-021-07242-7

75. Hung, S.-I., W.-H. Chung, S.-H. Jee, W.-C. Chen, Y.-T. Chang, W.-R. Lee, S.-L. Hu, M.-T. Wu, G.-S. Chen, T.-W. Wong, P.-F. Hsiao, W.-H. Chen, H.-Y. Shih, W.-H. Fang, C.-Y. Wei, Y.-H. Lou, Y.-L. Huang, J.-J. Lin, and Y.-T. Chen. 2006. "Genetic susceptibility to carbamazepine-induced cutaneous adverse drug reactions." *Pharmacogenet Genomics* 16 (4):297–306. doi: 10.1097/01.fpc.0000199500.46842.4a

76. Chonlaphat, S., M. Biswas, L. Palita, and S. Montinee. 2023. "Clinical pharmacogenomics implementation in Thailand: A dream come true." *Pharmacogenomics* 24 (6):297–301. doi: 10.2217/pgs-2023-0071

77. Cohen, C., H. Annick, and A. Khromava. 2018. "Comment on "central demyelinating diseases after vaccination against hepatitis B virus: A disproportionality analysis within the VAERS database." *Drug Saf.* doi: https://doi.org/10.1007/s40264-018-0733-4

78. Shi, X., D. Zhu, S. Nicholas, B. Hong, X. Man, and P. He. 2020. "Is traditional Chinese medicine "mainstream" in China? Trends in traditional Chinese medicine health resources and their utilization in traditional Chinese medicine hospitals from 2004 to 2016." *Evid Based Complement Alter Med (eCAM)* 2020:1–8. doi: 10.1155/2020/9313491

79. Walji, R., H. Boon, J. Barnes, Z. Austin, G. Baker, and S. Welsh. 2009. "Adverse event reporting for herbal medicines: A result of market forces." *Healthc Policy | Politiques de Santé* 4 (4):77–90. doi: 10.12927/hcpol.2009.20820

80. Sipos, M., A. Farcas, D. Leucuta, C. Bucsa, M. Huruba, and C. Mogosan. 2021. "Second-generation cephalosporins-associated drug-induced liver disease: A study in VigiBase with a focus on the elderly." *Pharmaceuticals* 14 (5):441. doi: 10.3390/ph14050441

81. Wolde, M. 2017. "Effects of aflatoxin contamination of grains in Ethiopia." *Int J Agric Sci* 7 (4):1298–1308.

82. Papay, J., N. Yuen, G. Powell, M. Mockenhaupt, and T. Bogenrieder. 2011. "Spontaneous adverse event reports of Stevens-Johnson syndrome/toxic epidermal necrolysis: Detecting associations with medications." *Pharmacoepidemiol Drug Saf* 21 (3):289–296. doi: 10.1002/pds.2276

83. Viswanathan, M., C. D. Patnode, N. D. Berkman, E. B. Bass, S. Chang, L. Hartling, M. H. Murad, J. R. Treadwell, and R. L. Kane. 2018. "Recommendations for assessing the risk of bias in systematic reviews of healthcare interventions." *J Clin Epidemiol* 97:26–34.

84. Toki, T., and S. Ono. 2020. "Assessment of factors associated with completeness of spontaneous adverse event reporting in the United States: A comparison between consumer reports and healthcare professional reports." *J Clin Pharm Ther* 45 (3):472–476.

85. Raschi, E., E. Poluzzi, F. Salvo, A. Pariente, F. De Ponti, G. Marchesini, and U. Moretti. 2018. "Pharmacovigilance of sodium-glucose co-transporter-2 inhibitors: What a clinician should know on disproportionality analysis of spontaneous reporting systems." *Nutr Metab Cardiovasc Dis* 28 (6):533–542.

86. Tamim, H, A. A. T. Monfared, and J. LeLorier. 2007. "Application of lag-time into exposure definitions to control for protopathic bias." *Pharmacoepidemiol Drug Saf* 16 (3):250–258.

87. Fox, G. J., A. Benedetti, C. D. Mitnick, M. Pai, D. Menzies, and Collaborative Group for Meta-Analysis of Individual Patient Data in MDR-TB. 2016. "Propensity score-based approaches to confounding by indication in individual patient data meta-analysis: non-standardized treatment for multidrug resistant tuberculosis." *PLoS One* 11 (3):e0151724.

88. Li, Y., Y. Lee, R. A. Wolfe, H. Morgenstern, J. Zhang, F. K. Port, and B. M. Robinson. 2015. "On a preference-based instrumental variable approach in reducing unmeasured confounding-by-indication." *Stat Med* 34 (7):1150–1168.

89. Bénard-Laribière, A., P. Noize, E. Pambrun, F. Bazin, H. Verdoux, M. Tournier, B. Bégaud, and A. Pariente. 2016. "Comorbidities and concurrent medications increasing the risk of adverse drug reactions: Prevalence in French benzodiazepine users." *Eur J Clin Pharmacol* 72:869–876.

90. Sandberg, L., H. Taavola, Y. Aoki, R. Chandler, and G. N. Norén. 2020. "Risk factor considerations in statistical signal detection: Using subgroup disproportionality to uncover risk groups for adverse drug reactions in VigiBase." *Drug Saf* 43:999–1009.

91. Almenoff, J., J. M. Tonning, A. L. Gould, A. Szarfman, M. Hauben, R. Ouellet-Hellstrom, R. Ball, K. Hornbuckle, L. Walsh, and C. Yee. 2005. "Perspectives on the use of data mining in pharmacovigilance." *Drug Saf* 28:981–1007.

92. Seabroke, S., G. Candore, K. Juhlin, N. Quarcoo, A. Wisniewski, R. Arani, J. Painter, P. Tregunno, G. N. Norén, and J. Slattery. 2016. "Performance of stratified and subgrouped disproportionality analyses in spontaneous databases." *Drug Saf* 39:355–364.

93. Zhao, Z., R. Liu, L. Wang, L. Li, C. Song, and P. Zhang. 2022. "A computational framework for identifying age risks in drug-adverse event pairs." In *AMIA Annual Symposium Proceedings*.

94. Ferrarotto, F. 2009. "Signaling Potential Gender Effect in a Spontaneous Reporting System: Cardiac Effects Associated with the Use of Antibiotics."

6 Role of Medical Coding in Signal Detection

Dipika Bansal, Beema T Yoosuf, and Muhammed Favas KT

6.1 INTRODUCTION

Pharmacovigilance (PV) is a scientific discipline that assures the safe, rational, and ethical use of medications, thereby optimizing patient safety and public healthcare. Spontaneous reporting systems (SRSs) serve as a critical source of drug safety information. They provide a unique perspective by offering insights into the most serious and rare adverse events (AEs). Data mining techniques enable the systematic analysis and mitigation of risks associated with medication use after approval using these systems, eventually leading to the compilation of drug safety signals.

The enormous drug safety data in SRSs presents a challenge to maintain reliable information or accurate data retrieval [1, 2]. A terminology is required to facilitate the recognition of medical conditions, representing unique concepts and offering the homogeneous grouping of related terms to support signal generation. Subsequently, it ought to promote intuitive or statistical identification of AEs with a threshold frequency or disproportionate incidence, as well as the identification of key occurrences that are frequently drug-related and facilitate the evaluation of new syndromes [3]. Thus, medical coding is an integral part of signal detection in the domains of PV and drug safety.

6.2 MEDICAL CODING

Medical coding is the translation of procedures, healthcare diagnoses, equipment, medical services, and other healthcare information into numeric codes. Coding is the translation of medical terms and concepts into a computer-interpretable format that indexes them uniquely [4]. These codes serve as a universal terminology for doctors, insurance companies, hospitals, drug safety authorities, government agencies, and other health-related organizations. Furthermore, medical coding is the act of integrating the universal medical dictionaries for translating the reported AEs, drugs, or even diseases into a standard term for the generation of drug safety signals. Subsequently, once the drug-related information has been accurately coded, the safety signals can be efficiently generated.

Coding dictionary brings hierarchical logical order to the plethora of descriptive terms that healthcare professionals (HCPs) as well as patients use to describe medical conditions and the broad spectrum of drugs. This can be accomplished through a process of data structuring, which involves organizing information in a way that is logical and meaningful for both humans and computers. By condensing complex descriptions into standardized codes or terms, it becomes feasible to accurately and succinctly record data within an electronic database. This, in turn, facilitates easy navigation and retrieval of information related to comparable medical conditions related with a specific pharmaceutical product. The data can then be presented in a summarized format or numerical tables for enhanced clarity and accessibility [1, 4, 5].

6.3 MEDICAL DICTIONARIES

Medical dictionaries are specialized lexicons that cover terms intended for the application in the pre- and post-approval phases of the drug regulatory process, such as diagnoses, signs and symptoms, adverse drug reactions (ADRs), therapeutic indications, the drug names, medical procedures, and medical history [6]. These terms can be used to record drug, disease, and AEs in the tabulation

DOI: 10.1201/9781032629940-6

of the safety data to generate drug safety signals as well as in constructing standard product information and documentation.

In PV, the features of the coding dictionary have a huge influence on the data. If the dictionary has relatively limited terms, compromises need to be made while coding the information. Furthermore, this may result in the meaningless handling of clinical data, for instance, cardiopulmonary failure and cardiac failure congestive may simultaneously become "heart failure." Likewise, if the relationships in the lexicon are not entirely accurate, a case that was originally described as "psychological problems" may end up being categorized as "psychotic" within the database.

Another advantage is that the choice of the dictionary allows for a more precise documentation of the AE. If a dictionary consists of 20 distinct kinds of diarrhea, it might be challenging to explain a basic research question concerning biguanide-associated diarrhea. The answer might be there were three reports of inflammatory diarrhea, four reports of bloody diarrhea, two reports of painful diarrhea, and five reports of watery diarrhea. When there isn't any appropriate group term for "diarrhoea" then the real AEs associated with the drug might be unnoticed. Such splitting without any homogeneous grouping might impede one's ability to detect new safety signals during the post-marketing safety surveillance. For this reason, including such dictionaries while coding medical data will facilitate the elimination of significant mistakes when generating drug safety signals.

6.3.1 DRUG DICTIONARIES

The huge range of available drugs renders a herculean task to maintain reliable information in the SRS. While considering the entirety of countless drug compositions, pharmaceutical formulations, routes of intake, therapeutic classes, brand names, indications, and doses, the process turns extremely complex. The expanding global pharma market, along with rising awareness of the necessity of patient well-being, has intensified the prominence of a conventional pharmaceutical database for the purpose of preclinical and post-marketing drug safety. Several types of standard classifications have been developed, and the most commonly employed drug dictionaries are addressed below.

6.3.1.1 Anatomical Therapeutic Chemical Classification

"The Anatomical Therapeutic Chemical (ATC) classification system, recommended by the World Health Organization (WHO), is the most widely recognized classification system for drugs." The ATC classification organizes active compounds into distinct groups based on their effects on organs or systems, as well as their therapeutic, pharmacological, and chemical properties [7, 8]. This classification system follows a five-level hierarchy.

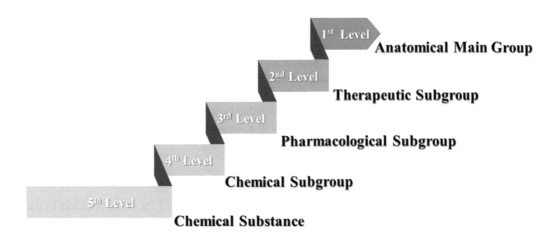

FIGURE 6.1   Anatomical Therapeutic Chemical (ATC) Classification of drugs at five different levels.

TABLE 6.1

ATC System Main Groups

A	Alimentary tract and metabolism
B	Blood and blood-forming organs
C	Cardiovascular system
D	Dermatological
G	Genitourinary system and sex hormones
H	Systemic hormonal preparations, excluding sex hormones and insulins
J	Anti-infective for systemic use
L	Antineoplastic and immunomodulating agents
M	Musculoskeletal system
N	Nervous system
P	Antiparasitic products, insecticides, and repellents
R	Respiratory system
S	Sensory organs
V	Various

At the top level, there are 14 primary anatomical or pharmacological groups (Table 6.1). Each primary group is further divided into two levels, which can be either pharmacological or therapeutic. The third and fourth levels encompass chemical, pharmacological, or therapeutic subgroups, while the fifth level identifies the specific chemical substance. In cases where pharmacological subgroups are more relevant than therapeutic or chemical subgroups, the third, second, and fourth levels are often used to provide detailed categorization (Table 6.2) [7, 9].

6.3.1.2 WHO Drug Dictionaries

"WHO DD is a global medicinal information dictionary that comprises medicinal products and active ingredients indicated for human use, such as active chemical substances, biotherapeutics, vaccines, nutritional supplements, herbal remedies, radiopharmaceuticals, and diagnostic agents." WHO DD is a database created in 1968 by the WHO Programme for International Drug Monitoring (PIDM) in response to the thalidomide tragedy [1, 10]. It contains information about drugs, including brand names, active ingredients, dosage forms, doses, countries of sale, and marketing authorization holders [11]. The Uppsala Monitoring Centre (UMC) manages and updates the WHO DD to meet international standards and user needs, ensuring its relevance and accuracy in monitoring drug safety throughout pre- and post-approval stages [11].

TABLE 6.2

Comprehensive Classification of Metformin Depicted in the Code Structure

A	Alimentary tract and metabolism (1st level: Anatomical Main Group)
A10	Drugs used in diabetes (2nd level: Therapeutic Subgroup)
A10B	Blood glucose-lowering drugs, excluding insulins (3rd level: Pharmacological Subgroup)
A10BA	Biguanides (4th level: Chemical Subgroup)
A10BA02	Metformin (5th level: Chemical Substance)

6.3.1.2.1   WHODrug

The dictionary was developed with a numerical code that exclusively identifies each distinct ingredient, resulting in a connection between drugs with diverse names in various nations with the same ingredient [9, 12, 13]. Medicines in WHODrug are classified into standardized drug groups using the ATC system, allowing for the grouping of medicinal products that have one or more common characteristics. The embedded data framework and drug classification in WHODrug support numerous means of aggregating drugs for the identification and analysis of potential AEs. The various levels of information in WHODrug are used to analyze the association between a drug or drug class with an AE [9].

6.3.1.2.2   Data Structure

WHODrug comprises drugs with distinct trade names associated with the active constituents, pharmaceutical formulation, strength, Marketing Authorization Holder (MAH), and marketing country, together with a categorization in accordance with the WHO ATC system. Additionally, it includes a number of umbrella terms for drugs that are not connected to specific active constituents.

All the medicines are assigned a unique identifier, a medicinal product identification (MPID) number, to distinguish them from one another, despite whether they differ in active ingredients, and to link related drugs in a logical way. For example, for different trade names with the same active constituent or in combination, it uses an alphanumeric code, i.e., the drug code. The alphanumeric drug code is an 11-character sequence composed of three components. The first component is six characters long, the drug record number; the second is two characters long (Sequence 1) and stands for active ingredient variation; and the third is three characters long (Sequence 2) and represents the name of the drug, which can be a brand name, a generic name, or an uncertain name [9, 13]. The example of a drug code for the medication using WHODrug is illustrated in Figure 6.2.

6.3.1.2.3   WHODrug Enhanced

WHODrug Enhanced (WHODE) is now formally withdrawn and will not be further produced. The dictionary has been available to users since 2005 and was last released on 1 September 2020. WHODrug Global is the exclusive dictionary available from March 1, 2021. WHODrug Global is the most inclusive and systematized drug dictionary and contains all the details found in WHODE plus herbal medications [14].

6.3.1.2.4   WHODrug Global

The WHODrug Global package has been released since March 1, 2017. The goal of WHODrug Global is to standardize and simplify the release of the WHODrug product family. UMC aims to facilitate

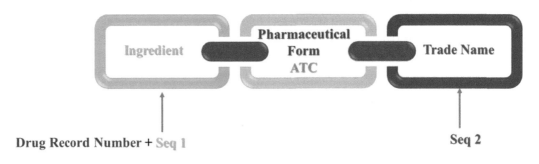

FIGURE 6.2   Drug code for medication using WHODrug.

the convenient use of WHODrug data by merging all WHODrug dictionaries into one, and the data is continuously updated with new releases twice a year on 1st March and 1st September [12].

WHODrug Global is the universal standardized reference for medicinal product information maintained by UMC. This provides a consistent drug vocabulary with precise terminology for classifying concomitant drugs owing to its unique drug code structure and extensive coverage. The dictionary is used to locate drug names and analyze medication-related information from approximately 150 countries, including active ingredients and ATC system of products [14].

6.3.1.2.5 Role in Safety Signal Detection

Medications reported in the post-marketing database as individual case safety reports (ICSRs) are coded with WHODrug that consists of ATC classification, facilitating interpretation and evaluation of drug safety signals. Furthermore, the structure of the dictionary allows for the identification and consolidation of information at various levels. This applies to instances where medications are documented as concurrent or interacting, as well as cases involving medications speculated as eliciting ADRs.

Within safety databases, ICSRs offer the opportunity to investigate and identify AE signals that may involve previously unknown or inadequately documented relationships between AEs and drugs. This signal detection process encompasses various applications, such as linking drugs to ADRs in case reports, identifying substandard drugs, analyzing AE reports in social media, investigating age-related ADRs, and examining country-specific reporting trends [11, 13, 15, 16].

In PV, disproportionality analysis is a well-known tool to enhance signal detection within drug safety databases. This analysis relies on disproportionality metrics incorporating the frequentist-based data mining algorithms (DMAs) such as the "proportional reporting ratio (PRR) and reporting odds ratio (ROR)" and Bayesian-based DMAs such as "Multi-Item Gamma Poisson Shrinker (MGPS) and Bayesian Confidence Propagation Neural Network (BCPNN)" that are computed based on the ATC system. When executing disproportionality analysis at the ATC code level, it involves calculating the reporting rate of a specific AE within a particular ATC code and comparing it to the reporting rate of the same event across all ICSRs in the SRSs, excluding those containing drugs from the ATC code being studied. Subsequently, the ATC system enables the study of drug utilization and classifying drugs based on their intended purposes, therapeutic properties, and chemical and pharmacological characteristics [1].

Moreover, employing WHODrug for drug coding enhances patient care and safety concerning medication usage. It also contributes to public health initiatives by furnishing reliable and well-balanced data to effectively evaluate the risk-benefit profile of pharmaceuticals.

6.4 DISEASE CLASSIFICATION

It is imperative to have a unique disease classification system in order to display and report the results in a systematic manner while striving to comprehend the disease pattern of a community. Statistical classification is a method of organizing individual items based on quantitative data related to one or more intrinsic features of the disease. Disease classification adheres to specific principles, with some diseases having their own distinct categories, while others are grouped together if they share common traits. It is important that there is no overlap between these categories. For example, a drug may be given for two indications in different doses, routes, and duration leading to the differences in AE frequency and distribution.

The WHO family International Classification of Diseases (ICD) provides a globally recognized system for classifying a wide range of health conditions [17]. These unique codes are essential for various purposes, including coding medical history, clinical trial diagnoses, and recording AEs. While ICD codes play a vital role in promoting effective treatment, prevention, and drug safety, they were not originally designed for these purposes, and their descriptions and groupings may not always be optimal for these functions. Nonetheless, they are crucial for achieving global comparability in public health and clinical research [1].

6.4.1 INTERNATIONAL CLASSIFICATION OF DISEASES

The systematic classification of diseases has its origins in the nineteenth century, and WHO became the custodian of the ICD with its sixth revision in 1948. However, now the ICD serves as the globally recognized standard for clinicians, policymakers, and patients to navigate, understand, and compare healthcare systems and services [17, 18]. Approximately 30 years ago, the ICD 10th revision (ICD-10) was released, now used in around 120 countries and available in 43 languages [19].

6.4.2 INTERNATIONAL CLASSIFICATION OF DISEASES 10TH REVISION

The ICD serves as a versatile tool with a wide range of global applications, offering essential insights into the prevalence, origins, and outcomes of human illnesses and fatalities through the data that is meticulously documented and encoded using the ICD. The primary objective of the ICD is to classify diseases, health-related conditions, external causes of disease, and injuries to facilitate the collection of valuable data on mortality and morbidity. The clinical terms coded using ICD play a fundamental role in recording medical information and gathering disease statistics across various levels of healthcare. This data and the resulting statistics support various aspects, including payment systems, service planning, drug safety, and extensive health research [18].

"ICD-10 promotes international comparability in the collection, classification, processing, and presentation of mortality statistics." The ICD is periodically updated to incorporate advancements in medical research. The ICD-10 differs from the ICD 9th revision (ICD-9) in several ways. ICD-10 is available in three volumes compared to ICD-9's two volumes. It involves changes in category titles, resulting in a more than twofold increase in the total number of categories. These changes reflect evolving knowledge about diseases and their causes, as well as the addition of discrete categories for particular diseases and complications that are of growing interest.

The adaptation from ICD-9 to ICD-10 brought significant changes in coding structure, shifting from numeric to alphanumeric codes. While the core rules for computing mortality data remain identical, adjustments were made to accommodate variations in mortality statistics. Tabulation lists for presenting mortality data were updated to align with the new classification. Evaluating the impact of these changes on cause-of-death statistics is crucial for understanding mortality trends accurately. Additionally, ICD-9 had limitations with its 13,000 three- to five-digit codes and no provision for new codes or indicating laterality. In contrast, ICD-10 offers a more extensive range of 68,000 three- to seven-digit codes and the potential for further expansion [20–23].

ICD-10 comprises several components, including tabular lists with lethal cause titles and codes along with inclusion and exclusion phrases for these titles in Volume 1. Volume 2 contains narratives, recommendations, and coding rules. Additionally, there is an alphabetical index to diseases, the features and external reasons of injury, and a tabulation of medications and chemicals in Volume 3 [20].

6.4.3 STRUCTURE OF ICD-10 CODES

Generally, ICD-10 codes are permitted to be up to seven characters long and have a format as follows: XXX.XXXX.X (category.anatomic site/severity.extension). The general disease or category is indicated by the first set of digits before the first decimal point. After the first decimal point, etiology, localized area, intensity, or clinical information is illustrated by the next three digits. Lastly, some conditions may have a second decimal point, followed by a final digit that may designate a prior or subsequent encounter, a condition's laterality, or the number of weeks gestation (in the instance of pregnancy) [23]. An example for ICD-10 codes for the types of diabetes is illustrated in Figure 6.3.

Disease coding, especially for diabetes and its complications, follows a structured system. For the majority of people with diabetes, a diagnosis code with a complication code is suitable, and the

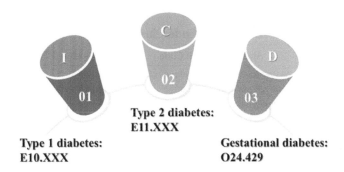

FIGURE 6.3 International Classification of Diseases 10th Revision (ICD-10) codes for the common forms of diabetes.

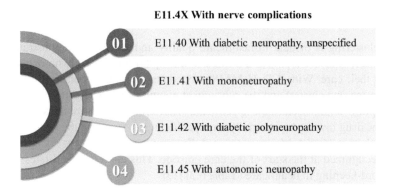

FIGURE 6.4 Examples of codes for type 2 diabetes with nerve complications.

digits after the decimal point remain the same regardless of diabetes type (e.g., type 1 or type 2). Each numerical code after the decimal point (ranging from 1 to 8) represents a different complication, and the second and third digits subdivide that complication. For instance, the code E11.65 is commonly used for type 2 diabetes with hyperglycemia. In contrast, E10.65 and E10.649 are common codes for type 1 diabetes with hyperglycemia and hypoglycemia without coma, respectively. More specific codes that further refine the categorization of complications are given in Figure 6.4.

Overall, this structured approach to disease coding simplifies data management. The ICD has been the cornerstone for comparing causes of death and morbidity worldwide for over a century. The 72nd World Health Assembly endorsed ICD-11, the most recent edition, in 2019, and it was implemented from January 1, 2022 [24].

6.4.4 INTERNATIONAL CLASSIFICATION OF DISEASES 11TH REVISION

"ICD-11 is a different and more powerful health information system, based on formal ontology, designed to be implemented in modern information technology infrastructures, and flexible enough for future modification and use with other classifications and terminologies." It is structured in such a way that clinically pertinent features of cases and information summaries for numerous intents, yet flexible for use in both advanced and simplified modes, and it includes built-in support for multiple languages. Furthermore, disease-related data coded according to ICD-11 is analogous to data coded to ICD-10 [19].

TABLE 6.3

Examples of ICD-11 MMS Code Clusters

Codes	Meaning of Each Code	Clustered Codes
Example 2: Proliferative diabetic retinopathy in type 2 diabetes		
5A11	Type 2 diabetes mellitus	5A11 / 9B71.01
9B71.01	Proliferative diabetic retinopathy	

6.4.5 ICD-11 FRAMEWORK

A crucial aspect of modernizing the ICD for the digital era is its foundation on a computable knowledge framework, setting ICD-11 (Figure 6.5) apart from earlier versions. This knowledge framework makes ICD-11 interoperable within digital health information settings. While it can still function in paper-based systems, the tools and capabilities are expected to strongly incentivize electronic adoption for most users. Additionally, the initial derived classification, "ICD-11th Revision, Mortality and Morbidity Statistics (ICD-11-MMS)," serves as the most direct successor to the previous version ICD-10.

The use of clustering is evident in healthcare quality and safety, as demonstrated by ICD-11-MMS. For instance, consider a patient admitted to a hospital for surgery who experiences a complication during their care. With ICD-11-MMS, it becomes possible to code both the initial reason for surgery (forming one cluster) and the subsequent complication. The cluster related to the complication can capture details such as the specific harm suffered (like severe nausea and constipation post-surgery), the drug involved (potentially any specific anesthetic), and the factors contributing to the problem (such as incorrect dosage or timing). Extension codes can also document whether the condition was recognized at the start of the care episode. This approach facilitates comprehensive and precise record-keeping in healthcare (Table 6.3) [19].

6.4.6 APPLICATIONS OF ICD-11

CD-11, as a classification system, facilitates deliberate documentation, assessment, interpretation, and comparison of mortality and morbidity data gathered from diverse geographical areas and different timeframes. It also ensures the semantic compatibility and usability of data collected for purposes beyond health statistics, including decision support, public health initiatives, research, guidelines, patient records, drug and device safety, and various other applications.

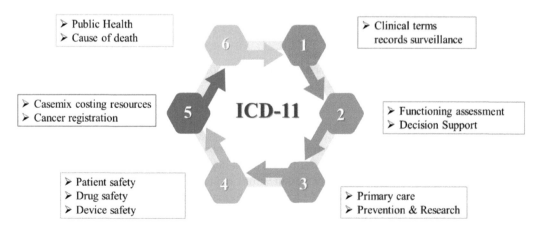

FIGURE 6.5   Purpose of International Classification of Diseases 11th Revision (ICD-11).

6.4.7 ROLE IN DRUG SAFETY

Developed with input from HCPs and organizations responsible for healthcare management, this classification is peculiarly tailored to facilitate coherent and valuable summarization of extensive disease data. This data can originate from either manual or electronic health records and can be coded using a structured terminology [19].

To develop effective strategies for reducing preventable AEs, it's essential to first understand the burden of disease in various healthcare settings and patient populations, as well as the common causes behind these events. This understanding serves as a foundation for prioritizing and streamlining the development and assessment of emerging prevention strategies related to healthcare service utilization and costs. Identifying modifiable risk factors that can be addressed through well-designed health system interventions is crucial. These factors may encompass drug, drug class, treatment protocols, prescription patterns, patient demographics, provider groups, healthcare settings, and care models. Successful interventions may benefit from ongoing refinement to improve feasibility and performance. Ultimately, it is vital to evaluate the impact of these strategies on health outcomes and costs, comparing them with other healthcare interventions to guide efficient resource allocation and optimize the value of healthcare expenditures.

Population-level administrative health data, when coupled with medication dispensing information, can serve as a valuable resource for this kind of research. Utilizing this data source, adverse drug event information becomes readily available and standardized across a large population, making it possible to analyze trends over extended timeframes, study prescription patterns, and make comparisons between different healthcare settings. Nonetheless, there is currently no consensus among health researchers regarding a reliable method for identifying AEs within these data sources. As a result, there is significant variability in the approaches employed to identify such events [25].

Additionally, the coding rules enhance the utility of mortality statistics by prioritizing specific categories, merging conditions, and methodically choosing one cause of death from a documented list of conditions. This chosen cause for tabulation is referred to as the underlying cause of death, while the remaining reported causes are categorized as non-underlying causes of death. The collective information involving both the causes constitutes the multiple causes of death [20].

6.5 ADVERSE DRUG REACTION DICTIONARIES

Signal generation is an integral part of the early detection of suspected ADRs in the domain of PV. Spontaneous reports are a tremendous source of information that aids in the generation of drug safety signals. The act of signal generation is based on the gathering and analysis of spontaneous reports submitted by HCPs or patients. The broad spectrum of available drugs and related ADRs make collecting precise information from spontaneous reports, a tremendous task. In order to streamline the recording and analysis of ADRs in PV, spontaneous reports often employ a controlled vocabulary for coding these AEs, typically the "**Med**ical **D**ictionary for **R**egulatory **A**ctivities" (MedDRA).

6.5.1 MEDICAL DICTIONARY FOR REGULATORY ACTIVITIES TERMINOLOGY

"MedDRA is a clinically validated international medical terminology used by regulatory authorities and the regulated biopharmaceutical industry. The terminology is used through the entire regulatory process, from pre-marketing to post-marketing, and for data entry, retrieval, evaluation, and presentation." Prior to the formation of MedDRA, there was a lack of internationally standardized medical terminology for regulatory consideration of medical products [26].

6.5.1.1 Evolution of MedDRA

As late as the early 1990s, there was no standard international terminology for reporting AEs. The "Coding Terminology Symbols for Thesaurus of Adverse Reaction Terms (COSTART)" and

ICD-9-CM were extensively adopted in the US for coding AEs. The combination of "World Health Organization's Adverse Reaction Terminology (WHO-ART)" and ICD-9 was commonly used in Europe and other parts of the world. Additionally, a few biopharmaceutical companies created a few custom dictionaries to meet their specific requirements.

Most of these older terminologies lacked general specificity, limiting their use for purposes other than coding AEs. Furthermore, it was not unusual that pre-marketing clinical trial AEs are categorized and summarized using one vocabulary, whereas marketed product AEs are encoded using a different terminology. This created a substantial challenge to integrate AE data throughout the pre- and post-approval phase, and it necessitated organizations to devote resources that maintaining numerous vocabularies for reporting and assessing AEs.

In 1993, a collaborative effort involving EU regulatory authorities and industry experts led to the evaluation and transformation of the UK medical terminology, resulting in the creation of MedDRA. By October 1994, the International Conference on Harmonization (ICH) adopted MedDRA Version 1.0 as the universal standard for regulatory terminology. To further refine MedDRA, an ICH M1 Expert Working Group was established. In February 1996, Version 1.0 underwent alpha testing by pharmaceutical companies and regulatory agencies, ultimately gaining approval as Version 2.0 in July 1997, with the updated name "Medical Dictionary for Regulatory Activities." In May 1998, the ICH MedDRA Management Committee was established, and by November of the same year, the Maintenance and Support Services Organization (MSSO)was engaged to oversee the maintenance and support of MedDRA. Furthermore, in January 1999, the Japanese Maintenance Organization (JMO) was founded, and in March of that year, the MSSO introduced MedDRA Version 2.1, which is complemented with a Japanese version by the JMO.MedDRA undergoes biannual updates (March and September), with the latest version being MedDRA Version 26.1 as of September 2023. This timeline of significant milestones leading to the initial release of MedDRA is illustrated in Figure 6.6 [25, 27, 28].

6.5.1.2   MedDRA Hierarchy

MedDRA terminology is widely utilized for documenting and reporting AEs in drug surveillance, covering both pre- and post-approval phases. It is also endorsed by the ICH for the electronic transmission of ICSRs [3, 4]. This hierarchical framework of MedDRA aggregates logically precise terms used for categorizing broader medical categories, enhancing data retrieval through clinically relevant groupings of terms [29].

The hierarchy establishes various levels of both higher (superordination) and lower order terms (subordination). A higher-level term serves as a broad grouping category that can be applied to

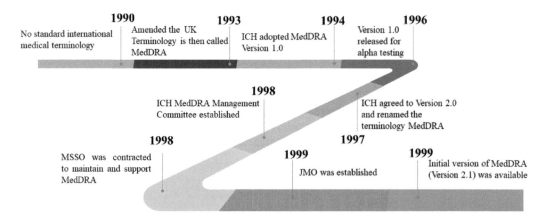

FIGURE 6.6   Milestones led to the initial release of MedDRA.

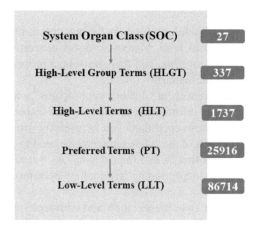

FIGURE 6.7 MedDRA hierarchy.

every lower level descriptor associated with it. Consequently, hierarchical levels illustrate vertical connections within the terminology. Hierarchies play a vital role in facilitating flexible data fetching and ensuring the transparent presentation of information. The five-level framework of the terminological system allows for data retrieval by particular or broad groupings, depending on the level of desired specificity [28].

MedDRA comprises "27 vertical axes, the system organ classes (SOCs)," for example, endocrine disorders, eye disorders, and product issues. Each SOC consists of four descending levels in the hierarchy, increasing in number, which are "high-level group terms (HLGTs), high-level terms (HLTs), preferred terms (PTs), and lowest level terms (LLTs)" [3, 29].

For MedDRA to facilitate the harmonization of coded data sharing, it is essential that users maintain consistency when assigning phrases to reporters' words ("Verbatim") of symptoms, indications, medical conditions, and AEs. LLTs, which are extremely specific ("**granular**") and detailed terms, play a crucial role in accurately capturing the verbatim. LLTs are typically analogously linked to their parent terms known as PTs. Additionally, PTs are comparably specialized and numerous.

"The structure of MedDRA is **multiaxial**, so that a PT may appear in more than one SOC. This enables terms to be grouped in a variety of ways that are medically appropriate. Each PT is assigned one primary SOC; all other SOC assignments for that PT are called secondary." Utilizing a single primary SOC helps avoid duplication of AEs when generating data across all SOCs. It is worth noting that MedDRA might not encompass all potential secondary SOC assignments for a specific PT. Nevertheless, through the change request procedure, new or revised SOC assignments can be developed [26, 29, 30].

6.5.1.3 Standardized MedDRA Queries

"Standardized MedDRA Queries (SMQs) are groupings of MedDRA terms, ordinarily at the Preferred Term (PT) level that relate to a defined medical condition or area of interest." SMQs are designed to assist in recognizing and retrieving ICSRs that may be potentially pertinent. The terms mentioned can include signs, symptoms, syndromes, diagnoses, laboratory findings, and other test data associated with the medical condition. Only LLTs linked to a PT in an SMQ are included; all others are excluded [30].

SMQs are the tools used to retrieve specific cases from a database coded with MedDRA. They offer flexibility through narrow and broad sub-search options, stratified grouping, and search algorithms. Users can also request modifications to SMQs to better align with their specific needs or changing requirements.

6.5.1.4   MedDRA in Signal Detection

To distinguish the various clinical conditions, terms in the nomenclature must express distinct medical concepts in a logical hierarchical level. Moreover, signal generation frequently requires that comparable conditions be detected together, and there must be a proper grouping of concepts that are medically connected in a certain manner.

For instance, if we identify two reports of "diabetic nephropathy" with a specific drug, additionally, we look into the vicinity of any information about reports of diabetic complications, renal and microalbuminuria, so that all of these conditions can be usefully grouped together. Rather than counting the number of safety reports of the single aliments, we need to count the number of reports on comparable, connected ailments that fit into the category. As a result, it is critical that the categories be homogeneous in nature, incorporating terminology pertaining to related or comparable clinical disorders.

MedDRA provides a pattern of case presentation and enumeration to enhance the statistical identification of potential signals. This presentation format allows for visual representation and enumeration of occurrences of similar event terms, with the ability to highlight if these numbers exceed an arbitrary threshold. Furthermore, the signal generation employs statistical methods for identifying disproportionality using the frequentist or Bayesian approach, and it is important to note the number of co-occurrences of AE reports associated with a drug. However, such approaches rely on the distinction of different conditions depending on the vocabulary used. Additionally, it is imperative to identify reports that exclusively contain indications, symptoms, and investigative findings that could potentially indicate new syndromes within the database [3].

Although the detailed vocabulary of MedDRA streamlines data entry by minimizing the need for interpretation, it does have implications for data retrieval, sorting, and presentation in drug discovery and development, PV, and risk mitigation. MedDRA's coherent hierarchical structure enhances data retrieval by organizing highly specific coding terms into broader medical categories (HLTs and HLGTs). Its multiaxial approach allows for flexible data retrieval through primary and secondary routes. However, the complexity of MedDRA often requires additional support to optimize the search results.

The multiaxial feature of MedDRA significantly influences data retrieval and signal detection. In essence, multiaxial terminology means that a concept can be represented in more than one SOC. For instance, taking the term "Diabetic Nephropathy" as an example, it is primarily categorized under the SOC "Renal and urinary disorders," but it also has secondary associations with the SOCs "Endocrine disorders" and "Metabolism and nutrition disorders" (Figure 6.8). Additionally, the "granular" element of MedDRA enables the appropriate capturing of the reporter's terms.

MedDRA is used in ICSRs to report AEs terminology which enables the aggregation of those reported phrases into medically logical stratified groupings to facilitate safety data analysis. Furthermore, it is used to include AE data in reports, compute the likelihood of comparable AEs, and capture and analyze linked data such as therapeutic indications, investigations, and medical histories [29].

Moreover, preliminary signal detection begins with MedDRA and the SMQs. SMQs were developed to navigate the high granularity and peculiar characteristics of MedDRA and to increase the possibility of identifying all terms related to a certain medical condition of interest. Disproportionality analysis using PRR or ROR revealed that the performance of the method varies across different levels of the terminology. In most instances, the PT level is the optimum option for indexing a drug safety signal [31].

In the post-marketing phase, specific SMQs or a selected group of SMQs can be employed to retrieve pertinent cases for subsequent medical review when there is a suspicion of an emerging safety signal. For example, if a pharmaceutical company suspects a potential signal of pancreatitis related to a new antidiabetic medication, they can apply the SMQ "Acute pancreatitis" to the data. Alternatively, the entire set of SMQs can be used for safety signal detection within the database.

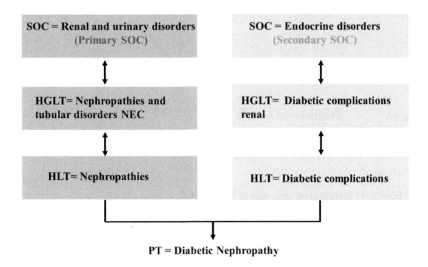

FIGURE 6.8 Multiaxial terminology.

Users may opt for more focused sub-search SMQs, such as narrow terms or specific hierarchical levels of SMQs, to reduce the dilution of the signal [30].

6.5.1.5 Advantages of MedDRA

ICH aims to create and sustain MedDRA, a globally standardized medical terminology that serves as a common language for regulatory communication and the assessment of data related to medications for human use. Consequently, MedDRA is tailored to support the registration, documentation, and safety surveillance of medications throughout their entire development process, including clinical trials and post-approval phases [27].

MedDRA is a versatile tool for data retrieval, serving purposes such as summarizing clinical trial data, PV, and addressing medical inquiries. The methods and tools for data retrieval may vary depending on the intended use. While MedDRA offers significant advantages for coding AEs and signal detection through data mining, its structure requires users to search for terms within predefined categories.

Moreover, MedDRA supports multiple languages that facilitate the precision and accuracy by allowing users to work in their native language. This enhances accuracy and precision when assigning terms and facilitates easy data sharing across multinational contexts. Detailed documentation is provided with each MedDRA release in numerous languages to support users effectively [29].

6.5.2 World Health Organization Adverse Reaction Terminology

WHO-ART, developed by the "WHO Collaborating Center for International Drug Monitoring," serves to code ADRs. To enhance signal detection within the WHO database, a potentially efficient method involves calculating the semantic distance between WHO-ART phrases to group related clinical conditions effectively.

6.5.2.1 Structure

WHO-ART is structured hierarchically with four levels and maintains an open-ended nature, allowing for the addition of new terms as needed. These levels include "System-Organ Classes (SOC), High-Level Terms (HLT), Preferred Terms (PT), and Included Terms".

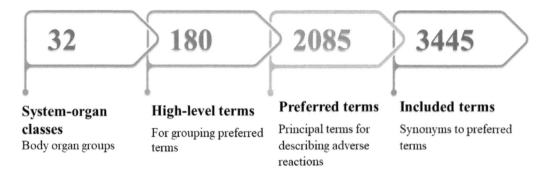

| 32 | 180 | 2085 | 3445 |

System-organ classes
Body organ groups

High-level terms
For grouping preferred terms

Preferred terms
Principal terms for describing adverse reactions

Included terms
Synonyms to preferred terms

FIGURE 6.9   Four-level hierarchical structure of WHO-ART.

HLTs are grouped together based on related or similar medical conditions for easier information retrieval. SOCs represent collections of adverse reaction PTs linked to the same organ system and are occasionally used in output contexts. Each PT can be associated with up to three distinct system-organ classes, and this allocation remains fixed for specific reports. The first SOC listed for each PT is considered the most relevant. The included terms are closely related to PTs and aid in identifying the appropriate PT for accurate coding of reported AEs. PTs are the primary terms employed to describe drug-related AEs.

The WHO-ART, a legacy tool for coding AEs, is no longer actively maintained. Developed and managed by UMC for over three decades, it served as a terminology for encoding ADRs, covering a wide range of medical terms needed for reporting AEs. Prior to the adoption of MedDRA in 2008, WHO-ART was the primary terminology for ADRs in VigiBase. Today, most users are encouraged to use MedDRA for coding. Despite being officially deprecated since 2015, individuals with specific needs for WHO-ART should contact UMC for assistance [31, 32].

6.3.1   CODING SYMBOLS FOR A THESAURUS OF ADVERSE REACTION TERMS

The COSTART, developed by the FDA, codes and retrieves post-marketing ADRs. It categorizes terms based on body systems, pathophysiology hierarchies, and includes a fetal/neonatal category. However, MedDRA Terminology has replaced COSTART [1, 33].

6.6   CONCLUSION

The most extensively used medical dictionaries consider the coding of drug names, diseases, and AEs. However, the use of coding dictionaries improves the quality of safety data gathered on PV databases, enhances effective drug safety signal analysis by offering clinically hierarchical logical groups of terms, and promotes electronic data transfer by providing a globally accepted standard. Subsequently, it promotes the intuitive or statistical identification of AEs with threshold frequency or significant events that are frequently drug-related and facilitate the generation of new safety signals. Moreover, medical coding lays the foundation for the implementation of artificial intelligence (AI) and machine learning (ML) into PV. These technologies promise increased efficiency in sifting the huge datasets, potentially reducing the need for extensive manpower. Furthermore, it ensures organized data and paves the way for future advancements, with AI and ML poised to revolutionize the field by making safety analysis more efficient and insightful.

ACKNOWLEDGMENT

We would like to express our sincere gratitude to Indian Council of Medical Research (ICMR) for their financial assistance.

REFERENCES

1. Brown EG. Dictionaries and coding in pharmacovigilance. In: Waller JTaP, editor. Stephens' Detection of New Adverse Drug Reactions. 5th ed. Chichester, West Sussex, England: John Wiley & Sons, Ltd.; 2004. pp. 533–557.
2. Souvignet J, Declerck G, Asfari H, Jaulent MC, Bousquet C. OntoADR a semantic resource describing adverse drug reactions to support searching, coding, and information retrieval. J Biomed Inform. 2016;63:100–107.
3. Brown EG. Effects of coding dictionary on signal generation: A consideration of use of MedDRA compared with WHO-ART. Drug Saf. 2002;25(6):445–452.
4. Fernando B, Kalra D, Morrison Z, Byrne E, Sheikh A. Benefits and risks of structuring and/or coding the presenting patient history in the electronic health record: Systematic review. BMJ Qual Saf. 2012;21(4):337–346.
5. Carpenter I, Ram MB, Croft GP, Williams JG. Medical records and record-keeping standards. Clin Med (Lond). 2007;7(4):328–331.
6. Brown EG, Wood L, Wood S. The medical dictionary for regulatory activities (MedDRA). Drug Saf. 1999;20(2):109–117.
7. Anatomical Therapeutic Chemical (ATC) Classification: World Health Organization (WHO). Available from: https://www.who.int/tools/atc-ddd-toolkit/atc-classification (Accessed on August 2023).
8. Chen L, Zeng WM, Cai YD, Feng KY, Chou KC. Predicting anatomical therapeutic chemical (ATC) classification of drugs by integrating chemical-chemical interactions and similarities. PLoS One. 2012;7(4):e35254.
9. Guidelines for ATC Classification and DDD Assignment 2023 2023, 26th edition. Available from: https://www.whocc.no/filearchive/publications/2023_guidelines_web.pdf (Accessed on August 2023).
10. Vargesson N. Thalidomide-induced teratogenesis: History and mechanisms. Birth Defects Res C Embryo Today. 2015;105(2):140–156.
11. Lagerlund O, Strese S, Fladvad M, Lindquist M. WHODrug: A global, validated and updated dictionary for medicinal information. Ther Innov Regul Sci. 2020;54(5):1116–1122.
12. What's New in WHODrug. 2019. Available from: https://who-umc.org/media/164657/what-s-new-in-whodrug-march-1-2019.pdf (Accessed on August 2023).
13. Juhlin K, Karimi G, Andér M, Camilli S, Dheda M, Har TS, et al. Using VigiBase to identify substandard medicines: Detection capacity and key prerequisites. Drug Saf. 2015;38(4):373–382.
14. WHODrug Global. 2021. Available from: https://who-umc.org/whodrug/whodrug-global/ (Accessed on August 2023).
15. van Stekelenborg J, Ellenius J, Maskell S, Bergvall T, Caster O, Dasgupta N, et al. Recommendations for the use of social media in pharmacovigilance: Lessons from IMI WEB-RADR. Drug Saf. 2019;42(12):1393–1407.
16. Wakao R, Taavola H, Sandberg L, Iwasa E, Soejima S, Chandler R, et al. Data-driven identification of adverse event reporting patterns for Japan in VigiBase, the WHO global database of individual case safety reports. Drug Saf. 2019;42(12):1487–1498.
17. Jakob R. Disease classification. In: Quah SR, editor. International Encyclopedia of Public Health. 2nd ed. Oxford: Academic Press; 2017. pp. 332–337.
18. Topaz M, Shafran-Topaz L, Bowles KH. ICD-9 to ICD-10: Evolution, revolution, and current debates in the United States. Perspect Health Inf Manag. 2013;10(Spring):1d.
19. Harrison JE, Weber S, Jakob R, Chute CG. ICD-11: An international classification of diseases for the twenty-first century. BMC Med Inform Decis Mak. 2021;21(Suppl 6):206.
20. International Classification of Diseases, Tenth Revision (ICD-10): Centers for Disease Control and Prevention, National Center for Health Statistics. Available from: https://www.cdc.gov/nchs/icd/icd10.htm (Accessed on August 2023).
21. International Classification of Diseases, 10th Revision (ICD-10): Centers for Disease Control and Prevention, National Center for Health Statistics. Available from: https://www.cdc.gov/nchs/data/dvs/icd10fct.pdf (Accessed on August 2023).
22. Anderson RN, Miniño AM, Hoyert DL, Rosenberg HM. Comparability of cause of death between ICD-9 and ICD-10: Preliminary estimates. Natl Vital Stat Rep. 2001;49(2):1–32.
23. Dugan J, Shubrook J. International Classification of Diseases, 10th Revision, Coding for Diabetes. Clin Diabetes. 2017;35(4):232–238.
24. International Statistical Classification of Diseases and Related Health Problems (ICD). Available from: https://www.who.int/standards/classifications/classification-of-diseases (Accessed on August 2023).

25. Hohl CM, Karpov A, Reddekopp L, Doyle-Waters M, Stausberg J. ICD-10 codes used to identify adverse drug events in administrative data: A systematic review. J Am Med Inform Assoc. 2014;21(3):547–557.
26. Babre D. Medical coding in clinical trials. Perspect Clin Res. 2010;1(1):29–32.
27. Medical Dictionary for Regulatory Activities (MedDRA). Available from: https://www.meddra.org/about-meddra/evolving-meddra (Accessed on August 2023).
28. Introductory Guide MedDRA, Version 26.0, March 2023. Available from: https://admin.meddra.org/sites/default/files/guidance/file/intguide_26_0_English.pdf.
29. MedDRA® Data Retrieval and Presentation: Points to Consider ICH-Endorsed Guide for MedDRA Users on Data Output, Release 3.21, March 2023. Available from: https://admin.meddra.org/sites/default/files/guidance/file/000861_datretptc_r3_23_mar2023.pdf.
30. Introductory Guide for Standardised MedDRA Queries (SMQs), Version 26.1, September 2023. Available from: https://admin.meddra.org/sites/default/files/guidance/file/SMQ_intguide_26_1_English.pdf (Accessed on September 2023).
31. WHO-ART Legacy Service. Available from: https://who-umc.org/vigibase/vigibase-services/who-art/ (Accessed on August 2023).
32. Iavindrasana J, Bousquet C, Degoulet P, Jaulent MC. Clustering WHO-ART terms using semantic distance and machine learning algorithms. AMIA Annu Symp Proc. 2006;2006:369–373.
33. Coding Symbols for a Thesaurus of Adverse Reaction Terms (COSTART). Available from: https://www.nlm.nih.gov/research/umls/sourcereleasedocs/current/CST/index.html#:~:text=The%20Coding%20Symbols%20for%20a,and%20Drug%20Administration%20(FDA).&text=COSTART%20is%20used%20for%20coding,drug%20and%20biologic%20experience%20reports (Accessed on August 2023).

7 Regulatory Aspects in Signal Detection and Assessment

James Buchanan

7.1 INTRODUCTION

A wealth of information is available from various regulatory authorities pertaining to the requirements for expedited and periodic reporting of safety information. The methods by which sponsors may conduct safety signal detection and evaluation, however, have not been the subject of detailed guidance by the regulatory authorities despite the clear expectation that sponsors and marketing authorization holders regularly review safety data for indications of product risks. Coupled with the multitude of ways in which safety signals can manifest, the apparent paucity of regulatory guidance has caused some degree of confusion within the industry as to how best to approach the task of signal detection and assessment. Nevertheless, there are a number of resources provided by regulatory authorities, which can serve as the basis for a consistent, methodical process by which safety data can be reviewed for safety signals and how these signals can be evaluated for evidence indicative of product risk. This chapter reviews the recent available material, first in the context of safety data derived from clinical studies and then from post-marketing data sources both in terms of signal detection and causality assessment.

7.2 CLINICAL DEVELOPMENT IN SIGNAL DETECTION AND ASSESSMENT

7.2.1 SIGNAL DETECTION

A discussion of signal detection should be based on a common understanding of what constitutes a safety signal. The clearest definition is perhaps that articulated by the Council for International Organizations of Medical Sciences (CIOMS) Working Group VI [1]: "an event with an unknown causal relationship to treatment that is recognized as worthy of further exploration and continued surveillance". This definition is equally applicable in both the clinical development and post-marketing environments as it is agnostic to the source of the data.

Regulatory authorities have had extensive experience evaluating product dossiers submitted for marketing authorization as well as safety risks that subsequently became apparent in the post-marketing period, some of which led to product withdrawals. The lessons learned from the experiences resulted in several useful guidance documents from FDA.

In 2005, as a result of a commitment from the Prescription Drug User Fee Act (PDUFA) III of 2002, the FDA released the Pre-Marketing Risk Assessment guidance [2]. The document is intended to provide "guidance to industry on good risk assessment practices during the development of prescription drug products, including biological drug products". The guidance covers a variety of topics relevant to risk identification during clinical development but importantly focuses the safety reviewer's attention on several areas that frequently have resulted in denial of marketing approval or in changes to post-marketing labeling or even marketing authorization withdrawal. These include the following:

- Use of pharmacokinetic (PK) information to detect evidence of a relationship between the appearance of adverse events to measures of product exposure, including valid biomarkers related to a safety concern, e.g., assessments of QT/QTc effects for novel antihistamines or creatine phosphokinase (CPK) as a marker for rhabdomyolysis in the evaluation of new HMG-CoA (3-hydroxy-3-methylglutaryl-CoA) reductase inhibitors

- Drug-induced prolongation of QTc
- Drug-induced liver toxicity
- Drug-induced nephrotoxicity
- Drug-induced bone marrow toxicity
- Drug–drug interactions
- Polymorphic metabolism
- Immunogenicity associated with biological products
- Gene-based biological products: Taking into account the genetic stability of products meant for long-term transfections, the transfection of nontarget cells, and the transmissibility of infection to close contacts
- Cell-based products: Assessments of adverse events related to distribution, migration, and growth beyond the initial intended administration site

These areas represent organ systems and areas wherein safety signals of particular clinical concern may arise, but the discussion generally does not necessarily describe what might constitute a safety signal within these areas. Some thresholds are provided by which a safety signal may be identified; for example, in the case of drug-induced QTc prolongation, an absolute QTc >450, 480, or 500 ms or a relative increase in QTc >30 or 60 ms from baseline and, in the case of drug-induced liver toxicity, an ALT/AST (alanine aminotransferase/aspartate aminotransferase) value >3× upper limit of normal with total bilirubin >2× upper limit of normal. However, in other areas what finding would represent a safety signal is not so clearly defined.

The guidance document does provide some general considerations for what circumstances could represent a safety signal. One example is identifying temporal associations between drug administration and the appearance of an adverse event. The onset of an adverse event close in time to the exposure to the product can be a piece of evidence not only pointing to a safety signal but also supportive of a possible causal association. An examination of temporal relationships can also shed light on adaptation and tolerance. Assessment of risk over time, such as a hazard rate curve, can elucidate changes in risk over time. When the size of the study population changes over time, a Kaplan–Meier curve can be a useful method to detect temporal associations.

The effect of dose on the occurrence of an adverse event should be examined when possible. This is part of the concept of the exposure-response relationship. When medications are taken for extended periods of time, event rates based on cumulative dosage analysis may be helpful. Exposure can also be evaluated in the context of PK parameters, such as C_{max} (maximal serum concentration) and AUC (area under the curve).

This guidance additionally highlights the importance of considering the reasons that subjects discontinue the study drug since the occurrence of adverse events or tolerability issues could limit a subject's willingness to continue treatment. Abnormal laboratory results, vital signs, or ECG findings that are not classified as adverse events may cause some patients to withdraw. Of particular concern are vague reasons, such as "withdrew consent", "patient preference", or "physician decision", for which the underlying reason could be an adverse event. The sponsor is encouraged to design case report forms carefully to avoid situations where safety issues are obfuscated.

Interestingly, while this guidance does not directly speak to the comparison of the relative frequency of adverse events between treatment groups, it does caution the safety reviewer to not simply utilize event frequency but rather consider expressing frequency in terms of person-time, particularly when the duration of treatment varies.

With respect to the concern for drug-induced liver injury (DILI), in 2007, the FDA released a document entitled *Guidance for Industry: Drug-Induced Liver Injury – Premarketing Clinical Evaluation* [3]. This focused guidance was warranted as DILI has been the most frequent cause of safety-related marketing authorization withdrawals. This guidance specifically identifies several circumstances that indicate a hepatic safety signal:

- An excess of aminotransferase elevations to >3× ULN (upper limit of normal) in a drug-treated group compared to a control group. However, the guidance cautions that "There are no good data analyses at this time on how great this excess should be compared to control (e.g., 2-fold, 3-fold) to suggest an increased risk of DILI".
- Marked elevations of aminotransferases to 5×, 10×, or 20× ULN in smaller numbers of subjects in the test drug group and not seen (or seen much less frequently) in the control group.
- One or more cases of ALT/AST >3× ULN, total bilirubin >2× ULN in the absence of evidence of cholestasis, i.e., serum alkaline phosphatase >2× ULN, in the test drug group compared to placebo. The use of absolute values of alkaline phosphatase to distinguish hepatocellular patterns from cholestatic ones has fallen out of favor to more recent methods, such as the R value [4, 5] or nR value, the latter of which has led to the proposal for the nR modified Hy's law [6, 7].

DILI, notably in Hy's law cases, is a diagnosis of exclusion. It is critical to consider other possible etiologies for the laboratory abnormalities before concluding that the drug is causally related.

With regard to the issue of QTc prolongation described in the Pre-Marketing Risk Assessment guidance, the FDA issued the guidance "E14 Clinical Evaluation of QT/QTc Interval Prolongation and Proarrhythmic Potential for Non-Antiarrhythmic Drugs" [8] to provide more details. This guidance, based on the International Conference on Harmonisation (ICH) E14, addresses the evaluation of the effect of drugs on the QT/QTc interval as well as the collection of cardiovascular adverse events. In this case the guidance does provide some specific findings that would represent a safety signal:

- In a "thorough QT/QTc study", evidence of QTc prolongation of around 5 ms is evidenced by an upper bound of the 95% confidence interval around the mean effect on QTc of 10 ms.
- In a general clinical trial, instances of marked QT/QTc prolongation (e.g., >450, 480, or 500 ms), or increase in QTc from baseline (e.g., >30 or 60 ms).
- The occurrence of serious arrythmias (e.g., Torsades de Pointes, ventricular tachycardia, ventricular fibrillation/flutter) or other adverse events that may indicate an underlying arrhythmia (e.g., sudden death, syncope, and seizures). An imbalance in the frequency of these events between study groups can signal a potential proarrhythmic effect of the investigational agent.

The document advises the sponsor to pay particular attention to subjects who discontinue the study due to QTc prolongation or who require a dose reduction.

In the guidance document "In Vitro Drug Interaction Studies – Cytochrome P450 Enzyme- and Transporter-Mediated Drug Interactions" [9], the FDA pointed to additional areas in which safety signals may arise. Drugs may be metabolized by a number of enzymes, including the cytochrome P-450 (CYP450) family and glucuronosyl- and sulfotransferases. A drug may inhibit or induce these enzymes which can effect not only the PKs of that drug but also other drugs metabolized by these systems. The clinical consequence of such altered PKs may be an increased or decreased pharmacological effect that in turn may produce adverse consequences.

The guidance also highlights an often-underappreciated area – interactions with drug transporters. Membrane transporters are expressed in many organs where they can have relevant effects on the PKs and pharmacodynamics of drugs. Understanding whether a drug is a substrate or inhibitor of key transporters can predict toxicity resulting from altered tissue distribution. Since the substrates for these transporters include naturally occurring compounds in the body, drugs that interact with these transporters can alter the distribution of these compounds. For example, drugs that inhibit the renal transporters organic anion transporter (OAT2), organic cation transporter (OCT2), and multidrug and toxin extruder (MATE) can give rise to apparent elevations in serum creatinine. The finding of increased creatinine levels in study subjects administered the transporter-inhibiting drug may be misinterpreted as an indication of acute renal injury when, in fact, it is a benign consequence of the inhibition of the renal transporters of creatinine.

Similar to the Pre-Marketing Risk Assessment guidance, this document directs the safety clinician to areas that may produce safety signals but does not stipulate what constitutes a safety signal. For example, the degree of CYP450 inhibition that would elicit adverse effects from a co-administered drug depends on the PKs and pharmacodynamics of that particular drug and cannot be prespecified.

The FDA crafted a guidance document for their internal medical reviewers to assist with the evaluation of safety data from a marketing application. This was released publicly in 2005 as the "Reviewer Guidance, Conducting a Clinical Safety Review of a New Product Application and Preparing a Report on the Review". It is no longer hosted on the FDA website, but an archived version is available [10].

The material provides the FDA medical reviewer with an overview of approaches for the evaluation of safety data. When reviewing the frequency of adverse events between treatment groups, focusing on events with the greatest differences between groups can highlight a safety signal. For example, events occur at an incidence of ≥5% and for which the incidence is at least twice, or some other percentage greater than the placebo incidence. In addition to the application of this risk difference, the risk ratio can also be utilized. The medical review of the tafamidis application sought to identify safety signals by identifying common treatment-emergent adverse events (TEAEs) that occurred in >5% of patients treated with tafamidis with a risk difference of ≥2% and a risk ratio of ≥1.2 compared to the placebo group [11]. Other factors to consider include adverse events with a relationship to dose, time to onset, propensity of the adverse event to occur in an identifiable demographic group, or association with concomitant drugs or disease suggesting a drug–drug or drug–disease interaction. In some cases, the frequency of an adverse event may not differ measurably between treatment groups, but a difference may appear when only severe events are considered. Laboratory data should be examined both in terms of measures of central tendency, which entail examining the extent to which various analytes change from baseline, and an exploration for laboratory outliers. The reviewer is also encouraged to explore the extent to which adverse events and laboratory anomalies lead to study drug discontinuation, particularly comparing such discontinuation rates between treatment groups. The areas of potential DILI and QT prolongation are similar to those described in the Pre-Marketing Risk Assessment guidance.

In September 2022, the FDA unveiled their suggested formats for Standard Safety Tables and Figures and recommendations for aggregating MedDRA Preferred Terms into FDA Medical Queries (FMQ) [12]. Among the Standard Safety Tables and Figures is a display of adverse event terms and frequency ordered by the familiar System Organ Class and Preferred Term format but displaying the terms by descending risk difference values. As described in the Medical Reviewer guidance, highlighting the AE (adverse event) terms with the greatest degree of risk difference is a method to focus on events most likely to represent safety signals.

An important contribution to safety analytics is the concept of performing analyses based on aggregation of adverse event terms that represent a common medical concept. When the relative frequencies of an adverse event are evaluated using single preferred terms, safety signals may be missed. For example, if the frequency of the individual term "anxiety" is compared between treatment groups, there may be no apparent difference. However, the concept of anxiety can be defined by the preferred terms "anxiety", "panic disorder", "panic attack", "nervousness", and "generalized anxiety disorder". A comparison of the frequency of the appearance of this aggregation of terms may now indicate a difference between groups where the individual term comparison failed to do so. The opposite effect may occur wherein a difference in frequency is seen based on an individual preferred term but disappears when the aggregate grouping is used. In both cases, the aggregation of terms representing a medical condition provides a more reliable assessment of the frequency of that condition.

Standard MedDRA Queries (SMQ) were an early attempt to address this concept. While useful for casting a broad net across a dataset to identify cases that may represent a topic of interest, SMQs tend to lack sufficient specificity for routine analyses. Because the terms that comprise an FMQ are specific for that medical concept, FDA estimates that a case identified using an FMQ has at least a 90% probability of actually representing that medical concept. Some companies have been using internally generated aggregate term lists for data analyses, and combining a clinical SOC (system organ class) term with an Investigations SOC term (e.g., hypokalemia and blood potassium decreased) has been a common approach.

With a move toward aggregate term analysis, we also move from an analysis of adverse "events" to an analysis of adverse "effects". An adverse effect can comprise not only multiple individual adverse event terms but also other data elements such as laboratory values, vital signs, concomitant medications, medical history, ECG results, and other test findings. The concept of an adverse effect analysis results in a more holistic approach to defining a product's risks. Toward that end, the FMQs also include what FDA terms "algorithmic FMQs". As an example, consider an algorithmic FMQ that comprises four components: (1) two preferred terms pertaining to a medical concept, (2) a particular laboratory analyte whose value exceeds the upper limit of normal, (3) the use of a particular concomitant medication within three days of the onset of a preferred term, and (4) the appearance of the preferred term in a patient with a particular medical history. When all four conditions are met, the medical concept of interest has been identified for that case. The concept can be illustrated using the actual algorithmic FMQ for rhabdomyolysis. This FMQ is defined by these four components: (1) any "rhabdomyolysis" FMQ preferred term, (2) urine myoglobin above the upper limit of normal, (3) a CPK value >5× the upper limit of normal in the absence of either a baseline CPK value above the upper limit of normal or a CPK/CPK-MB value >0.05 within three days of the onset of the preferred term event, and (4) the presence of the preferred terms "myalgia" and "muscular weakness" and either "myoglobin urine present" or "chromaturia" with onset dates within seven days of each other.

The FMQ lists, including the components of various algorithmic FMQs, will evolve over time as FDA weighs industry comments and experience gained from review of sponsor marketing applications. These lists represent an important step forward toward improved detection of safety signals and communication of safety findings with regulatory authorities. They also provide an opportunity for better standardization of such aggregate term lists across the industry.

Finally, in a preview of some post-marketing methods of signal detection, the use of designated medical event (DME) [13] and important medical event (IME) [14] lists can be of equal value in clinical trial datasets. The DME list presents those events that are unlikely to occur spontaneously and are often the result of drug exposure. The finding of a report of a DME event may be considered to represent a safety signal worthy of further evaluation. The IME list includes terms that should always be considered serious regardless of how they might be reported. Cases that include adverse event terms on the IME list may also be considered safety signals, particularly those with a fatal outcome. Of note, the DME terms are represented within the IME list.

7.2.2 Signal Assessment

The previous discussion has identified several regulatory guidance documents that help with focusing the safety reviewer's efforts on areas in which safety signals traditionally have arisen. Some additionally provide some details as to what finding would constitute a safety signal, e.g., a laboratory finding of a potential Hy's law case, an adverse event frequency difference of >5% of patients with a risk difference of ≥2%, and a risk ratio of ≥1.2 compared to the placebo group. Once a safety signal is identified, most guidance documents do not assist the safety reviewer with the task of evaluating whether there is convincing evidence of a causal association between the safety finding and the study drug. Two references, however, do provide useful advice.

Although not a regulatory body, the CIOMS does provide guidance that often informs regulatory positions. The CIOMS Working Group VI report [1], in Appendix 7, described causality criteria and threshold considerations for inclusion of safety data in the development core safety information (DCSI) document. The DCSI is a document maintained by the sponsor, which describes the "core" adverse events believed to be attributable to the investigational drug, i.e., identified risks. The CIOMS report provided criteria that may be applied to (1) individual cases, (2) multiple cases (i.e., aggregate safety data), and (3) previous knowledge of the drug or class. The criteria from the latter two groupings that are germane to this discussion are summarized below.

Evidence from multiple cases

- Higher incidence on average when compared to a placebo or active comparator (whether statistically significant or not)
- Positive dose-response (fixed or escalating dose studies)
- Greater frequency of event-specific patient discontinuations relative to comparator(s)
- In active groups compared to comparator group(s), earlier onset and/or greater severity
- Consistency of pattern of presenting symptoms
- Consistency of time to onset
- Consistent trends across studies
- Consistent pattern of clinical presentation and latency

Previous knowledge of AE or drug/class, including metabolites

- Rarity of the event in comparable untreated populations or indications
- Event is commonly drug-related (e.g., neutropenia, Stevens-Johnson syndrome)
- PK evidence (e.g., interactions)
- Known mechanism
- Recognized class effect
- Similar findings in animal or *in vitro* models
- Closeness of drug characteristics to those of other drugs known to cause the adverse event

The FDA released the draft guidance "Safety Reporting Requirements and Safety Assessment for IND and BA/BE Studies" [15] in June 2021. Within this guidance is a discussion as to what evidence would constitute a "reasonable possibility" that a drug is causally associated with an adverse event. The intent of these criteria is to assist the safety reviewer in deciding when an expedited report of a suspected unexpected serious adverse reaction (SUSAR) should be submitted to the FDA because the sponsor believes there is a reasonable possibility that the study drug is causally associated with the adverse event. Although this assessment is being performed on a single case, the criteria provided are equally applicable to an aggregate assessment. In fact, the types of criteria described fit surprisingly well within the Bradford Hill Criteria of the causality framework. These criteria include the following:

- Extent of the increase in incidence seen in the test group compared to the control groups
- Evidence of a dose-response
- Temporal relationship (for example, early increase post-drug initiation, such as DILI occurring in the usual one- to six-month window, or malignancy events occurring after a lag period between the dates of exposure and date of event onset)
- Consistency of the increase in multiple trials
- Presence of a plausible mechanism of action

- Nonclinical evidence (from toxicology or pharmacology animal studies, genetic studies such as knock-out or knock-in mouse models, or human genetic data) to support the finding
- Pharmacology of the drug (including results from receptor, transporter, or enzyme binding or activation studies, and animal models) and known class effects
- Pattern across the study population (for example, the event is observed more frequently in individuals likely to be susceptible to it (e.g., acute kidney injury in individuals with prior chronic kidney disease, myocardial infarctions in older individuals or those with existing coronary heart disease, and hyperkalemia in individuals on angiotensin-converting enzyme [ACE] inhibitors)
- Occurrence of other potentially related adverse events (e.g., occurrence of both strokes and transient ischemic attacks, unexpectedly large increase in creatine kinase, and events of rhabdomyolysis)

The Bradford Hill Criteria of Causality was developed by Sir Austin Bradford Hill [16] to provide a means to "pass from [an] observed association to a verdict of causation". Nine criteria were proposed and are described in further detail below:

Strength of the Association: Consider this a statistical description. Depending on the source of the data, this could be as simple as the difference in frequency rates between an exposed group and a control group (the risk difference), or it could include an actual statistical measure of the extent of the difference as described by an odds ratio or a hazard ratio. However, the lack of an apparent statistically significant difference should not necessarily be interpreted as the lack of supporting evidence as the dataset may be too small to make meaningful comparisons. Alternatively, a statistically significant finding may arise by chance following multiple comparisons when the study was not designed with this comparison as the primary endpoint.

Consistency: Has the finding been consistently observed in other settings? In clinical development this could mean the finding has been replicated across multiple clinical trials or multiple patient populations. In the post-marketing environment, it could mean the finding was observed across multiple post-marketing regional datasets, or within a dataset the finding was observed across multiple regions (see an example of this in the section on EudraVigilance below).

Specificity: A specific drug results in a specific effect. Such one-to-one relationships are, admittedly, uncommon. Drugs can typically produce a variety of untoward effects, and medical conditions often have multiple etiologies. However, when such a specific effect is observed, it can be compelling evidence.

Temporality: The effect must occur after exposure to the putative cause, and the closer the event occurs in time after the exposure, the stronger the temporal relationship. Positive dechallenge and rechallenge information can also be considered to be part of a temporal relationship.

Biological gradient: In the context of drug development, this can be envisioned as a dose-response relationship. As mentioned in previous discussions above, in addition to actual dose, the relationship may extend to cumulative dose and PK parameters, such as C_{max} and AUC.

Plausibility: Is there a plausible mechanism to explain how the drug might cause the adverse effect? However, the lack of a plausible mechanism might reflect inadequacies in pharmacological knowledge and does not, per se, suggest the lack of an association.

Coherence: This criterion is generally difficult to apply to drug evaluations, and its application is perhaps the most misunderstood in publications applying it to a drug/event evaluation. Consider coherence to be the inverse of plausibility. In the plausibility criterion, we start with the known effects of the drug and ask if it could reasonably produce the observed effect. In the case of coherence, we start with the known etiologies of the event and ask if any of these are consistent with the known pharmacology of the drug.

Experiment: This is another criterion that is commonly misunderstood. Clinical experimental evidence is already accounted for in the Strength of Association criterion. Experimental evidence in the context of this criterion would be nonclinical studies, including safety pharmacology and toxicology studies, or *in vitro* studies such as metabolizing enzyme or transporter interactions or ion channel inhibition.

Analogy: This final criterion evaluates the extent to which similar compounds produce similar effects. Consider evidence for class effects.

Taken together, the CIOMS and FDA criteria, when considered within the Bradford Hill framework, offer a robust method to evaluate the strength of evidence for or against a causal relationship between a drug and an adverse event. This process does not rely on a scoring system, such as other algorithms, and, thus, does not suffer from the fundamentally arbitrary nature of these scoring methods. Although this process provides a structured approach, it still allows for the application of clinical judgment. A guide to implementing the regulatory recommendations within the Bradford Hill framework is detailed in Figure 7.1.

7.3   POST-MARKETING SIGNAL DETECTION AND ASSESSMENT

Post-marketing signal detection involves the evaluation of spontaneous reports. Not all the signal detection methods used during clinical development can be applied to the post-marketing environment. Individual case review can continue to use the DME and IME lists. However, without a clearly known frequency rate, data mining techniques must be used in large post-marketing datasets.

Coincident with the release of the FDA's Pre-Marketing Risk Assessment guidance was the publication of the Good Pharmacovigilance Practices (GVP) and Pharmacoepidemiology guidance [17]. This latter document was intended to provide guidance on safety signal detection and interpretation in the post-marketing environment. A safety signal is defined as a worry about an excess of unfavorable events relative to what would be anticipated to be connected to the use of a product, requiring additional research to ascertain whether the product was the cause of the event or not. According to the guidance a single post-marketing case may represent a safety signal if it is well documented, if the report describes a positive rechallenge, or if the event is extremely rare in the absence of drug use. Other circumstances that may represent a safety signal include the following:

- New unlabeled adverse events, especially if serious
- An apparent increase in the severity of a labeled event
- Occurrence of serious events thought to be extremely rare in the general population
- New product-product, product-device, product-food, or product-dietary supplement interactions
- Identification of a previously unrecognized at-risk population (e.g., populations with specific racial or genetic predispositions or co-morbidities)
- Confusion about a product's name, labeling, packaging, or use
- Concerns arising from the way a product is used (e.g., adverse events seen at higher than labeled doses or in populations not recommended for treatment)
- Concerns arising from potential inadequacies of a currently implemented risk minimization action plan (e.g., reports of serious adverse events that appear to reflect failure of a RiskMAP[1] goal)
- Other concerns identified by the sponsor or FDA

The previously mentioned FDA guidance on QT prolongation included advice that the available post-marketing adverse event data should be examined for evidence of QT/QTc interval prolongation and TdP and for adverse events possibly related to QT/QTc interval prolongation, such as cardiac arrest, sudden cardiac death, and ventricular arrhythmias (e.g., ventricular tachycardia and ventricular fibrillation). While the other events that are reported more frequently would be of particular concern if reported in a population at low risk for them (e.g., young men experiencing sudden death), well-characterized episodes of TdP have a high probability of being linked to drug use.

Strength of the Association

- AE frequency difference between treatment groups as measured by an odds ratio or hazard ratio (consider FDA FMQs in addition to single preferred terms)
- TEAEs that occurred in >5% of patients treated with study drug with a risk difference of ≥2% and a risk ratio of ≥1.2 compared to the control group
- High signal of disproportional reporting score

Consistency

- Consistently higher incidence vs placebo or active comparator across studies
- Consistency of pattern of presenting symptoms
- Consistency of time to onset

Temporality

- Onset shortly after drug administration
- Earlier onset and/or greater severity in active vs comparator group
- Positive dechallenge / positive rechallenge

Biological Gradient

- Positive dose-response relationship (including cumulative dose)
- AE frequency associated with pharmacokinetic variables (e.g., C_{max}, AUC)

Plausibility

- Plausible mechanism of action
- Known pharmacology including results from receptor, transporter, or enzyme binding or activation studies

Coherence

- Rarity of the event in untreated patients
- Matches terms on the Designated Medical Event list

Experiment

- Nonclinical evidence to support the finding, including *in vivo* and *in vitro* studies
- Positive drug-drug or drug-transporter interaction studies

Analogy

- Recognized class effect
- Occurrence of other potentially related adverse events (e.g., occurrence of both strokes and transient ischemic attacks)
- Event is commonly drug-related (e.g., neutropenia, Stevens-Johnson Syndrome)

FIGURE 7.1 Steps in the application of regulatory recommendations for signal detection and assessment.

Generally, when a case indicates a safety concern, the sponsor should seek other similar cases in the safety database to assemble a case series. In the evaluation of a case series, the guidance document offers several factors to consider in an assessment of causal association:

- Occurrence of the adverse event in the expected time (e.g., type 1 allergic reactions occurring within days of therapy, cancers developing after years of therapy)
- Absence of symptoms related to the event prior to exposure
- Evidence of positive dechallenge or positive rechallenge
- Consistency of the event with the established pharmacological/toxicological effects of the product, or for vaccines, consistency with established infectious or immunologic mechanisms of injury
- Consistency of the event with the known effects of other products in the class
- Existence of other supporting evidence from preclinical studies, clinical trials, and/or pharmacoepidemiologic studies
- Absence of alternative explanations for the event (e.g., no concomitant medications that could contribute to the event; no co- or pre-morbid medical conditions)

The FDA recognized that data mining techniques may also augment other signal detection methods. The various data mining methodologies produce a score that quantifies the disproportionality between the observed number of adverse event reports and expected values for a given product-event combination. The FDA cautions that these methods are inherently exploratory or hypothesis generating. Although the outputs are often referred to as signals of disproportionate reporting, it is important to understand that these are statistical values, not clinical signals. Clinical judgment is required to determine which of these data mining scores warrant further evaluation.

The FDA separately provided additional criteria for a causality assessment. This was described as factors that the FDA takes into account when evaluating a sponsor's assessment of a signal.

- Strength of the association (e.g., relative risk of the adverse event associated with the product)
- Temporal relationship of product use and the event
- Consistency of findings across available data sources
- Evidence of a dose-response for the effect
- Biologic plausibility

It is noteworthy that these criteria are similar in many respects to the causality criteria discussed previously provided by the CIOMS and FDA and the structure of the Bradford Hill causality framework.

A longstanding resource published by the MHRA (Medicines and Healthcare Products Regulatory Agency) is the Good Pharmacovigilance Practice Guide [18]. Signal detection activities are described but principally in a general manner. The guide speaks to individual case review yet cautions that "a single report of a suspected adverse reaction can only rarely be considered as a signal in itself". With respect to the review of multiple case reports, the guide notes that data mining methods may be employed in the search for safety signals. These methods can be accompanied by looking for changes in the number of cases of an adverse event reported over time; for example, in a quarter-to-quarter comparison. However, specific criteria to either detect safety signals or evaluate them are not covered. The material presented primarily describes operational considerations.

The MHRA guide refers the reader to additional material on post-authorization evaluation of pharmacovigilance data found in Volume 9A, Part 1, Section B produced by the European Commission [19].[2] This guidance primarily covers operational aspects of pharmacovigilance. Brief mention is made of identifying signals as a new unexpected hazard or a change in the severity, frequency, or characteristics of an expected adverse event.

In 2012 the European Medicines Agency (EMA) released Module IX of the GVP guidances [20]. Overall, this series of guidances provided an extensive description of the type of pharmacovigilance practices expected of marketing authorization holders. Importantly, Module IX described the signal management process. Although the name implies a discussion of how to identify a safety signal, the actual content focuses more on operational aspects. Subsequently, in 2017 the EMA published Addendum I to Module IX titled "Methodological aspects of signal detection from spontaneous reports of suspected adverse reactions" [21]. The addendum addresses signals of disproportionate reporting that result from various data mining algorithms and the thresholds that can be applied. In order to limit false positives, the guidance suggests that it is better to raise the threshold for the number of individual case safety reports (ICSRs) than the value for the disproportionality statistic. For the disproportionality statistic, rather than using the point estimate, a formal lower confidence bound is advocated. The rationale is that when the statistic is based on a few ICSRs, the lower bound falls further below the point estimate and makes an SDR score above the threshold less likely. However, what the preferred number of ICSR reports and what the SDR score threshold should be are not addressed.

A further method to investigate post-marketing spontaneous reports is to examine temporal changes in reporting frequency. The simplest method is to measure the number of ICSRs pertaining to a particular event over a fixed period of time and compare it to a subsequent period of time. Alternatively, SDR scores can be derived for particular periods of time and compared to subsequent periods to determine if the SDR score is increasing or decreasing.

Event terms reported in ICSRs can be compared to the DME and IME lists. Cases with terms matching any of those on these lists may be considered a safety signal, particularly IME terms with a fatal outcome.

The addendum further describes signal detection approaches in special populations, specifically pediatric and geriatric populations. An SDR score that is higher in one of these subgroups than in the general population should be highlighted for additional consideration. Given the lower number of patient reports in these subgroups, a lower threshold for the number of ICSRs should be used. The specific SDR ratios and ICSR number thresholds to be employed are not defined; however, more specific details are described in the discussion of EudraVigilance monitoring methods below.

EudraVigilance, the database of safety reports from clinical trials and post-marketing experience, is maintained by the EU regulatory network. In 2016 the EMA published a guidance titled "Screening for Adverse Reactions in EudraVigilance" [22]. The publication coincided with the intent to allow access to marketing authorization holders. The document cautions that SDR scores that highlight a particular drug-event pair should be further evaluated before considering the finding as a safety signal. This is an important concept. As noted previously in the discussion of FDA guidance, the SDR score is a statistical finding, not necessarily a clinically relevant safety signal. The preferred disproportionality method is the reporting odds ratio (ROR), and the applicable threshold is the lower bound of the 95% confidence interval (ROR025). The EMA suggests that the criteria for a safety signal are met when the lower bound of the 95% confidence interval of the ROR is ≥ 1, the number of ICSR cases is at least three for drugs requiring additional monitoring or at least five for all other drugs, the event is on the IME list, and subgrouping by geographical region of the report shows the regional SDR score exceeds the threshold in more than one region.

Additional criteria to indicate a safety signal include those ICSRs that contain an event on the DME list or a fatal event on the IME list, regardless of the SDR score.

With respect to pediatric and geriatric populations, this guidance further clarifies what would be considered a safety signal for further review in these groups. The relative ROR score is calculated for these groups. For example, among pediatric patients the relative ROR score is the ROR score for an event in the pediatric group divided by the ROR score for the event in the rest of the population. A signal of disproportionate reporting in the pediatric subgroup is thus defined by the lower bound of the 95% confidence interval of the ROR score among pediatric patients ≥ 1, the lower bound of the 95% confidence interval of the pediatric ROR score divided by the lower bound of the 95%

confidence interval of the non-pediatric ROR score ≥1, the number of ICSR cases is ≥2 for drugs requiring additional monitoring or ≥3 for all other drugs, and the event is on the IME list.

A similar approach is taken for geriatric patients whereby the relative ROR score is the ROR score for an event in the geriatric group divided by the ROR score for the event in the rest of the population. A signal of disproportionate reporting is defined in this group as the lower bound of the 95% confidence interval of the ROR score among geriatric patients is ≥1, the lower bound of the 95% confidence interval of the geriatric ROR score divided by the lower bound of the 95% confidence interval of the non-geriatric ROR score is ≥1, the number of ICSR cases is ≥3 for drugs requiring additional monitoring or ≥5 for all other drugs, and the event is on the IME list. Note that the number of ICSR case thresholds is different from the pediatric group criteria.

Additionally, cases showing a positive rechallenge should be identified for further review.

What is missing from this discussion is what constitutes recommended evaluations to assess for a causal relationship between the event and the drug. Referring back to GVP Module IX, Section IX.B.3 pertains to factors to consider in the evaluation process. For example, cases showing a compatible temporal association, positive de- or rechallenge, lack of potential alternative causes, assessed as possibly related by the reporting healthcare professional, with supportive results of relevant investigations, consistency of the evidence across cases, dose-reaction relationship, and a possible mechanism based on a biological and pharmacological plausibility. The reader can appreciate some of the similarities with the previously described criteria from CIOMS and FDA within the Bradford Hill framework. Thus, while adverse event information arising from clinical development and post-marketing sources represent different datasets, and some of the tools to identify safety signals may differ between these two circumstances, the methods by which causality can be assessed can be similar providing a consistent approach to expand the understanding of the risk profile of a product.

Approaches to signal detection and evaluation have been proffered by a number of sources. Focusing on those provided by regulatory authorities can both provide insight into how these agencies approach safety data and offer a path for the sponsor to evaluate their data in a manner consistent with the regulatory authorities.

NOTES

1. Since the publication of this guidance document in 2005, RiskMAPs have been replaced by the Risk Evaluation and Mitigation Strategy (REMS) program.
2. https://ec.europa.eu/docsroom/documents/2809/attachments/1/translations/en/renditions/pdf

REFERENCES

1. Council for International Organizations of Medical Sciences. "Management of safety information from clinical trials: Report of CIOMS Working Group VI." In *Management of Safety Information from Clinical Trials: Report of CIOMS Working Group VI*. 2005.
2. FDA Guidance for Industry, Premarketing Risk Assessment. 2005. https://www.fda.gov/media/71650/download (accessed August 8, 2023).
3. FDA Guidance for Industry, Drug-Induced Liver Injury: Premarketing Clinical Evaluation. 2007. https://www.fda.gov/media/116737/download (accessed August 8, 2023).
4. Kullak-Ublick, Gerd A., Raul J. Andrade, Michael Merz, Peter End, Andreas Benesic, Alexander L. Gerbes, and Guruprasad P. Aithal. "Drug-induced liver injury: Recent advances in diagnosis and risk assessment." *Gut* 66, no. 6 (2017): 1154–1164.
5. Leise, Michael D., John J. Poterucha, and Jayant A. Talwalkar. "Drug-induced liver injury." *Mayo Clinic Proceedings* 89 (2014): 95–106.
6. Robles-Diaz, Mercedes, M. Isabel Lucena, Neil Kaplowitz, Camilla Stephens, Inmaculada Medina-Cáliz, Andres González-Jimenez, Eugenia Ulzurrun et al. "Use of Hy's law and a new composite algorithm to predict acute liver failure in patients with drug-induced liver injury." *Gastroenterology* 147, no. 1 (2014): 109–118.

7. Hayashi, Paul H., Don C. Rockey, Robert J. Fontana, Hans L. Tillmann, Neil Kaplowitz, Huiman X. Barnhart, Jiezhan Gu et al. "Death and liver transplantation within 2 years of onset of drug-induced liver injury." *Hepatology* 66, no. 4 (2017): 1275–1285.

8. FDA Guidance for Industry, E14 Clinical Evaluation of QT/QTc Interval Prolongation and Proarrhythmic Potential for Non-Antiarrhythmic Drugs. 2005. https://www.fda.gov/files/drugs/published/E14-Clinical-Evaluation-of-QT-QTc-Interval-Prolongation-and-Proarrhythmic-Potential-for-Non-Antiarrhythmic-Drugs.pdf (accessed August 8, 2023).

9. FDA Guidance for Industry In Vitro Drug Interaction Studies—Cytochrome P450 Enzyme- and Transporter-Mediated Drug Interactions. 2020. https://www.fda.gov/media/134582/download (accessed August 8, 2023).

10. FDA Reviewer Guidance Conducting a Clinical Safety Review of a New Product Application and Preparing a Report on the Review. 2005. https://purl.fdlp.gov/GPO/LPS116705

11. Center for Drug Evaluation and Research, Clinical Review, Application Number 211996Orig1s000 212161Orig1s000. 2019. https://www.accessdata.fda.gov/drugsatfda_docs/nda/2019/211996Orig1s000,%20212161Orig1s000MedR.pdf (accessed August 8, 2023).

12. FDA Advancing Premarket Safety Analytics. 2022. https://healthpolicy.duke.edu/events/advancing-premarket-safety-analytics (accessed August 8, 2023).

13. Designated Medical Event (DME) List. 2020. https://www.ema.europa.eu/en/documents/other/designated-medical-event-dme-list_en.xlsx (accessed August 9, 2023).

14. MedDRA Important Medical Event Terms List – Version 26.0. 2023. https://www.ema.europa.eu/en/documents/other/meddra-important-medical-event-terms-list-version-260_en.xlsx (accessed August 23, 2023) [Note: the IME list is specific to each MedDRA version].

15. FDA Guidance for Industry Sponsor Responsibilities—Safety Reporting Requirements and Safety Assessment for IND and Bioavailability/Bioequivalence Studies. 2021. https://www.fda.gov/media/150356/download (accessed August 9, 2023).

16. Hill, Austin Bradford. "The environment and disease: Association or causation?" *Proceedings of the Royal Society of Medicine* 58 (1965): 295–300.

17. FDA Guidance for Industry Good Pharmacovigilance Practices and Pharmacoepidemiologic Assessment. 2005. https://www.fda.gov/files/drugs/published/Good-Pharmacovigilance-Practices-and-Pharmacoepidemiologic-Assessment-March-2005.pdf (accessed August 9, 2023).

18. Medicines and Healthcare products Regulatory Agency. 2009. *Good Pharmacovigilance Practice Guide*. London, Pharmaceutical Press.

19. Volume 9 Pharmacovigilance. 2014. https://ec.europa.eu/docsroom/documents/2809/attachments/1/translations/en/renditions/pdf (accessed August 9, 2023).

20. Guideline on Good Pharmacovigilance Practices (GVP) Module IX – Signal Management. 2012. https://www.ema.europa.eu/en/documents/scientific-guideline/guideline-good-pharmacovigilance-practices-module-ix-signal-management-superseded_en.pdf (accessed August 9, 2023). Superseded by Guideline on good pharmacovigilance practices (GVP) Module IX – Signal management (Rev 1) 2017. https://www.ema.europa.eu/en/documents/scientific-guideline/guideline-good-pharmacovigilance-practices-module-ix-signal-management-superseded_en.pdf (accessed August 9, 2023).

21. Guideline on Good Pharmacovigilance Practices (GVP) Module IX Addendum I – Methodological Aspects of Signal Detection from Spontaneous Reports of Suspected Adverse Reactions. 2017. https://www.ema.europa.eu/en/documents/scientific-guideline/guideline-good-pharmacovigilance-practices-gvp-module-ix-addendum-i-methodological-aspects-signal_en.pdf (accessed August 9, 2023).

22. Screening for Adverse Reactions in EudraVigilance. 2016. https://www.ema.europa.eu/en/documents/other/screening-adverse-reactions-eudravigilance_en.pdf (accessed August 9, 2023).

8 Disproportionality Methods for Signal Detection in Pharmacovigilance

Krishna Undela and Christy Thomas

8.1 SIGNAL DETECTION IN PHARMACOVIGILANCE

Signal detection is an important part of pharmacovigilance, which is the science of monitoring the safety of medications after they are marketed. It is the process of data mining to identify the safety signals of a medication. A safety signal is an indication that a medication may have an adverse effect that was not previously known or suspected. In this context, a signal refers to a higher-than-expected number of adverse events with a specific medication, generating a 'disproportionate' rate of reporting compared to other events recorded in the pharmacovigilance database (1). By detecting safety signals early, pharmacovigilance can help prevent harm to patients. The renowned organizations in pharmacovigilance defined a 'signal' as follows:

> A signal is 'reported information on a possible causal relationship between an adverse event and a medication, the relationship being unknown or incompletely documented previously'. Depending upon the seriousness of the event and the quality of the information available, more than a single report is required to generate a signal. A signal could be a labeled or expected adverse reaction, or it could be an unlabeled or unexpected reaction. It could be labeled and still exhibit itself as a signal in the events, such as an increase in their frequency of occurrence (2).
>
> *World Health Organization*

> 'A safety signal is information on a new or known adverse event that may be caused by a medicine and requires further investigation'. Safety signals can be detected from a wide range of sources, such as spontaneous reports, clinical studies, and scientific literature (3).
>
> *European Medicines Agency*

> 'A signal is essentially a hypothesis of a risk with a medicine with data and arguments that support it, derived from data from one or more of many possible sources' (4).
>
> *Uppsala Monitoring Centre*

Though the above definitions are different from each other, they have something in common. A signal is 'information' derived from 'different sources' on a 'potential harm' that may be caused by a 'medicine' and it requires 'further evaluation'.

8.2 METHODS OF SIGNAL DETECTION

The signal detection methods in pharmacovigilance are mainly divided into qualitative and quantitative. Qualitative methods involve the review of data by experts to identify patterns or trends that suggest a possible safety signal. These methods are often used to identify signals from spontaneous reports, which are often incomplete or inaccurate. Quantitative methods use statistical techniques to identify signals in data. These methods are more objective than qualitative methods, but they can

DOI: 10.1201/9781032629940-8

also be more sensitive to false positive. This chapter contains a detailed description of the most commonly used quantitative method for signal detection, i.e. disproportionality analysis.

8.3 DISPROPORTIONALITY ANALYSIS

Disproportionality analysis is a statistical method used to identify potential safety signals from spontaneous reports of adverse events. The disproportionality analysis compares the observed number of reports of a specific medication-adverse event combination to the number that would be expected if there was no association between the medication and the adverse event. The expected number of reports is calculated based on the background rates of the medication and the adverse event combination in the population (5). The number obtained from the disproportionality analysis can be interpreted as the number of times more likely the observed number of reports is than the expected number. For example, an outcome of 1 indicates that there is no disproportionality between the observed and expected numbers of reports. A value greater than 1 indicates that there is a positive disproportionality, meaning that the observed number of reports is more than expected. A value less than 1 indicates that there is a negative disproportionality, meaning that the observed number of reports is less than expected.

The different methods for studying the disproportionality of adverse event reports include frequentist methods such as proportional reporting ratio (PRR) and reporting odds ratio (ROR); and Bayesian methods such as the Multi-Item Gamma Poisson Shrinker (MGPS) method and the Bayesian Confidence Propagation Neural Network (BCPNN) method (1). This chapter focuses on the calculation and interpretation of frequentist methods of disproportionality analysis, as described in Figure 8.1.

8.4 PROPORTIONAL REPORTING RATIO (PRR)

The PRR is a statistical method used to detect signals in pharmacovigilance databases (6). Based on this method, a signal will be identified when the event of interest 'E' is reported relatively more frequently for the medication of interest 'M' than with other medications. This relative increase in the event of interest 'E' reporting for the medication of interest 'M' in a pharmacovigilance database is reflected in the below 2×2 contingency table (7). The computation of the PRR is presented in Table 8.1.

The descriptions of codings for the values in Table 8.1 are as follows:

- The value A indicates the number of individual cases with the medication of interest 'M' involving an event of interest 'E'.
- The value B indicates the number of individual cases related to the medication of interest 'M', involving any other adverse events but 'E'.
- The value C indicates the number of individual cases involving the event of interest 'E' in relation to any other medications but 'M'.
- The value D indicates the number of individual cases involving any other adverse events but 'E' and any other medications but 'M'.

TABLE 8.1

The Computation of the Proportional Reporting Ratio (PRR)

	Event of Interest (E)	All Other Events	Total
Medication of Interest (M)	A	B	A+B
All Other Medications	C	D	C+D
Total	A+C	B+D	N = A+B+C+D

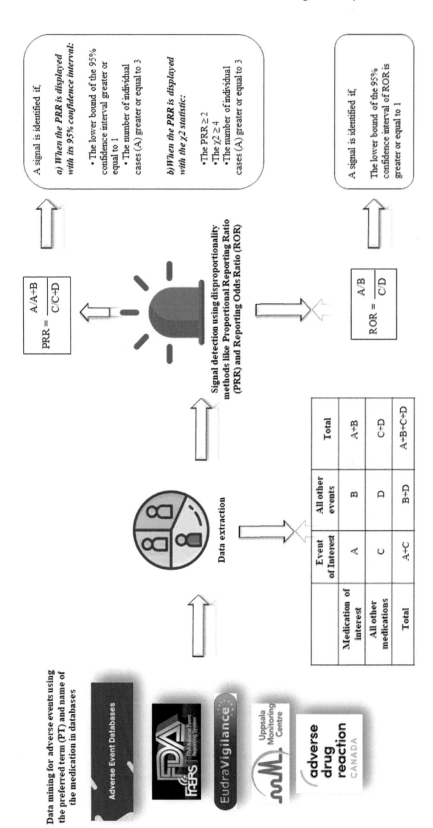

FIGURE 8.1   Methodology of disproportionality analysis.

Using the values given in Table 8.1, the PRR is computed as follows (6):

$$PRR = \frac{A/A + B}{C/C + D}$$

It is understood from the above formula that the PRR compares the ratio of reporting an adverse event for a medication of interest to the ratio of reporting the adverse event for all other medications in the database.

8.4.1 COMPUTATION OF 95% CONFIDENCE INTERVAL OF THE PRR

The 95% confidence interval (CI) takes into account the uncertainty in the estimate of the PRR. It is a range of values that is likely to contain the true value of the PRR, with a 95% probability. The lower and upper limits of 95% CI will be calculated by using the following formulas (1):

$$\text{Lower } 95\% \text{ CI} = e^{\ln(PRR)-1.96\sqrt{1/A+1/C-1/(A+B)-1/(C+D)}}$$

$$\text{Upper } 95\% \text{ CI} = e^{\ln(PRR)+1.96\sqrt{1/A+1/C-1/(A+B)-1/(C+D)}}$$

8.4.2 COMPUTATION OF CHI-SQUARE (χ^2) VALUE OF THE PRR

The Chi-square is a statistic that is traditionally used in disproportionality analyses. The Chi-square value is used as an alternative measure of association between the medication of interest 'M' and the event of interest 'E' based on the following calculation (European Medicines Agency 2006):

$$\chi^2 = \left[(AD - BC)^2 (A+B+C+D)\right] / \left[(A+B)(C+D)(A+C)(B+D)\right]$$

8.4.3 INTERPRETATION OF PRR IN SIGNAL DETECTION

The following criteria are followed to detect the signals in pharmacovigilance by using the PRR:

a. When the PRR is displayed with its 95% CI:
   - The lower bound of the 95% CI greater or equal to 1
   - The number of individual cases (A) greater or equal to 3
b. When the PRR is displayed with the χ^2 statistic:
   - The PRR ≥ 2
   - The $\chi^2 \geq 4$
   - The number of individual cases (A) greater or equal to 3

8.5 REPORTING ODDS RATIO (ROR)

The reporting odds ratio (ROR) is a similar measure to that of the PRR and is also used for the purpose of identifying the signals in pharmacovigilance. It is measured using the following formula and considering the values from the same 2×2 contingency table (Table 8.1) (8):

$$ROR = \frac{A/B}{C/D}$$

The above formula indicates that the ROR compares the odds of an adverse event being reported for a medication of interest to the odds of the adverse event being reported for all other medications in the database.

The lower and upper limits of 95% CI of ROR are calculated by using the following formulas (1):

$$\text{Lower } 95\% \text{ CI} = e^{\ln(\text{ROR})-1.96\sqrt{1/A+1/C-1/(A+B)-1/(C+D)}}$$

$$\text{Upper } 95\% \text{ CI} = e^{\ln(\text{ROR})+1.96\sqrt{1/A+1/C-1/(A+B)-1/(C+D)}}$$

8.5.1 Interpretation of ROR in Signal Detection

The signal between the medication of interest and event of interest will be confirmed when the lower bound of the 95% CI of ROR is greater or equal to 1 (8).

8.6 THE RELATION BETWEEN PRR AND ROR

The PRR and ROR are dimensionless numbers that can be interpreted as the number of times more likely the observed number of reports is than the expected number, assuming no association between the medication and the event. The PRR is identical to the calculation of a relative risk (RR) from a cohort study, and the ROR is identical to the calculation of an odds ratio (OR) from a case-control study (8). It is well recognized that these measures will give very similar results providing, as is virtually always the case in this context, 'A' is a small proportion of 'A+B' and 'C' is a small proportion of 'C+D'. Effectively, this is a similar argument used to show that the OR in a case-control study approximates the RR in a cohort study. The ROR is more sensitive to rare events than the PRR. However, the PRR is less affected by biases in the data, such as over-reporting certain adverse events.

8.7 ADVANTAGES OF DISPROPORTIONALITY ANALYSIS

Because clinical trials are known to be limited in their ability to identify rare adverse events, post-marketing safety data are an invaluable resource for the early detection of new safety signals. In this context, several advantages of disproportionality analysis are as follows.

8.7.1 Detecting Rare Adverse Events

Disproportionality analysis is particularly effective in identifying rare adverse events that may not be evident in small-scale clinical trials. By analyzing large volumes of real-world data, it can detect potential safety signals that might have been missed by traditional methods. This early detection can help ensure timely intervention and risk mitigation for patients who may be experiencing these rare events (9).

8.7.2 Identifying Unexpected Associations

Disproportionality analysis can uncover unexpected associations between medications and adverse events. This means it can identify potential safety concerns that would not have been predicted based on prior knowledge or existing research. By analyzing patterns and trends in the data, disproportionality analysis can highlight potential risks or interactions that may require further investigation.

8.7.3 Understanding Safety Profiles across Diverse Populations

Another advantage of disproportionality analysis is its ability to analyze data from diverse patient populations. This allows for a more comprehensive understanding of the safety profile of medications in different demographic groups, including those that may be underrepresented in clinical trials. By considering a broader range of patient data, this analysis can identify subpopulations that may be at higher risk for adverse events and inform targeted interventions or prescribing guidelines.

8.8 LIMITATIONS OF DISPROPORTIONALITY ANALYSIS

Disproportionality analysis is a valuable tool for signal detection in pharmacovigilance. However, it is important to note that it is not a perfect method. It can be affected by several factors, such as the quality of the data, the choice of statistical methods, and the interpretation of the results. Moreover, it is sensitive to rare events, can be affected by biases in the data, and can be difficult to distinguish between real signals and false positives (10, 11). Despite these limitations, disproportionality analysis is a valuable tool for signal detection in pharmacovigilance. By using disproportionality analysis in conjunction with other methods, such as qualitative analysis, it is possible to improve the chances of detecting real safety signals early.

It is important to note that while disproportionality analysis has its limitations, as mentioned above, it still provides valuable insights into medication safety. By utilizing this method alongside other pharmacovigilance techniques and considering its results within the broader context of scientific evidence, regulators and healthcare professionals can make informed decisions to ensure the safe and effective use of medications.

8.9 CASE STUDY

This case study aims to detect the signal between the hypothetical medication 'Exagliflozin' and an adverse event 'Pancreatitis'. Table 8.2 contains the assumed numbers of adverse event reports in a pharmacovigilance database for the calculation of PRR and ROR.

TABLE 8.2

The Assumed Numbers of Adverse Event Reports in a Pharmacovigilance Database for the Calculation of PRR and ROR

	Pancreatitis	All Other Events	Total
Exagliflozin	300 (A)	19,700 (B)	20,000 (A+B)
All Other Medications	29,700 (C)	9,950,300 (D)	9,980,000 (C+D)
Total	30,000 (A+C)	9,970,000 (B+D)	10,000,000 (A+B+C+D)

PROPORTIONAL REPORTING RATIO (PRR)

$$PRR = \frac{A/A+B}{C/C+D}$$

$$PRR = \frac{300/20000}{29700/9980000}$$

$$PRR = 5.0$$

$$Lower\,95\%\,CI = e^{\ln(PRR)-1.96\sqrt{1/A+1/C-1/(A+B)-1/(C+D)}}$$

$$Lower\,95\%\,CI = e^{\ln(5)-1.96\sqrt{1/300+1/29700-1/(20000)-1/(9980000)}} = 4.5$$

$$Upper\,95\%\,CI = e^{\ln(PRR)+1.96\sqrt{1/A+1/C-1/(A+B)-1/(C+D)}}$$

$$Upper\,95\%\,CI = e^{\ln(5)+1.96\sqrt{1/300+1/29700-1/(20000)-1/(9980000)}} = 5.6$$

$$95\%\,CI = 4.5 - 5.6$$

$$\chi2 = \left[(AD - BC)^2(A + B + C + D)\right] / \left[(A + B)(C + D)(A + C)(B + D)\right]$$

$$\chi2 = \left[(2985090000 - 585090000)^2(300 + 19700 + 29700 + 9950300)\right] / \left[(20000)(9980000)(30000)(9970000)\right]$$

$$\chi2 = 964.8$$

Interpretation: PRR identified a statistically significant signal between the medication 'Exagliflozin' and the adverse event 'Pancreatitis'. The results indicate that the reports of pancreatitis were 5 times more (based on the PRR value) with Exagliflozin compared to all other medications in the database.

REPORTING ODDS RATIO (ROR)

$$ROR = \frac{A/B}{C/D}$$

$$ROR = \frac{300/19700}{29700/9950300}$$

$$ROR = 5.1$$

$$Lower\,95\%\,CI = e^{\ln(ROR)-1.96\sqrt{1/A+1/C-1/(A+B)-1/(C+D)}}$$

$$Lower\,95\%\,CI = e^{\ln(5.1)-1.96\sqrt{1/300+1/29700-1/(20000)-1/(9980000)}} = 4.6$$

$$Upper\,95\%\,CI = e^{\ln(ROR)+1.96\sqrt{1/A+1/C-1/(A+B)-1/(C+D)}}$$

$$Upper\,95\%\,CI = e^{\ln(5.1)+1.96\sqrt{1/300+1/29700-1/(20000)-1/(9980000)}} = 5.7$$

$$95\%\,CI = 4.6 - 5.7$$

Interpretation: ROR identified a statistically significant signal between the medication 'Exagliflozin' and the adverse event 'Pancreatitis'. The results indicate that the reports of pancreatitis were 4.5 times more (based on the lower bound of the 95% CI of ROR) with Exagliflozin compared to all other medications in the database.

8.10   OVERVIEW OF RECENTLY PUBLISHED STUDIES

An overview of recently published studies that used disproportionality methods for signal detection in pharmacovigilance is presented in Table 8.3.

TABLE 8.3

Overview of Recently Published Studies

Study ID	Databases Used	Outcome Measures	Medication(s)		Identified Signals
Battini V 2023	FAERS, VigiBase	PRR, ROR, EBGM, ERAM	Sulfonylureas	Gliclazide	Therapy failure in depressed cases
			Thiazolidinediones	Troglitazone	
			Dipeptidyl peptidase-4 inhibitors	Saxagliptin	
				Vildagliptin	
			Glucagon-like peptide-1 analogs	Dulaglutide	
				Liraglutide	
				Semaglutide	
			Sodium-glucose cotransporter-2 inhibitors	Dapagliflozin	
Dermoncourt A 2023	VigiBase	ROR (>4)	Anakinra		Musculoskeletal and connective tissue disorders
			Canakinumab		Musculoskeletal and connective tissue disorders, immune system disorders
			Abatacept		Musculoskeletal and connective tissue disorders
			Rituximab		Immune system disorders
			Belimumab		Neonatal infections
Goldman A 2022	FAERS	ROR, IC_{025}	Evolocumab		Composite hyperglycemic adverse events, mild hyperglycemia
			Statins		Composite hyperglycemic adverse events, diabetes mellitus
Huang S 2023	FAERS	PRR, ROR	Vemurafenib + Cobimetinib		Chorioretinopathy, detachment of retinal pigment epithelium, retinal detachment, serous retinal detachment, cataract, eye disorder, periorbital edema, ocular toxicity, retinopathy, serous retinopathy
			Dabrafenib + Trametinib		Chorioretinopathy, detachment of retinal pigment epithelium, macular fibrosis, papilledema, retinal detachment, serous retinal detachment, vitreous detachment, vitreous floaters, iris adhesions, eye edema, retinal disorders, eye contusion, ocular toxicity, photophobia, retinal vein occlusion, retinopathy, serous retinopathy
			Encorafenib + Binimetinib		Chorioretinopathy, detachment of retinal pigment epithelium, macular detachment, retinal detachment, serous retinal detachment, subretinal fluid, vitreous floaters, eye disorders, retinal disorders, ocular toxicity, retinopathy, serous retinopathy

(Continued)

TABLE 8.3 (Continued)
Overview of Recently Published Studies

Study ID	Databases Used	Outcome Measures	Medication(s)	Identified Signals
Kim YS 2022	FAERS, VigiBase, EudraVigilance	ROR	Sacubitril/Valsartan, lisinopril, enalapril, atenolol, valsartan, eplerenone, losartan, metoprolol, carvedilol, furosemide, spironolactone, bumetanide, ivabradine	Hypotension, renal dysfunction, hyperkalemia, angioedema
Palapra H 2022	FAERS, VigiBase, CARD	PRR, ROR, IC_{025}	Canagliflozin Dapagliflozin Empagliflozin	Pancreatitis
Raschi E 2017	FAERS, VigiBase, EudraVigilance	ROR	Canagliflozin	Infections and infestations, metabolism and nutrition disorders, renal and urinary disorders, reproductive system and breast disorders, skin and subcutaneous tissue disorders, social circumstances
			Dapagliflozin	Congenital, familial, and genetic disorders, infections and infestations, metabolism and nutrition disorders, renal and urinary disorders, skin and subcutaneous tissue disorders
			Empagliflozin	Infections and infestations, metabolism and nutrition disorders, renal and urinary disorders, reproductive system and breast disorders, skin and subcutaneous tissue disorders
Schaefer A 2020	FAERS, VigiBase	PRR (>5)	Pembrolizumab	Hypophysitis, adrenal insufficiency, vitiligo, pemphigoid, rash maculopapular, autoimmune colitis, colitis, thyroiditis, hypothyroidism, hyperthyroidism, thyroid disorder, autoimmune hepatitis, hepatitis, hepatotoxicity, pneumonitis, myasthenia gravis, Type 1 diabetes mellitus, diabetic ketoacidosis, uveitis, myocarditis, myositis, tubulointerstitial nephritis
Shu Y 2022	FAERS	ROR	Semaglutide	Eructation, obstructive pancreatitis, pancreatitis, early satiety, vomiting projectile, flatulence, gastro-esophageal reflux disease, pancreatic carcinoma and 37 other gastrointestinal adverse events

(Continued)

TABLE 8.3 *(Continued)*

Overview of Recently Published Studies

Study ID	Databases Used	Outcome Measures	Medication(s)	Identified Signals
Shu Y 2023	FAERS	ROR	Quetiapine	Cardiac disorders, dizziness, tachycardia, palpitations, cardio-respiratory arrest, syncope, sinus tachycardia, cardiac arrest, arrhythmia, bradycardia, myocarditis, atrial septal defect, cardiogenic shock, torsade de pointes, cardiomyopathy, ventricular extrasystoles, dizziness postural, cyanosis, ventricular tachycardia, cardiovascular disorder, ventricular septal defect, ventricular fibrillation, long QT syndrome, tricuspid valve incompetence, congestive cardiomyopathy, Brugada syndrome, supraventricular tachycardia, patent ductus arteriosus, ventricular arrhythmia, tachyarrhythmia, extrasystoles, bundle branch block right

Abbreviations: CARD: Canadian Adverse Reaction Database, EBGM: Empirical Bayes Geometric Mean, ERAM: Empirical Bayes Regression Adjusted Mean, FAERS: FDA Adverse Event Reporting System, IC: Information Component, PRR: Proportional Reporting Ratio, ROR: Reporting Odds Ratio.

REFERENCES

1. Faillie, J.L. 2019. Case–non-case studies: Principle, methods, bias and interpretation. *Therapies* 74 (2): 225–32.
2. Edwards, IR, Biriell, C. 1994. Harmonisation in pharmacovigilance. *Drug Saf* 10 (2): 93–102.
3. European Medicines Agency. Safety signal. https://www.ema.europa.eu/en/glossary/safety-signal (accessed November 15, 2023).
4. Uppsala Monitoring Centre. What is a signal? https://who-umc.org/signal-work/what-is-a-signal/ (accessed November 15, 2023).
5. Bate, A, Evans, S.J.W. 2009. Quantitative signal detection using spontaneous ADR reporting. *Pharmacoepidemiol Drug Saf* 18 (6): 427–36.
6. Evans, SJ, Waller, PC, Davis, S. 2001. Use of proportional reporting ratios (PRRs) for signal generation from spontaneous adverse drug reaction reports. *Pharmacoepidemiol Drug Saf* 10 (6): 483–6.
7. Van Puijenbroek, E.P., Bate, A, Leufkens, H.G.M., Lindquist, M., Orre, R, Egberts, A.C.G. 2002. A comparison of measures of disproportionality for signal detection in spontaneous reporting systems for adverse drug reactions. *Pharmacoepidemiol Drug Saf* 11 (1): 3–10.
8. Rothman, K.J., Lanes, S, Sacks, S.T. 2004. The reporting odds ratio and its advantages over the proportional reporting ratio. *Pharmacoepidemiol Drug Saf* 13 (8): 519–23.
9. Montastruc, JL, Sommet, A, Bagheri, H, Lapeyre-Mestre, M. 2011. Benefits and strengths of the disproportionality analysis for identification of adverse drug reactions in a pharmacovigilance database. *Br J Clin Pharmacol* 72 (6): 905–8.
10. Caster, O, Aoki, Y, Gattepaille, LM, Grundmark, B. 2020. Disproportionality analysis for pharmacovigilance signal detection in small databases or subsets: Recommendations for limiting false-positive associations. *Drug Saf* 43 (5): 479–87.
11. Council for International Organizations of Medical Sciences. 2010. Practical aspects of signal detection in pharmacovigilance : report of CIOMS Working Group VIII. https://cioms.ch/wp-content/uploads/2018/03/WG8-Signal-Detection.pdf (accessed November 15, 2023).

9 Bayesian Methods of Signal Detection

Dipika Bansal, Muhammed Favas KT, and Beema T Yoosuf

9.1 INTRODUCTION

Despite significant efforts to investigate drug safety before introducing new compounds to the market, certain adverse drug reactions (ADRs) remain unidentified until after the drug's release. This persistent challenge underscores the complexity of the pharmaceutical landscape. While comprehensive pre-market testing and clinical trials are carried out to evaluate a drug's safety and efficacy, there are various factors contributing to the continuous development of previously unrecognized ADRs. One major factor is the rarity of some adverse reactions. In clinical trials, which involve a limited number of participants, it may be statistically unlikely to encounter exceedingly rare ADRs. Moreover, certain subpopulations, like pregnant women or juveniles, are often excluded from clinical trials due to ethical and safety concerns. Consequently, ADRs specific to these groups may only surface once the drug is administered more widely. Another critical factor is drug interactions. Patients often take multiple medications simultaneously, and the interactions between these drugs can lead to unexpected ADRs. These interactions can be challenging to predict, even with extensive pre-market testing. Furthermore, some ADRs have a delayed onset, making them difficult to identify during initial testing. These delayed reactions may only become apparent after a drug has been on the market for an extended period, as the cumulative exposure reveals previously concealed risks.

To address these post-market challenges, regulatory bodies and drug monitoring centres have established computational data mining techniques. These methods supplement traditional approaches by sifting through vast spontaneous reporting databases and searching for patterns and associations that might indicate potential ADRs. However, it is crucial to emphasize that these outcomes from data mining are not conclusive evidence of causation. Instead, they serve as valuable starting points for further investigation. Screening individual case safety reports (ICSRs) or spontaneous reports is a primary strategy for post-marketing drug safety surveillance. This approach involves analysing real-world data collected from healthcare providers, patients, and other sources. Global initiatives have enhanced this process by aggregating information from around the world, increasing the chances of early detection of drug safety issues. Nevertheless, managing the enormous datasets involved in spontaneous reporting systems (SRSs) presents its own challenges. To extract meaningful insights, quantitative statistical approaches are essential. These statistical methods help identify statistically significant signals amid the noise, allowing researchers to prioritize potential ADRs for further evaluation.

9.1.1 SPONTANEOUS REPORTING SYSTEMS

In the post-marketing setting, SRSs are the principal source of data for safety surveillance which includes "VigiBase by World Health Organisation (WHO)" (1), "FDA Adverse Event Monitoring System (FAERS) by the U.S. Food and Drug Administration (FDA)" (2), and "EudraVigilance by European Medicines Agency" (3). SRSs are voluntary reports of suspected ADRs communicated by patients, healthcare providers, consumers, and/or manufacturers. Typically, the information comprises the patient's demographic information, the drug(s) being taken, the suspected ADR, and the outcome of the ADR (4, 5). Essential details in SRSs often include demographic data, report dates, drug/biologic information, report sources, suspected ADRs, and patient outcomes.

The purpose of designing these systems was to provide data that allows for the examination of potential drug safety issues, which would otherwise be impossible to detect within the limited time-frame of a clinical drug development stage. When a new drug enters the phase following approval for drug use, the possibility of interactions with concomitants, foods, biologics, and medical devices develops, which may alter the drug safety profile. Monitoring pharmaceutical safety following the marketing is vital for public health. It ensures ongoing evaluation of drug benefits and risks, safeguarding patients by responding to emerging concerns. Signal detection is crucial in pharmacovigilance as it provides the first indication that a medication is potentially harmful.

9.1.2 Signal Generation

The WHO has defined "signal" as "Reported information on a possible causal relationship between an adverse event and a drug, the relationship being unknown or incompletely documented", with the caveat that "usually more than one case is required" (6). Due to the voluntary nature of the reporting procedure, several issues arise. One severe issue is over-reporting, which can occur as a result of publicity or a drug warning. There may be underreporting, due to the nature of the ADRs and their severity, or inadequate knowledge of the reporting procedure. Multiple reports of ADRs can occur due to the duplication of identical events, stemming from either different individuals or recurrent reports from the same person.

Indeed, once a suspicion arises ("signal generation"), it must be confirmed by the collection of new evidence ("signal strengthening"). The third step ("signal quantification") entails using epidemiological approaches to confirm and quantify the drug-ADR association. Once a signal has been validated, regulatory agencies employ a range of measures, the choice of which hinges on the signal's perceived importance. This prioritization of ADRs hinges on factors including their impact on public health, the seriousness of the ADR, and the extent of disproportionality in reporting. Potential regulatory actions encompass continuous tracking and assessment of how a signal evolves and changes over time, altering labelling or usage guidelines, restricting distribution, or, in severe cases, removing the product from the market. The latter decision is typically made when the risk-benefit balance becomes unfavourable, posing significant risks to public health that outweigh the benefits associated with the drug's use.

9.1.2.1 Data Mining Algorithms

Data mining is a powerful technique that may be utilized to address a variety of challenges. It involves uncovering patterns within vast datasets by employing advanced technologies like machine learning, statistical analysis, and database systems. This interdisciplinary field, bridging computer science and statistics, strives to extract valuable insights from data and restructure it into a coherent format for further analysis. In the realm of pharmacovigilance, data mining algorithms (DMAs) have become indispensable tools. They enable professionals to quantitatively detect signals related to ADRs, aiding in the timely identification of possible safety issues and contributing to more proactive and informed decision-making in drug regulation and public health (7).

Traditional quantitative signal detection methodologies primarily relied on sales or prescription reporting rates. In contrast, modern approaches use "measures of disproportionality" to assess unexpectedness (8). In this scenario, "unexpectedness" indicates that the observed frequency of a drug-ADR pairing exceeds what would be anticipated by chance. The primary goal of "quantitative signal detection" is to identify genuine signals as early as feasible while minimizing false positives. Each combination of drugs and adverse events is assessed and quantified, with a predetermined threshold determining whether it warrants further review. Various techniques for gauging unexpectedness exist, but they all depend on comparing observed occurrences with expected values, ensuring a systematic and data-driven approach to signal detection.

TABLE 9.1

The 2×2 Contingency Table for the Calculation of Disproportionality

	Drug(s) of Interest	All Other Drugs in the Database	Total Σ
Adverse drug reaction(s) of interest	a (Co-occurrences)	b	$a + b$
All other adverse drug reaction	c	d	$c + d$
Total Σ	$a + c$	$b + d$	$a + b + c + d$

Notes: a – Number of reports in the SRSs containing both the suspected drug and the suspected ADR (co-occurrences); b – Number of reports in the SRSs containing the suspected ADR with other drugs; c – Number of reports in the SRSs containing the suspected drug with other ADR; d – Number of reports in the SRSs containing other medications and other ADR.

9.1.2.2 Framework for Calculating Disproportionality

A 2×2 contingency table containing the categorical variables, drug consumption and ADR occurrence, acts as the foundation for all disproportionality analysis (DPA). Drugs might be listed as either suspected or concurrent medication. The frequency of a specific combination (co-occurrences) can be determined by considering all reports linked with a drug or only those in which the drug is suspected of causing the adverse event (Table 9.1).

This option is not inherent in any approaches and warrants more investigation. Of course, all medicines must be examined while seeking out interactions.

9.1.3 MEASUREMENTS OF DISPROPORTIONALITY

Quantitative signal detection methods in DMAs can be broadly categorized into two: frequentist and Bayesian approaches. The frequentist approach, exemplified by the "Reporting Odds Ratio (ROR)" and the "Proportional Reporting Ratio (PRR)", was pioneered by Evans et al. (9). These methods calculate statistical ratios to assess the association between drugs and adverse events based on observed and expected values. In contrast, Bayesian approaches like the "Empirical Bayes Geometric Mean (EBGM)" (4) and "Bayesian Confidence Propagation Neural Network (BCPNN)" (10) employ Bayes' theorem to estimate the likelihood of an ADR occurring when a specific drug is used, incorporating prior information and observed data, thus offering a probabilistic perspective to signal detection.

9.1.3.1 Proportional Reporting Rate

PRR determines the proportion of specific ADR or cluster of ADRs for given drug(s) of interest, with all other drugs in the SRSs serving as the comparator. If the suspected drug and ADR are not associated, the anticipated level of PRR should be 1, where PRR > 1 indicates that the report occurred more frequently than expected. One disadvantage is that PRR may be undefined if there are no instances of a condition outside the target population (b = 0). It may also produce abnormally high results for rare events (when the values for a, b, and c are too low).

$$PRR = a/(a+c)/b/(b+d)$$
$$= \frac{a(b+d)}{b(a+c)}$$

The confidence interval (CI) can be estimated using the standard epidemiology 2×2 table defined by Rothman et al. (11).

$$95\% \, CI = exp(ln(PRR) \pm 1.96 \, SE), \text{where } SE = sqrt\left((1/a)+(1/b)+(1/c)+(1/d)\right)$$

9.1.3.2 Reporting Odds Ratio

ROR is another disproportionality metric where the anticipated frequency of a particular drug-ADR pair is independent of the observed frequency. ROR has been offered as a way of addressing the non-selective underreporting of certain medications or AEs. One consideration with ROR is that events particularly linked to medicine may be difficult to detect because the metric can be unbounded when there are no cases in the comparator group. This circumstance may occur in unusually notable cases of rare disorders that should be investigated further.

$$ROR = (a/c)/(b/d)$$

$$= \frac{ad}{bc}$$

The CI can be estimated as

$$95\% \, CI = exp\left(ln(ROR) \pm 1.96 \, SE\right), \text{where } SE = sqrt\left((1/a) + (1/b) + (1/c) + (1/d)\right)$$

9.1.3.3 Chi-Squared – χ^2

A measure of marginal expectation can be computed through a Chi-squared test with one degree of freedom. The accuracy of this test is dependable when the expected cell counts in a contingency table are sufficiently substantial. However, when these expected counts become small, typically below 5, the Chi-squared test may yield inaccurate outcomes. In such cases, if necessary, Yates' correction can be applied as an adjustment to mitigate the impact of low cell numbers and enhance the accuracy of the test results.

$$\chi^2 = \frac{[n(|ad - bc| - n/2)^2]}{[(a+b)(c+d)(a+c)(b+d)]}$$

Chi-squared with one degree of freedom and Yates' correction can be calculated using the above formula (12), where n indicates the total number of cases.

9.1.4 BAYESIAN APPROACHES

For any ICSR in the SRSs, there is a certain "prior probability" which is the likelihood that a particular ADR will be dealt with on it. If that case report mentions a particular medication, the likelihood that the ADR has been linked could be different, the "posterior probability". According to Bate A. et al., "If the posterior probability is higher than the prior probability, then the presence of the drug on the report has enhanced the chance of the ADR being present, and the drug – ADR pair are present together in the SRS more often than expected" (10).

Bayes' law states:

$$P(A/D) = \frac{P(D, A)}{P(D)}$$

"$P(A/D)$ = Posterior probability"
"$P(A)$ = Prior probability, that a specific ADR being present on a safety report"
"$P(D)$ = Prior probability, that a specific drug being present on a safety report"
"$P(D, A)$ = Coincident probability, that both specific drug and ADR co-occur on the same report"

From Table 9.1, it can be written as

$$"P(A) = (a+c)/n"$$
$$"P(D) = (a+b)/n"$$
$$"P(D, A) = a/n"$$

9.1.4.1 Empirical Bayes Gamma Mixture

DuMouchel describes Empirical Bayes Gamma Mixture (EBGM) as a method for evaluating drug-ADR combinations with "interestingly large counts in a large frequency table, having millions of cells, most of which have an observed frequency of 0 or 1" (13). In quantitative signal detection, the total occurrences of a specific drug-ADR combination are expected to conform to a Poisson distribution, where the parameter μ is determined as λE ($\mu = \lambda E$). Here, E represents the anticipated co-occurrences for both the drug and the ADR if there were no actual relationship among them. To accommodate the increased relative risk associated with a drug-ADR combination, the parameter λ is assumed to have a prior distribution represented as a combination of gamma distributions. This approach helps capture the variability and uncertainty in λ. This statistical framework helps assess the significance of the observed drug-ADR pairings in pharmacovigilance analysis.

The empirical Bayes aspect of this approach emerges by determining the prior distribution characteristics in a way that maximizes the "product of negative binomial marginal densities" for the co-occurrence counts of each drug-ADR combination. In the context of investigating possible drug-ADR associations, key components of the posterior distribution for λ, such as the expected value (λ_{EB}) and the 5% lower posterior quantile (λ_{EB05}), are frequently employed. These statistical parameters play a pivotal role in assessing the likelihood and significance of drug-ADR pairs in pharmacovigilance analysis. The U.S. FDA has adopted a signal detection threshold of $\lambda_{EB05} \geq 2$ for adverse event reports in the SRSs. This criterion ensures that, regardless of the total co-occurrence count, a specific drug-ADR pair is described at least two times as often as one would expect if there were no real connection between that particular drug and the suspected adverse event. For instance, a λ_{EB05} of 5 means that a drug-ADR combination has occurred five times more frequently than anticipated based on random chance. It is important to note that a high rate of reporting does not actually indicate a higher likelihood of the ADR occurring or imply a cause-and-effect relationship between the drug and the reported ADR, as other factors like reporting bias and coincidental events can influence reporting rates in pharmacovigilance.

In the context of EBGM, the conventional stratification parameters typically include age, gender, and reporting year. Stratification serves as a technique to minimize the chances of identifying potential drug-ADR combinations that may be influenced by independent associations involving a drug, a strata parameter (such as age or gender), and an adverse event, all sharing the same strata parameter. By stratifying the analyses based on the reporting year, the intention is to mitigate the risk of detecting signals that could be attributed to temporal trends rather than actual drug-ADR relationships. This approach helps ensure that the signals identified are more likely to be genuinely linked with the drug and ADR rather than being driven by factors related to reporting patterns or "trendiness" associated with specific drugs or adverse events during certain time periods (14).

9.1.4.2 Bayesian Confidence Propagation Neural Network

The BCPNN is a type of feed-forward neural network that leverages Bayes' law principles for learning and inference. Similar to other neural network architectures, one of BCPNN's key advantages lies in its self-organizing capability. Additionally, it can be efficiently implemented on parallel computing platforms, which enhances its computational efficiency and scalability for handling complex tasks in various domains, including machine learning and statistical analysis. This neural network also offers an efficient computational paradigm that works effectively with sequential machines. A further benefit of BCPNN is the ease with which the weights can be interpreted as probabilistic elements. The information recorded in the BCPNN as weights is employed here to determine drug-ADR dependence. The Bayesian neural network possesses the computational capability to examine all potential connections and detect potential signals effectively. Its transparency makes it easy to interpret the findings, and its robustness allows it to generate valid, meaningful insights even when data is missing, which is common in SRS reports with several empty fields. This quality is particularly advantageous as it simplifies validation and confirmation processes, making it easy to reproduce results. Moreover, this Bayesian network streamlines the analysis by requiring only a single pass over the data, which significantly enhances its efficiency and ease of training. To efficiently search through the SRS reports, a sparse matrix approach is employed. This approach leverages the factuality that only a small fraction of all conceivable drug-ADR co-occurrences have non-zero values in the SRSs, optimizing the network's performance while handling vast datasets in pharmacovigilance (15).

BCPNN employs Bayesian statistical methods within a neural network framework to systematically examine all described drug-ADR co-occurrences, aiming to uncover previously unrecognized associations. This technique has gained prominence and is now a regular tool for signal detection in WHO-VigiBase, a global pharmacovigilance database. It is also under investigation for potential use with observational claims data. In BCPNN, the robustness of the association between a drug and an ADR is quantified using the information component (IC), a logarithmic metric. The IC can be conceptualized as the logarithmic ratio of the observed co-occurrence rate to the expected co-occurrence rate of a drug and an ADR, assuming no relationship between them (null hypothesis). This statistical approach provides a robust method for identifying and evaluating potential signals in pharmacovigilance data (10).

$$IC = log_2\left(\frac{P(D,A)}{P(D)\,P(A)}\right)$$

$$= log_2\left(\frac{P(A\mid D)}{P(A)}\right)$$

where $P(A\mid D)$ is the conditional probability of A given D.

An IC value that is greater than zero for the co-occurrence of a drug and ADR signifies a stronger relationship than expected, implying that the drug has a notable impact on the likelihood of the ADR occurring in a specific case report. Conversely, IC values close to zero indicate a lack of association, indicating that the drug's intake does not significantly affect the probability of the ADR in that ICSR, or that prior and posterior probabilities remain unchanged (16).

The IC is determined through Bayesian analysis and serves as an approximation of the unknown likelihood of P(D, A), P(D), and P(A). To make this calculation more manageable, a Dirichlet prior distribution is modelled for the probability parameters. This choice is made because it aligns with the multinomial distribution of the data, allowing for straightforward mathematical expressions for the posterior probabilities of P(D, A), P(D), and P(A). It is important to note that there is not a convenient closed-form formula available for the posterior distribution of the IC. However, recent research has demonstrated that employing Monte Carlo (MC) simulations with the established closed-form expressions for P(D, A), P(D), and P(A) can be advantageous in gaining deeper insights into the characteristics of the posterior IC distribution. This approach helps us gain a more comprehensive understanding of the IC's behaviour and properties.

Some general concerns regarding the observed to expected ratios have to be taken into consideration. Since observed to expected ratios have essentially no upper limit, they are useful metrics of association, particularly for adverse events with low expected frequencies. Another issue with the ratio is that a large observed number of co-occurrences may have a spillover effect on the expected count of co-occurrences. Specifically, when a suspected drug is widely used, and there are an unusually high number of reports linking that drug to a specific ADR, this can significantly influence the overall occurrence of that ADR. As a result, the observed-to-expected ratio may not accurately reflect the true extent of the relationship between the drug and the ADR. In other words, the ratio may underestimate the extent of the association between the drug and the ADR due to the sheer volume of reports related to the drug's widespread use. In typical safety datasets, the concerns mentioned above usually do not have a substantial influence on pairwise IC analysis when examining the link between drug intake and the occurrence of ADRs. However, when studying different types of events or working with smaller datasets, these issues can become more relevant. To mitigate the risk of obtaining misleading results in such cases, it can be a good practice to compare the anticipated IC values for significant relationships with the standard log odds ratios. This comparative analysis can provide a more robust assessment of the strength and significance of these associations, helping to ensure the accuracy and reliability of the findings. The comparative analysis of Bayesian methods of signal generation is compiled in Table 9.2.

9.1.4.2.1 Moderating Prior Distribution

The goal of IC analysis is to produce valuable leads about quantitative relationships in a relational database. It is critical in this context to avoid exposing an abundance of connections with weak

TABLE 9.2

Comparative Analysis of Bayesian Methods of Signal Generation: Advantages and Disadvantages

	BCPNN	EBGM
Advantages		
Signal Detection	Effective for detecting signals associated with rare adverse events or with sparse data	Effective for identifying potential safety signals through disproportionality analysis
Bayesian Framework	Utilizes Bayesian statistics, providing a probabilistic framework for signal analysis, accounting for data uncertainty	Employs Bayesian estimation to adjust observed-to-expected ratios, improving stability
Handling Uncertainty	Propagates confidence values through a neural network, allowing for consideration of uncertainty in associations between drugs and adverse events	Addresses uncertainty in rare event estimates through empirical Bayes shrinkage, yielding robust results
Flexibility	Suitable for various data sources and can handle different types of adverse event data	Widely used and well-established method with a proven track record in pharmacovigilance
Disadvantages		
Complexity	Requires a deeper understanding of neural networks and Bayesian methods, making it less accessible for some users	May require expertise in setting appropriate thresholds for signal detection
Computation Intensive	May require significant computational resources, especially for large databases	The model can be computationally intensive for extensive databases
Interpretability	Results can be complex and challenging to interpret, as BCPNN generates multiple posterior probabilities	Interpretation of results may not be straightforward for non-experts
Dependency on Parameters	Performance can depend on hyperparameters, which may need to be tuned for specific applications	Performance can be sensitive to parameter choices, especially the threshold for signal detection

data support while focusing on the estimated magnitude of the association. In this regard, the use of Bayesian dependency derivation with a conservative prior distribution has proven to be advantageous in situations where data is limited. It helps to temper or moderate the estimated strength of association between variables. Essentially, by incorporating a cautious or conservative prior distribution in the Bayesian analysis, the method takes into account the uncertainty associated with limited data and provides more cautious or restrained estimates of the association's magnitude. This approach can be particularly valuable in cases where data scarcity could otherwise lead to exaggerated or unreliable assessments of associations between variables (10, 13). Using credibility intervals in conjunction with Bayesian moderation presents a practical and effective approach that strikes a balance between methods that solely depend on statistical significance (which can be sensitive to weak associations when there is abundant data) and methods that rely on raw observed-to-expected ratios (which are less reliable when data support is limited). As data accumulates, the influence of the prior distribution diminishes, and the difference between Bayesian and classical (frequentist) estimates becomes minimal for combinations with substantial data support. To account for the assumption of independence, this prior distribution incorporates hyperparameters that correspond to the products of the relevant marginal frequency values. In essence, this approach allows for more robust and balanced assessments of associations, adapting to the amount of data available and considering the underlying assumption of independence when determining credibility intervals. The advantage of this approach is that the posterior IC distribution is always centred between 0 and the raw observed-to-expected log ratio, and it can be analysed regardless of whether one of the marginal values is 0, which is essential for computational stability.

9.1.4.2.2   Credibility Interval Estimates

The IC_{025} is a statistical metric that signifies the 2.5th percentile of the posterior IC distribution. It serves as the lower boundary of a two-sided 95% credibility interval associated with the IC. This 95% credibility interval is a standard method employed in IC analysis to assess and rank the strength of associations between variables. In essence, the IC_{025} provides a lower threshold for the range of possible IC values, helping to establish a CI around the estimated IC and offering a measure of uncertainty in the assessment of associations (10, 16). The adoption of a "lower credibility interval limit" accommodates uncertainty cautiously or conservatively. The objective is to decide on an impression such that its real value will be higher than what is estimated with a particular level of certainty (at the 97.5th percentile).

The credibility intervals for the IC are typically calculated using a normal approximation to the IC distribution (10, 16). However, MC experiments have revealed that although the distribution of IC tends to approach a normal distribution as sample sizes grow larger (asymptotically), this assumption leads to relatively inaccurate approximations when dealing with uncommon pairs of events. Notably, more than 80% of the reported drug-ADR pairs in the WHO VigiBase dataset link into this critically important categorization of uncommon events. Consequently, there is a clear need for improvement in the calculation of credibility intervals for such cases. One potential approach to address this issue is to employ brute force MC simulations to determine posterior percentiles. This method can yield extremely accurate estimates of credibility intervals; however, this advantage is accompanied by a rise in computational intricacy or complexity. Despite the computational demands, it offers a way to obtain highly precise results for cases involving uncommon pairs of events, ensuring more reliable assessments of IC values in these situations.

9.1.4.2.3   Stratified IC Analysis

Though the primary objective of dependency derivation is to formulate hypotheses or conjectures, and some degree of wrong leads is permissible in this setting, the fraction of false leads should be kept to a minimum. Detecting and controlling for potential confounders is one methodology that can improve specificity. As with other epidemiological applications, adjusting overall estimates could be utilized in cases where there is no indication of effect modifications; if not stratum-specific estimations have to be employed (17).

The adjusted IC estimations are weighted averages of the stratum-specific IC values, as proposed previously (18). This method, however, necessitates a proper identification of stratification parameters since it is highly responsive to the data-thinning effect. Indeed, preliminary experiments carried out using MC simulation show that this method of adjusting the IC generally results in wider credibility intervals than the unadjusted IC, this, along with a reduction in precision, is a technical drawback because it makes the calculation of accurate credibility intervals more difficult.

9.1.5 STATISTICAL SOFTWARE AVAILABLE

Several commercial software packages designed for analysing spontaneous reports offer tools for calculating various statistics used to detect potential signals in pharmacovigilance data. However, there are relatively few readily accessible packages that specifically focus on performing these calculations. One such example is SAS (Statistical Analysis System), which provides code for computing commonly employed metrics in pharmacovigilance, including the PRR, the ROR, BCPNN, and the EBGM metrics. SAS's code is flexible and authorize for the integration of stratification, and with some modifications, it can also be adapted to consider co-occurrences of drugs and/or events. Additionally, the code is capable of calculating posterior probabilities to assess whether a particular drug-event pair constitutes a signal, a concept originally described by Gould in pharmacovigilance analyses. This comprehensive set of tools within SAS can be valuable for safety analysts and researchers working in pharmacovigilance to identify and evaluate potential safety signals in drug-adverse event data (19).

The R package PhViD offers a versatile set of tools for pharmacovigilance analysis. It allows users to compute a range of metrics such as PRR, ROR, IC_{025}, and λ_{EB05} metrics. Additionally, it includes newer methods based on Fisher's exact test. Moreover, PhViD provides computations to approximate the False Discovery Rate (FDR) and False Negativity Rate (FNR), as described by Ahmed et al. (20). This comprehensive suite of calculations is particularly useful when the goal is to screen a SRS for possible drug-ADR associations without having a specific drug or adverse event identified beforehand. However, there are scenarios where there is a specific interest in exploring associations between a particular adverse event and a specific drug or drug class. In such cases, the code within PhViD can still be applied, but additional methods for computation can be employed to utilize the precision or particularity of the questions under consideration. These strategies can help manage the computational burden and provide more sharp insights into the targeted drug-event associations. The codes described are suitable when there is no predefined drug or ADR of interest, meaning the goal is to conduct a broad screening of the database to detect possible relationships or connections between drugs and ADRs. However, it is common in pharmacovigilance to have a specific interest in exploring a particular ADR associated with a specific drug or drug class, or even both. While the previously mentioned codes can still be used for such scenarios, it is essential to note that other computational approaches can come into play when dealing with specific questions of interest. These alternative strategies are especially valuable because they allow for a more targeted approach, focusing computational resources on the specific drug-event associations under investigation. This approach not only enhances the manageability of the computational burden but also ensures that the analysis is finely tuned to address the precise research objectives related to the identified drug or ADR of interest.

9.1.6 EXAMPLE

We evaluated the FAERS database from 2016 q2 to 2023 q1 for identifying venetoclax-related febrile neutropenia cases, and the signals that were generated. We identified 1,48,58,406 safety reports in total from FAERS, due to all medications. Venetoclax was responsible for 38,885 cases of all the ADRs (Table 9.3).

TABLE 9.3

The 2×2 Contingency Table for Venetoclax and Febrile Neutropenia

Groups	Venetoclax	Other Drugs	Sums
Febrile Neutropenia	2031	46,640	48,671
	a	b	$(a + b)$
Other Events	36,854	1,47,72,881	1,48,09,735
	c	d	$(c + d)$
Sums	38,885	1,48,19,521	1,48,58,406
	$(a + c)$	$(b + d)$	$N (a + b + c + d)$

Venetoclax is a medication known as a "selective and orally bioavailable small-molecule inhibitor of BCL-2". BCL-2 is a protein that plays a role in preventing cell death (apoptosis). Venetoclax is specifically indicated for the treatment of patients diagnosed with "chronic lymphocytic leukemia (CLL), small lymphocytic lymphoma (SLL), and acute myeloid leukemia (AML)". This therapy is employed as a focused treatment approach for managing haematological malignancies. It operates by inhibiting BCL-2, a protein that can support the survival and proliferation of malignant cells. Here we evaluated the risk of "febrile neutropenia" associated with the drug "venetoclax". After downloading the original FAERS quarterly data extracts, the drug file (DRUG) and reaction file (REAC) linked through PRIMARYID (a unique number for identifying a FAERS report) were imported into Microsoft Access 2019 to find the co-occurrences between the drug venetoclax and the ADR-febrile neutropenia. We used both generic and brand names as drug name keywords, including "venetoclax", "venclexta", and "venclyxto". The synonyms of the drugs were found using the Drugs@FDA. So, the drug was coded to a single uniform term. The ADR known as "febrile neutropenia" was categorized and recorded using MedDRA version 26.1 ("Medical Dictionary for Regulatory Activities") at the preferred term (PT) level. This categorization was achieved by employing standardized MedDRA queries (SMQs). SMQs are predefined sets of terms and criteria within the MedDRA terminology that are used to systematically identify and categorize adverse events or reactions associated with medications. In this case, the use of SMQs helped ensure consistent and standardized reporting of febrile neutropenia as an ADR in accordance with MedDRA guidelines. The MedDRA code for febrile neutropenia is 10016288 (PT). As MedDRA is a multi-axial representation of a medical concept, so there will be multiple system organ classes (SOCs). Here, there are two SOCs for the term "febrile neutropenia", the blood and lymphatic system disorders (10005329) as well as general disorders and administration site conditions (10018065). As SMQs are constructed at MedDRA PT level, they enable the standardized communication of safety information and consistent data retrieval. Finally, after mapping drug names and MedDRA coding of the ADRs, there were 2,031 febrile neutropenia events among the 38,885 ADRs. The R package PhViD was employed for the signal generation.

In the process of generating signals using the EBGM method, each observed count was treated as if it were sampled from a Poisson distribution. This was done with the assumption that the expected count follows the same Poisson distribution, based on the assumption that adverse events and drugs are independent of each other. In simpler terms, this means that when using EBGM to detect potential relationships among drugs and ADRs, the analysis considers each observed count of a particular drug-event co-occurrence as a random occurrence, following a Poisson distribution. The "expected count" here refers to what would be anticipated if there were no relationship or dependence between the specific drug and the ADR, assuming they occur independently (13). The criteria for potentially considering a couple (referring to a pair of a specific drug and specific ADR) as a signal were established with a minimum threshold of one notification. The prior hyperparameters were initialized with values of $\alpha 1 = 0.20$, $\beta 1 = 0.06$, $\alpha 2 = 1.40$, $\beta 2 = 1.80$, and $\omega = 0.10$ as per DuMouchel (13). The output of the signal detection method EBGM resulted in a

TABLE 9.4

Signal Values for the Venetoclax-Febrile Neutropenia Association

Disproportionality Indicators	Value	Interpretation
Rate	5.22309	The proportion of febrile neutropenia caused by venetoclax compared to all venetoclax-related ADRs (% $a / (a + c)$)
Chi-squared Yates (χ^2)	28618.4546	Values greater than 4 correspond to p < 0.05, and the 2×2 table does have a normal distribution
Proportional Reporting Rate	PRR = 16.596 LB95 (log (PRR)) = 2.765864	Drug and event are statistically significantly related if PRR $\geq$ 2, N $\geq$ 3
Reporting Odds Ratio	ROR = 17.45549 LB95 (log (ROR)) = 2.814066	Drug and event are statistically significantly related if LB95 > 1, N $\geq$ 2
Empirical Bayes Gamma Mixture	λ_{EB05} = 15.6216648	Drug and event are statistically significantly related if λ_{EB05} > 2, N > 0
Information Component	IC_{025} = 2.723839	Drug and event are statistically significantly related if IC_{025} > 0

posterior expectation of log2(λ) = 3.995 and a 5% quantile of the posterior distribution of λ_{EB05} = 15.622 (Table 9.4).

This implies that the drug venetoclax and the event febrile neutropenia are statistically significantly related (pointing to a putative ADR).

The Bayesian model known as BCPNN employs different statistical methods depending on whether MC simulations are performed or not. When MC simulations are not used, BCPNN utilizes the beta-binomial model as originally proposed by Bate et al. (10). In this scenario, the statistic of interest, often denoted as IC, is calculated using a normal approximation technique described in Bate et al. However, this approximation takes into account the accurate expectation and variance values described by Gould in 2003 (18). On the other hand, when the statistic of interest (IC) is determined using MC simulations, the model relies upon the Dirichlet-multinomial model which was recently proposed by Noren et al. (21). In both scenarios, whether using MC simulations or not, the minimum threshold for the count of notifications required for a drug-adverse event pair to be considered potentially significant is set to 1. This means that even a single reported case can trigger further investigation as a potential signal in pharmacovigilance analysis. The 2.5% quantile of the posterior distribution, IC_{025} was found to be 3.918 (Table 9.4), when no MC simulations were run. The IC_{025} was found to be 2.724 after 1,00,000 MC simulations.

9.2   CONCLUSION

Post-marketing drug safety datasets pose significant challenges due to their sheer size and inherent issues such as heterogeneity and selection bias. Despite these challenges, quantitative methods have emerged as valuable tools to assist clinical experts in sifting through these extensive data repositories. These methods serve as an initial screening process, helping identify potential concerns and associations that warrant further investigation by epidemiologists and healthcare professionals. However, it is important to recognize that not all signals generated from SRSs are indicative of genuine safety issues. Many signals are, in fact, noise. This noise can arise from various factors, including treatment indications, concurrent use of other pharmaceuticals, reporting bias, and more. Additionally, some reported adverse events might already be well-documented, clinically insignificant, or biologically implausible. In this context, quantitative methods act as a crucial filter, helping prioritize signals for deeper scrutiny while recognizing that thorough clinical and epidemiological evaluation is essential to confirm and understand the significance of these signals. This collaborative approach, combining quantitative analysis and expert evaluation, is vital for improving drug safety surveillance and ensuring the well-being of patients worldwide.

ACKNOWLEDGEMENT

We would like to express our sincere gratitude to the Indian Council of Medical Research (ICMR) for their financial assistance.

REFERENCES

1. Lindquist, Marie. "VigiBase, the WHO global ICSR database system: Basic facts." *Drug Information Journal* 42, no. 5 (2008): 409–419.
2. Sakaeda, Toshiyuki, Akiko Tamon, Kaori Kadoyama, and Yasushi Okuno. "Data mining of the public version of the FDA adverse event reporting system." *International Journal of Medical Sciences* 10, no. 7 (2013): 796.
3. Postigo, Rodrigo, Sabine Brosch, Jim Slattery, Anja van Haren, Jean-Michel Dogné, Xavier Kurz, Gianmario Candore, Francois Domergue, and Peter Arlett. "EudraVigilance medicines safety database: Publicly accessible data for research and public health protection." *Drug Safety* 41 (2018): 665–675.
4. Poluzzi, Elisabetta, Emanuel Raschi, Carlo Piccinni, and Fabrizio De Ponti. "Data mining techniques in pharmacovigilance: analysis of the publicly accessible FDA adverse event reporting system (AERS)." In *Data mining applications in engineering and medicine*. IntechOpen, 2012.
5. Bihan, Kévin, Bénédicte Lebrun-Vignes, Christian Funck-Brentano, and Joe-Elie Salem. "Uses of pharmacovigilance databases: An overview." *Therapies* 75, no. 6 (2020): 591–598.
6. Edwards, I. Ralph, and Jeffrey K. Aronson. "Adverse drug reactions: Definitions, diagnosis, and management." *Lancet* 356, no. 9237 (2000): 1255–1259.
7. Balakin, Konstantin V. *Pharmaceutical data mining: approaches and applications for drug discovery*. John Wiley & Sons, 2009.
8. Bate, Andrew, and Stephen J. W. Evans. "Quantitative signal detection using spontaneous ADR reporting." *Pharmacoepidemiology and Drug Safety* 18, no. 6 (2009): 427–436.
9. Evans, Stephen J. W., Patrick C. Waller, and S. Davis. "Use of proportional reporting ratios (PRRs) for signal generation from spontaneous adverse drug reaction reports." *Pharmacoepidemiology and Drug Safety* 10, no. 6 (2001): 483–486.
10. Bate, Andrew, Marie Lindquist, I. Ralph Edwards, Sten Olsson, Roland Orre, Anders Lansner, and R. Melhado De Freitas. "A Bayesian neural network method for adverse drug reaction signal generation." *European Journal of Clinical Pharmacology* 54 (1998): 315–321.
11. Rothman, Kenneth J., Stephan Lanes, and Susan T. Sacks. "The reporting odds ratio and its advantages over the proportional reporting ratio." *Pharmacoepidemiology and Drug Safety* 13, no. 8 (2004): 519–523.
12. Sheskin, David J. *Handbook of parametric and nonparametric statistical procedures*. Chapman and hall/CRC, 2003.
13. DuMouchel, William. "Bayesian data mining in large frequency tables, with an application to the FDA spontaneous reporting system." *The American Statistician* 53, no. 3 (1999): 177–190.
14. Gould, A. Lawrence, Theodore C. Lystig, Yun Lu, Haoda Fu, and Haijun Ma. "Methods and issues to consider for detection of safety signals from spontaneous reporting databases: A report of the DIA Bayesian safety signal detection working group." *Therapeutic Innovation & Regulatory Science* 49 (2015): 65–75.
15. Heckerman, David. "Bayesian networks for data mining." *Data Mining and Knowledge Discovery* 1 (1997): 79–119.
16. Orre, Roland, Anders Lansner, Andrew Bate, and Marie Lindquist. "Bayesian neural networks with confidence estimations applied to data mining." *Computational Statistics & Data Analysis* 34, no. 4 (2000): 473–493.
17. Mantel, Nathan, and William Haenszel. "Statistical aspects of the analysis of data from retrospective studies of disease." *Journal of the National Cancer Institute* 22, no. 4 (1959): 719–748.
18. Gould, Lawrence A. "Practical pharmacovigilance analysis strategies." *Pharmacoepidemiology and Drug Safety* 12, no. 7 (2003): 559–574.
19. Gould, A. Lawrence. "Accounting for multiplicity in the evaluation of "signals" obtained by data mining from spontaneous report adverse event databases." *Biometrical Journal: Journal of Mathematical Methods in Biosciences* 49, no. 1 (2007): 151–165.

20. Ahmed, Ismaïl, Cyril Dalmasso, Françoise Haramburu, Frantz Thiessard, Philippe Broët, and Pascale Tubert-Bitter. "False discovery rate estimation for frequentist pharmacovigilance signal detection methods." *Biometrics* 66, no. 1 (2010): 301–309.
21. Norén, G. Niklas, Andrew Bate, Roland Orre, and I. Ralph Edwards. "Extending the methods used to screen the WHO drug safety database towards analysis of complex associations and improved accuracy for rare events." *Statistics in Medicine* 25, no. 21 (2006): 3740–3757.

10 Subgroup and Sensitivity Analysis

Faiza Javed and Anoop Kumar

10.1 SUBGROUP AND SENSITIVITY ANALYSIS

Precision medicine is the most important discipline of medicine in the current era in which drugs are prescribed based on the genetic basis of individuals to get maximum benefits. Not only benefits but assessment of risks is also important to provide effective as well as safe medicines to the populations. Therefore, it is important to identify which patients are more prone to particular adverse drug reactions (ADRs) with particular medicines. Hence, subgroup analysis is important to identify and prevent particular ADRs. The subgroup analysis can be done based on gender, age groups, locations, comorbid conditions, and so on. There are many signals that were identified only in male or female groups only or in particular age groups or in particular locations only.

Sensitivity analysis is another important analysis in the detection of signals of drugs to check the effects of concomitant drugs on the outcome. It has been observed that signal strength has been significantly decreased after the exclusion of cases of concomitant drugs. This chapter explains in detail about subgroup and sensitivity analysis with a suitable case study using the FDA Adverse Event Reporting System (FAERS) database.

10.2 CASE STUDY: ASSOCIATION OF ANTI-DIABETIC DRUGS WITH DRUG-INDUCED LIVER INJURY (DILI)

Pharmacovigilance is known as the study of the drug's benefits and risks, which helps in pharmacotherapeutic decision-making, individually, regionally, nationally, or internationally [1]. In accordance with the World Health Organization (WHO), pharmacovigilance refers to actions that detect, assess, understand, and prevent adverse effects related to drugs [2]. Various unexpected ADRs are observed in the post-marketing phase of drugs which are important for patient safety. WHO defines signal as reports of suspected causal relationships between adverse events and drugs, for which a causal relationship has not been fully documented or is unknown. The seriousness of the event and the quality of the information often require more than one case report to generate a signal [3].

FAERS of the United States is a web-based tool for the public that contains human adverse events data reported to the FDA by healthcare providers, consumers, and pharmaceutical industries. It is a user-friendly and highly interactive database [4]. OpenVigil provides clean data related to ADRs of drugs from the FAERS database [5].

Signals of Disproportionate Reporting (SDR) are defined as the statistical association of drugs and ADRs which are also known as drug-event pairs [6]. Disproportionality analysis is used to generate a hypothesis for SDRs and assess individual case reports clinically from global databases such as FAERS [7]. Data mining algorithms (DMAs) such as Proportionality Reporting Ratio (PRR) can be defined as a measure of how common a drug's ADR is with a common event in a complete database, whereas ROR can be defined as an odd of a particular event with the drug of interest, in comparison with the odds of a same event occurring with all other drugs in the complete database [8].

The term diabetes mellitus (DM) refers to a range of conditions characterized by hyperglycemia as a result of a deficit in insulin production or activity. DM causes chronic hyperglycemia, which damages, inhibits, and destroys various organs, including blood vessels, heart, retina, kidney, and nervous system [9]. Globally, diabetes prevalence is on the rise. According to the International

DOI: 10.1201/9781032629940-10

Diabetes Federation (IDF) 2021 data, a total of 537 million adults aged between 20 and 79 years are living with diabetes, i.e., 1 in 10 probability. Diabetes is the main reason for 6.7 million deaths in 2021, i.e., 1 every 5 seconds [10]. A number of classes of drugs such as biguanides, sulfonylureas, meglitinides, alpha-glucosidase inhibitors (AGIs), glucagon-like peptide-1 (GLP-1) analogs, thiazolidinediones, dipeptidyl peptidase-4 (DPP4) inhibitors, and sodium-glucose cotransporter-2 (SGLT-2) inhibitors are used for DM, and these medicines are generally prescribed for a longer duration. Unfortunately, prolonged use of these medicines results in various ADRs with different classes such as sulfonylureas (glipizide, gliclazide, glimepiride, glyburide, and glibenclamide) associated with dizziness, insomnia, diarrhea, meglitinides (repaglinide) associated with weight gain, upper respiratory tract infection, cardiovascular ischemia, biguanide (metformin) associated with diarrhea, skin rash, vitamin B12 deficiency, thiazolidinediones (pioglitazone, rosiglitazone) associated with cardiac failure, bone fracture, myalgia, AGIs (miglitol, acarbose) associated with flatulence, diarrhea, abdominal pain, DPP4 inhibitors (sitagliptin, linagliptin, alogliptin) associated with nasopharyngitis, urinary tract infection, lymphocytopenia, SGLT-2 inhibitors (canagliflozin, dapagliflozin, empagliflozin) associated with dyslipidemia, hyperphosphatemia, hypovolemia, fungal vaginosis, diabetic ketoacidosis, and hypoglycemia [11].

Drug-induced liver injury (DILI) is a cause of hepatic dysfunction and acute liver failure. For a drug to be proven as the cause of liver disease, other plausible causes must be excluded and a clinical drug signature must be identified [12]. Acute liver failure is based on the fulfillment of three criteria which include hepatic encephalopathy, increased prothrombin time (>3 seconds), or international normalized ratio (INR) >1.5, a total bilirubin level >3.0 mg/Dl along with high alanine transaminase or aspartate transaminase levels (>500 U/L) [13]. The most common drugs associated with DILI are non-steroidal anti-inflammatory drugs (NSAIDs) such as diclofenac and nimesulide, followed by antibiotics such as amoxicillin-clavulanate (AMC), immunosuppressants, antiplatelet agents, and statins [14]. However, most of the anti-diabetic drugs are not known for DILI except troglitazone, the first thiazolidinedione drug approved for non-insulin-dependent DM [15]. It was withdrawn from the United Kingdom market [16] and the United States [17] due to cases of acute liver failure [18–21]. Emerging reports have indicated the association of some anti-diabetics (dapagliflozin, metformin, etc.) with DILI [22, 23]. However, the exact association is unclear so far. Thus, in the current investigation, we have performed a signal analysis of anti-diabetic drugs associated with DILI using DMAs.

10.2.1 MATERIAL AND METHODS

10.2.1.1 Data Sources

The Individual Case Safety Reports (ICSRs) were taken out from OpenVigil 2.1, which uses cleaned data (removal of duplicates and missing information) from the FDA database. It allows access to the US and internationally available data. The duration of the collection is from January 1, 2004 (2004Q1) to September 30, 2021 (2021Q3). The listedness of DILI with particular anti-diabetic drugs was done using the Summary of Product Characteristics (SmPC) of the innovator available at https://www.medicines.org.uk/emc#gref.

The following terms were used in the search query box of OpenVigil 2.1 MedDRA version [24]:

Drug: (name of the drug)
Adverse Event: Preferred Term (PT), DILI
Data presentation and statistics: Frequentist methods
Counting records: Case (entire case)

10.2.1.2　Subgroup Analysis

The subgroup analysis was done to identify the signal of DILI with particular anti-diabetic drugs in different gender and age groups. The age groups are categorized according to WHO into 0–11 as child, 12–17 as adolescent, 18–64 as adult, and ≥ 65 as elder.

10.2.1.3 Sensitivity Analysis

The sensitivity analysis was done to check the effect of concomitant drugs which are known for DILI on the outcome.

10.2.1.4 Statistical Analysis

The signal analysis was done using DMAs like PRR along with Chi-squared with Yates' correction and Reporting Odds Ratio (ROR) with a 95% confidence interval. PRR and ROR were calculated using standard formulas. A PRR ≥ 2 with an associated Chi-squared value ≥ 4, the lower limit of 95% confidence interval of ROR exceeds 1, and the count of co-occurrence ≥ 3 is considered a positive signal. The analysis was done using OpenVigil 2.1-Med-DRA [25].

10.2.2 Results

A total of 9920 cases of DILI were reported. Out of 9920, 726 were found to be associated with anti-diabetic drugs.

10.2.2.1 Cases of DILI Associated with Anti-Diabetic Drugs

The number of cases of DILI was found with all classes of anti-diabetic drugs except glibenclamide. The highest number of cases of DILI were found with metformin (386 cases) followed by gliclazide (59 cases), sitagliptin (52 cases), glimepiride (45 cases), and so on as shown in Figure 10.1.

10.2.2.2 Association of Anti-Diabetic Drugs with DILI

Among all anti-diabetic drugs, gliclazide, acarbose, alogliptin, and rapaglinide were found to be associated with DILI as indicated by values of PRR with Chi-squared and ROR. Among all identified drugs, acarbose has shown higher strength of signal (PRR 7.4 [4.8; 11.4] Chi-squared value 104.2 and ROR value 7.4 [4.8; 11.6]) followed by gliclazide, alogliptin, and repaglinide (Table 10.1).

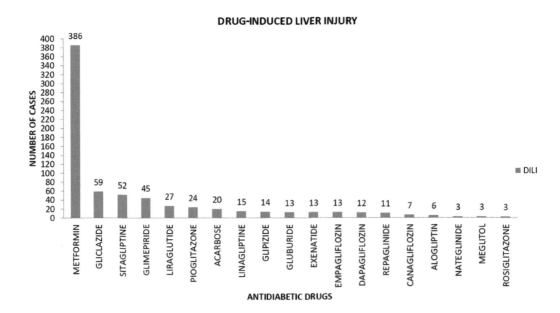

FIGURE 10.1 Number of cases of anti-diabetic drugs associated with DILI in OpenVigil 2.1-MedDRA-v24 tool (data 2004Q1–2021Q3).

TABLE 10.1

Disproportionality Analysis Results of Identified Anti-Diabetic Drugs Associated with DILI

DRUGS	Chi-Squared with Yates' Correction (Chi-Squared Value > 4)	Proportional Reporting Ratio (PRR) and 95% Confidence Interval (PRR $\geq$ 2)	Reporting Odds Ratio (ROR) and 95% Confidence Interval (ROR $\geq$ 2)
Gliclazide	192.5	5.1 (3.9; 6.6)	5.1 (4.0; 6.6)
Acarbose	104.2	7.4 (4.8; 11.4)	7.4 (4.8; 11.6)
Alogliptin	16.0	5.1 (2.3; 11.4)	5.1 (2.3; 11.5)
Rapaglinide	6.3	2.2 (1.2; 4.0)	2.2 (1.2; 4.0)

10.2.2.3 Subgroup Analysis

Subgroup analysis was performed with all anti-diabetic drugs to know the association of DILI in males or females and different age groups. The results of the subgroup analysis are discussed below.

10.2.2.4 Number of Cases Based on Gender

The number of cases of DILI with all anti-diabetic drugs was also categorized based on gender. The highest number of cases of DILI in males were observed with metformin (153) followed by sitagliptin (36), glimepiride (28), gliclazide (26), and so on, whereas, in females, the highest number of cases were observed with metformin (163), gliclazide (23), liraglutide (20), and so on as shown in Figure 10.2a.

10.2.2.5 Association of Anti-Diabetic Drugs with DILI According to Gender

The PRR along with the Chi-squared value ROR was calculated for all anti-diabetic drugs. Among all drugs, alogliptin, acarbose, and glimepiride have shown significant association with DILI in males, whereas in females, acarbose, repaglinide, metformin, and gliclazide have shown significant association with DILI as indicated by their ROR, PRR with Chi-squared values as compiled in Table 10.2.

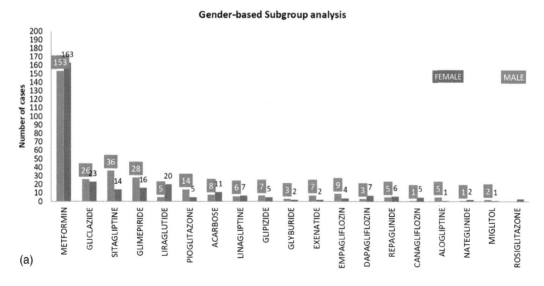

(a)

FIGURE 10.2 Subgroup analysis (a) number of cases of anti-diabetic drugs associated with DILI in OpenVigil 2.1-MedDRA-v24 tool (data 2004Q1–2021Q3) in males and females. *(Continued)*

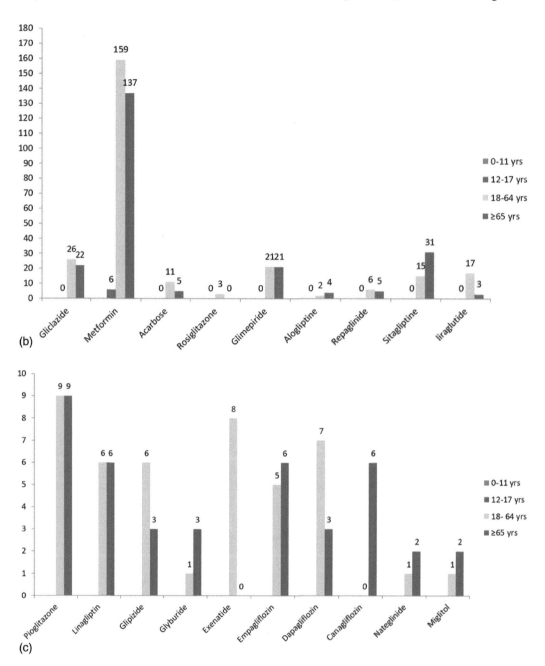

FIGURE 10.2 *(Continued)* (b) Number of cases of anti-diabetic drugs (glicazide, metformin, acarbose, rosiglitazone, glimepiride, alogliptin, repaglinide, sitagliptin, liraglutide) associated with DILI in OpenVigil 2.1-MedDRA-v24 tool (data 2004Q1–2021Q3) in different age groups. (c) Number of cases of anti-diabetic drugs (pioglitazone, linagliptin, glipizide, glyburide, extenatide, empagliflozin, dapagliflozin, canagliflozin, nateglinide, miglitol) with DILI in OpenVigil 2.1-MedDRA-v24 tool (data 2004Q1–2021Q3) in different age groups associated with DILI in OpenVigil 2.1-MedDRA-v24 tool (data 2004Q1–2021Q3) in different age groups.

TABLE 10.2

Disproportionality Analysis Results of Identified Anti-Diabetic Drugs Associated with DILI in Subgroups as per Gender and Different Age Groups

Drug	Characteristics	Drug-Induced Liver Injury		Chi-Squared with Yates' Correction (Chi-Squared Value > 4)	Proportional Reporting Ratio (PRR) and 95% Confidence Interval (PRR ≥ 2)	Reporting Odds Ratio (ROR) and 95% Confidence Interval (ROR ≥ 2)
		Number of Cases	Total Cases			
Gliclazide	Male	26	3498	31.9	1.6 (1.3; 1.9)	1.6 (1.3; 1.9)
	Female	23	4183	61.7	1.9 (1.6; 2.1)	1.9 (1.6; 2.2)
	0–11 years	0	175	6.0	0	0
	12–17 years	0	248	21.8	0	0
	18–64 years	26	4514	64.2	4.4 (2.9; 6.4)	4.4 (2.9; 6.5)
	65 years and above	22	2145	32.4	3.3 (2.2; 4.9)	3.3 (2.2; 5.0)
Acarbose	Male	8	3497	28.5	6.0 (3.0; 11.9)	6.0 (3.0; 12.1)
	Female	11	4346	87.1	10.7 (5.9; 19.2)	10.8 (5.9; 19.5)
	0–11 years	0	175	27.1	0	0
	12–17 years	0	248	15.3	0	0
	18–64 years	11	4514	64.9	8.4 (4.7; 15.2)	8.5 (4.7; 15.5)
	65 years and above	5	2145	5.5	3.2 (1.3; 7.7)	3.2 (1.3; 7.7)
Alogliptin	Male	5	3497	23.1	7.8 (3.2; 18.6)	7.8 (3.2; 18.9)
	Female	1	4346	0.06	2.9 (0.4; 20.3)	2.9 (0.4; 20.5)
	0–11 years	0	175	223.9	0	0
	12–17 years	0	248	0	0	0
	18–64 years	2	4514	1.9	3.9 (0.9; 15.5)	3.9 (0.9; 15.7)
	65 years and above	4	2145	14.9	6.9 (2.6; 18.5)	7.0 (2.6; 18.8)
Repaglinide	Male	5	3497	1.6	2.0 (0.8; 4.8)	2.0 (0.8; 4.8)
	Female	6	4346	7.7	3.3 (1.5; 7.4)	3.4 (1.5; 7.5)
	0–11 years	0	175	223.9	0	0
	12–17 years	0	248	9.3	0	0
	18–64 years	6	4514	7.2	3.2 (1.5; 7.2)	3.2 (1.5; 7.3)
	65 years and above	5	2145	0.3	1.5 (0.6; 3.6)	1.5 (0.6; 3.6)
Metformin	Male	153	3498	31.92	1.6 (1.3; 1.9)	1.6 (1.3; 1.9)
	Female	163	4346	61.7	1.9 (1.6; 2.1)	1.9 (1.6; 2.2)
	0–11 years	0	175	0.001	0	0
	12–17 years	6	248	7.3	3.3 (1.5; 7.4)	3.3 (1.5; 7.5)
	18–64 years	159	4514	12.9	1.3 (1.1; 1.6)	1.3 (1.1; 1.6)
	65 years and above	137	2145	45.0	1.8 (1.5; 2.1)	1.8 (1.5; 2.1)
Glimepiride	Male	28	3497	14.3	2.0 (1.4; 3.0)	2.0 (1.4; 3.0)
	Female	16	4346	1.9	1.5 (0.9; 2.4)	1.5 (0.9; 2.4)
	0–11 years	0	175	4.9	0	0
	12–17 years	0	248	11.7	0	0
	18–64 years	21	4514	2.9	1.5 (0.9; 2.2)	1.5 (0.9; 2.2)
	65 years and above	21	2145	5.0	1.7 (1.0; 2.6)	1.7 (1.0; 2.6)

10.2.2.6 Number of Cases in Different Age Groups

The number of cases of DILI with all anti-diabetic drugs was also categorized based on different age groups. The highest number of cases of DILI in the age group 18–65 years were observed with metformin (159) followed by gliclazide (26), glimepiride (21), liraglutide (17), and so on, whereas, in the age group >65 years, the highest number of cases were observed with metformin (137), sitagliptin (31), gliclazide (22), and so on as shown in Figure 10.2b, c.

Other age groups don't have a significant number of cases except metformin with six cases in the 12–17-year age group.

10.2.2.7 Association of Anti-Diabetic Drugs with DILI in Different Age Groups

The DMAs were calculated for all anti-diabetic drugs according to the age group. Among all drugs, gliclazide, acarbose, and repaglinide have shown significant association with DILI in the 18–65-year age group and also, in the >65-year age group, acarbose, alogliptin, and gliclazide have shown significant association with DILI. Apart from this, metformin in the 12–17-year age group shows a significant association as indicated by its ROR and PRR with Chi-squared values as compiled in Table 10.2.

10.2.2.8 Sensitivity Analysis

The cases of DILI with selected anti-diabetic drugs (metformin, glimepiride, repaglinide, gliclazide, acarbose, and alogliptin) were decreased after removing the cases associated with concomitant drugs as shown in Figure 10.3.

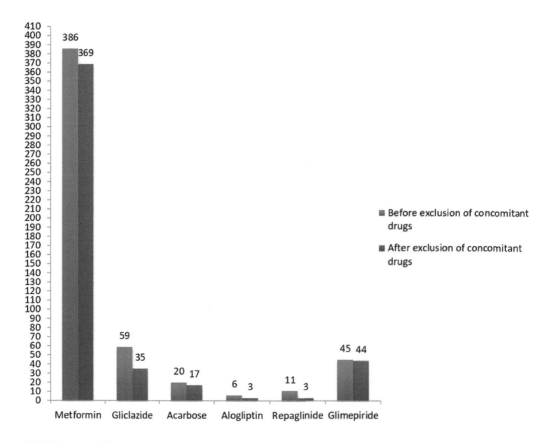

FIGURE 10.3 Number of cases before and after exclusion of concomitant drugs.

TABLE 10.3

Sensitivity Analysis Results after Exclusion of Cases of DILI Related to Concomitant Drugs

Drug of Interest	Concomitant Drug	Chi-Squared with Yates' Correction (Chi-Squared Value > 4)	Proportional Reporting Ratio (PRR) and 95% Confidence Interval (PRR ≥ 2)	Reporting Odds Ratio (ROR) and 95% Confidence Interval (ROR ≥ 2)
Metformin	Without azithromycin	112.1	1.7 (1.5; 1.8)	1.7 (1.5; 1.8)
	Without atenolol	106.2	1.7 (1.5; 1.8)	1.7 (1.5; 1.8)
	Without both drugs	104.4	1.6 (1.5; 1.8)	1.7 (1.5; 1.8)
Repaglinide	Without atorvastatin	3.4	1.9 (1.0; 3.8)	1.9 (1.0; 3.8)
	Without rabeprazole	5.6	2.1 (1.1; 3.8)	2.1 (1.1; 3.8)
	Without furosemide	2.5	1.8 (0.9; 3.7)	1.8 (0.9; 3.7)
	Without all drugs	0.8	1.6 (0.7; 3.5)	1.6 (0.7; 3.5)
Glimepiride	Without ibuprofen	14.7	1.7 (1.3; 2.2)	1.7 (1.3; 2.2)
Gliclazide	Without escitalopram	173.9	4.8 (3.7; 6.2)	4.8 (3.7; 6.2)
	Without indapamide	122.6	4.2 (3.2; 5.5)	4.2 (3.2; 5.6)
	Without amlodipine	114.6	4.3 (3.2; 5.7)	4.3 (3.2; 5.8)
	Without atorvastatin	121.3	4.4 (3.3; 6.0)	4.5 (3.3 ;6.0)
	Without all drugs	84.1	4.1 (3.0; 5.8)	4.2 (3.0; 5.8)
Acarbose	Without amlodipine	93.5	7.1 (4.5; 11.1)	7.1 (4.5; 11.2)
	Without pantoprazole sodium	75.8	6.3 (4.0 ;10.0)	6.3 (4.0; 10.1)
	Without both drugs	74.0	6.5 (4.0; 10.4)	6.5 (4.0; 10.5)
Alogliptin	Without lansoprazole	9.5	4.3 (1.7; 10.3)	4.3 (1.7; 10.4)
	Without ranitidine	14.4	4.7 (2.1; 10.6)	4.8 (2.1; 10.7)
	Without both drugs	4.8	3.4 (1.3; 9.2)	3.4 (1.3; 9.3)

The strength of the signal was decreased after the exclusion of cases of concomitant drugs, however, an association of DILI with gliclazide, acarbose, alogliptin, and repaglinide was still found to be significant whereas the association of DILI with metformin and glimepiride was found to be non-significant as indicated by disproportionality measures (Table 10.3).

10.2.2.9 Discussion

A wide variety of anti-diabetic drugs are associated with many adverse effects, but the association with DILI is not yet proven and documented. Various case reports are available in the literature regarding the adverse effects of anti-diabetic drugs on the hepatic system. Dourakis et al. have reported a case of an older woman aged 60 years highlighting the development of sclera's acute hepatitis-like illness after gliclazide therapy for six weeks followed by immediate withdrawal of the drug and patient recovery within four weeks [25]. Caksen et al. have reported a 14-year-old non-diabetic girl who suffered from hepatitis, hemiplegia along with dysphasia due to an overdose of gliclazide (20 mg/kg/day) while attempting suicide [26]. Chounta A. et al. have reported a case of a 65-year-old man with type 2 diabetes developed cholestatic liver injury after ingestion of glimepiride, the patient recovered within 50 days without further recurrences after withdrawal of glimepiride [27]. A 70-year-old woman after taking metformin for five weeks without any other concomitant drugs suffered from DILI has been reported [23]. A case report was also reported of an 83-year-old man with metformin-induced hepatotoxicity with constitutional syndrome [28]. Some other case reports also reported that metformin also causes mixed hepatic damage, i.e., hepatocellular and cholestatic damage [29–31].

In a rare case, elevated liver enzymes have been noted which is a case of acute hepatotoxicity in a 78-year-old woman while taking repaglinide [32]. In a case report published by Marcy et al., hepatotoxicity with second-generation TZDs has been reported which mainly involves rosiglitazone and pioglitazone taken by a 33-year-old black woman suffering from DM, hypertension, and congestive heart failure [33]. A voluntary withdrawal of the anti-diabetic drug triglitazone was taken on March 21, 2000, by Warner-Lambert Company due to more than 100 cases of hepatotoxicity through the Medicine Control Agency of the United Kingdom and FDA [34]. A case report of a 48-year-old woman was reported by Levine J.A. et al. to have jaundice with acute cholestatic liver injury with dapagliflozin use [22].

To the best of our knowledge, there was no disproportionality analysis done so far to check the association of available anti-diabetic drugs with DILI. However, recently, Katshuhiro O. et al. have performed a signal analysis of DPP-4 inhibitor associated with gynecomastia and the results have shown a positive signal for gynecomastia in older men with diabetes [35]. The results of the current investigation have shown a significant association of DILI with repaglinide, gliclazide, acarbose, and alogliptin among all anti-diabetic drugs.

10.2.2.10 Strengths and Limitations

The strength of the study is the use of OpenVigil 2.1-MedDRA-v24 to query the FAERS database which provides clean data (removal of duplicates and missing data) for the analysis and enough sample size to identify rare adverse reactions. The study has only found out the association of anti-diabetic drugs with DILI. Thus, further causality assessment is required.

10.3 CONCLUSION

Subgroup analysis is important to identify the strength of the identified signal in different genders, age groups, locations, and comorbid conditions. Sensitivity analysis is also important to check the effects of concomitant drugs on the outcome. Both these analyses are helpful to verify the identified signal.

REFERENCES

1. Meyboom RHB, Egberts ACG, Gribnau FWJ et al. Pharmacovigilance in perspective. Drug-Saf. 1999;21: 429–447. doi: 10.2165/00002018-199921060-00001
2. Härmark L, van Grootheest AC. Pharmacovigilance: methods, recent developments and future perspectives. Eur J Clin Pharmacol. 2008;64:743–752. doi:10.1007/s00228-008-0475-9
3. Practical Aspects of Signal Detection in Pharmacovigilance. https://cioms.ch/wp-content/uploads/2018/03/WG8-Signal-Detection.pdf Accessed July 2, 2022.
4. Center for Drug Evaluation and Research. (2021, October 23). FDA Adverse Event Reporting System (FAERS) Public Dashboard. U.S. Food and Drug Administration. https://www.fda.gov/drugs/questions-and-answers-fdas-adverse-event-reporting-system-faers/fda-adverse-event-reporting-system-faers-public-dashboard Accessed June 15,2022.
5. Böhm R, von Hehn L, Herdegen T, et al. OpenVigil FDA – inspection of U.S. American adverse drug events pharmacovigilance data and novel clinical applications. PLoS One. 2016;11(6):e0157753. doi:10.1371/journal.pone.0157753
6. Vieira JML, de Matos GC, da Silva FAB, Bracken LE, Peak M, Lima EDC. Serious adverse drug reactions and safety signals in children: a nationwide database study. Front Pharmacol. 2020;11:964. doi:10.3389/fphar.2020.00964
7. Caster O, Aoki Y, Gattepaille LM et al. Disproportionality analysis for pharmacovigilance signal detection in small databases or subsets: recommendations for limiting false-positive associations. Drug Saf. 2020;43:479–487. doi:10.1007/s40264-020-00911-w
8. Pharmacovigilance – Guidance material. https://allaboutpharmacovigilance.org/ Accessed June 25, 2022.
9. Alam U, Asghar O, Azmi S, Malik RA. General aspects of diabetes mellitus. Handb Clin Neurol. 2014;126:211–222. doi:10.1016/B978-0-444-53480-4.00015-1
10. IDF Diabetes Atlas. https://diabetesatlas.org/ Accessed June 26, 2022.
11. Ganesan K, Rana MBM, Sultan S. Oral hypoglycemic medications. In: StatPearls. Treasure Island (FL): StatPearls Publishing; May 8, 2022.
12. Abboud G., Kaplowitz N. Drug-induced liver injury. Drug Saf. 2007;30(4):277–294. doi:10.2165/00002018-200730040-00001

13. Chan KA, Truman A, Gurwitz JH, et al. A cohort study of the incidence of serious acute liver injury in diabetic patients treated with hypoglycemic agents. Arch Intern Med. 2003;163(6):728–734. doi:10.1001/archinte.163.6.728

14. Licata A. Adverse drug reactions and organ damage: The liver. Eur J Intern Med. 2016;28:9–16. doi:10.1016/j.ejim.2015.12.017

15. Ghazzi MN, Perez JE, Antonucci TK, Driscoll JH, Huang SM, Faja BW, Whitcomb RW. Cardiac and glycemic benefits of troglitazone treatment in NIDDM. The troglitazone study group. Diabetes. 1997;46(3):433–439. doi: 10.2337/diab.46.3.433. PMID: 9032099.

16. Wise J. Diabetes drug withdrawn after reports of hepatic events. BMJ. 1997;315(7122):1564. doi:10.1136/bmj.315.7122.1559n

17. Henney JE. Withdrawal of troglitazone and cisapride. JAMA. 2000;283(17):2228. doi:10.1001/jama.283.17.2228

18. Vella A, de Groen PC, Dinneen SF. Fatal hepatotoxicity associated with troglitazone. Ann Intern Med. 1998;129(12):1080. doi: 10.7326/0003-4819-129-12-199812150-00032. PMID: 9867776.

19. Neuschwander-Tetri BA, Isley WL, Oki JC, Ramrakhiani S, Quiason SG, Phillips NJ, Brunt EM. Troglitazone-induced hepatic failure leading to liver transplantation. A case report. Ann Intern Med. 1998;129(1):38–41. doi: 10.7326/0003-4819-129-1-199807010-00009. PMID: 9652998.

20. Herrine SK, Choudhary C. Severe hepatotoxicity associated with troglitazone. Ann Intern Med. 1999;130(2):163–164. doi: 10.7326/0003-4819-130-2-199901190-00021. PMID: 10068372.

21. Murphy EJ, Davern TJ, Shakil AO, Shick L, Masharani U, Chow H, Freise C, Lee WM, Bass NM. Troglitazone-induced fulminant hepatic failure. Acute liver failure study group. Dig Dis Sci. 2000;45(3):549–553. doi: 10.1023/a:1005405526283. PMID: 10749332.

22. Levine JA, Ann Lo A, Wallia A, Rogers M, VanWagner LB. Dapagliflozin-induced acute-on-chronic liver injury. ACG Case Rep J. 2016;3(4):e169. doi:10.14309/crj.2016.142

23. Zheng L. Metformin as a rare cause of drug-induced liver injury, a case report and literature review. Am J Ther. 2016;23(1):e315–e317. doi: 10.1097/MJT.0000000000000007. PMID: 24263160.

24. Evans SJ, Waller PC, Davis S. Use of proportional reporting ratios (PRRs) for signal generation from spontaneous adverse drug reaction reports. Pharmacoepidemiol Drug Saf. 2001;10(6):483–486. doi:10.1002/pds.677

25. Dourakis SP, Tzemanakis E, Sinani C, Kafiri G, Hadziyannis SJ. Gliclazide-induced acute hepatitis. Eur J Gastroenterol Hepatol. 2000;12(1):119–121. doi:10.1097/00042737-200012010-00021

26. Caksen H, Kendirci M, Tutuş A, Uzüm K, Kurtoğlu S. Gliclazide-induced hepatitis, hemiplegia and dysphasia in a suicide attempt. J Pediatr Endocrinol Metab. 2001;14(8):1157–1159. doi:10.1515/jpem-2001-0814

27. Chounta A, Zouridakis S, Ellinas C, et al. Cholestatic liver injury after glimepiride therapy. J Hepatol. 2005;42(6):944–946. doi:10.1016/j.jhep.2005.02.011

28. de la Poza Gómez G, Rivero Fernández M, Vázquez Romero M, Angueira Lapeña T, Arranz de la Mata G, Boixeda de Miquel D. Síndrome constitucional asociado a hepatotoxicidad por metformina [constitutional syndrome associated to metformin induced hepatotoxicity]. Gastroenterol Hepatol. 2008;31(10):643–645. doi:10.1016/S0210-5705(08)75812-6

29. Kutoh E. Possible metformin-induced hepatotoxicity. Am J Geriatr Pharmacother. 2005;3(4):270–273.

30. Nammour FE, Fayad NF, Peikin SR. Metformin-induced cholestatic hepatitis. Endocr Pract. 2003;9(4):307–309. doi:10.4158/EP.9.4.307

31. Saadi T, Waterman M, Yassin H, Baruch Y. Metformin-induced mixed hepatocellular and cholestatic hepatic injury: case report and literature review. Int J Gen Med. 2013;6:703–706. doi:10.2147/IJGM.S49657

32. Jaiswal S, Mehta R, Musuku M, Tran L, McNamee W Jr. Repaglinide induced acute hepototoxicity. JNMA J Nepal Med Assoc. 2009;48(174):162–164.

33. Marcy TR, Britton ML, Blevins SM. Second-generation thiazolidinediones and hepatotoxicity. Ann Pharmacother. 2004;38(9):1419–1423. doi:10.1345/aph.1E072

34. Chaudhry MU, Simmons DL. Case of the month. Hepatic and renal failure in a patient taking troglitazone and metformin. J Ark Med Soc. 2001;98(1):16–19.

35. Ohyama K, Tanaka H, Shindo J, Shibayama M, Iwata M, Hori Y. Association of gynecomastia with anti-diabetic medications in older adults: data mining from different national pharmacovigilance databases. Int J Clin Pharmacol Ther. 2022;60:24–31. doi:10.5414/CP204066

11 Causality Assessment of ICSRs

Vivekanandan Kalaiselvan and Rishi Kumar

11.1 INTRODUCTION

The components of Individual Case Safety Reports (ICSRs) are discussed below.

11.2 INITIAL REPORT INFORMATION

Initial report date is the date when the report was first recorded by the initial reporter. For example, it is the date available on the suspected Adverse Drug Reaction (ADR) reporting form/consumer reporting form by the reporter/consumer or by other means of ADR reporting.

11.2.1 PATIENT INFORMATION

This section captures the information about the patient details. Patient information consists of the Patient Age at the time of onset of the reaction/event and the Patient Initials, which are the initials of the patient's name. The initials should be entered in the following order: first alphabet letter of the name and surname without any sign or space between, and it may be of a maximum of three alphabet letters. For example, the initials of Ram Kumar Verma should be RKV.

11.2.1.1 Patient Sex
- *Sex of Patient*: This can be selected from Male, Female and Unknown (in the case of transgender, it can be mentioned in the case narrative).
- *Patient Age or Date of Birth*: It provides information about the patient category. It is important for the calculation of dose and category of the patient whether the patient belongs to the categories of infant, neonate, child, adolescent, adult, geriatric etc.

11.2.2 REPORT TITLE

The report title shall be written as reaction name: drug name. If there are multiple reactions and multiple drugs involved then the report title shall include the most serious reaction and most suspected drugs in the report title. In the case of FDC, the report title can be, e.g., Anaemia: Lamivudine + Zidovudine or Anaemia: Combivir. Combivir is the brand name of Lamivudine + Zidovudine.

11.2.3 CASE NARRATIVE

Case narrative and reporter's comments: Both the case narrative and reporter's comment fields are free-text fields which can expand to accommodate large texts. By writing the case narrative with the words and phrases as described by the initial reporter, it is possible to keep the original narrative and use it in combination with the structured fields for a better analysis.

11.2.3.1 Free Text
Medical and past drug history: It covers the medical history of the patient such as disease and conditions like pregnancy surgical procedures. It also covers the reporter's comments, additional information on the drug (e.g., is the ADR adequately labelled?) and the sender's comment.

DOI: 10.1201/9781032629940-11

11.2.4   Reaction Term – Entering an Event/Reaction

11.2.4.1   Time Onset

- *Onset date* – This is the date of the start of the reaction/event. You can enter an incomplete date; however, the minimum requirement is the year of occurrence.
- *End date* – It is the date at which the reaction/event was identified as one of the following: recovered/resolved and recovered/resolved with sequelae. You can enter an incomplete date; however, the minimum precision is the year.

11.2.5   Outcome

Informative values in the field of *ADR outcome*:

- Recovered/resolved
- Recovering/resolving
- Not recovered/not
- Resolved
- Fatal

11.2.5.1   Seriousness

- Seriousness is classified as either Yes or No
- Death, Life-threatening, Congenital anomaly, Hospitalization/Prolongation
- Of hospitalization, Disabling, Other medically important conditions

11.2.6   Drug Name

Suspected and concomitant drugs should be coded from MedDRA:

1. Brand Name
2. Active Pharmaceutical Ingredient Details
3. Reference (Authentic Source of Information, e.g., MAH website)

11.2.6.1   Drug Information

Drug information comprises pharmaceutical form, route of administration, authorization holder, dose and dose regimen.

11.2.6.2   Indications

A drug refers to the requirement for the use of that drug for treating a particular disease, e.g., diabetes is an indication for insulin.

11.2.7   Action Taken

Informative values from any field given action taken section:

- Drug withdrawn, Dose reduced, Dose does not change
- Not applicable and Unknown

11.3   CAUSALITY ASSESSMENT

Causality Assessment (relatedness of suspected drug and adverse reaction/event occurred) is the most important element that needs to be assessed based on the available data. There are many methods available for the causality assessment. We show a method which is widely accepted and easy to understand.

11.3.1 QUERY

In case of non-availability of information, which is required for causality assessment of the case, efforts should be made to collect such information from the primary reporter.

11.3.2 METHODS OF CAUSALITY ASSESSMENT

Hutchison defined causality assessment as a "method for eliciting a state of information about a particular drug-event connection as input and delivering as output a degree of belief about the truth of the proposition that the drug caused the event to occur". Factors that need to be considered for Causality Assessment are as follows:

- *Prior reports of reaction*: The availability of data related to a particular reaction will be an additional advantage for the evaluator of Causality Assessment. In this part, it is expected to have the adverse drug reaction-related data available in the hard copy or soft copy to check whether the event/reaction is already reported for the suspected drug.
- *Temporal relationship*: There should be a time relationship between the drug intake and the occurrence of the event in the patient.
- *De-challenge*: It means the withdrawal of the drug from the schedule of treatment of patient. It may be either positive or negative. The positive de-challenge is the stoppage of reaction/event after the withdrawal of suspected medicine. The negative de-challenge is when the reaction/event continues after the withdrawal of the suspected medicine.
- *Re-challenge*: The re-challenge is considered to be not ethical in case of serious adverse reactions/events unless the benefit outweighs the risk associated with the use of suspected medicine.
- *Dose-response relationship*: If the reaction/event subsided after reducing or increasing the dose of the medicine, it shall also be considered while doing the Causality assessment.
- *Alternative aetiologies*: It is also needed to be understood while performing causality assessment.
- *Past history of reaction to same or similar medication*: If patients have a past history of reaction with the same or similar medication, it needs to be assessed while doing the causality assessment.

11.3.2.1 WHO-UMC Method of Causality Assessment

The WHO-UMC method of Causality Assessment is the most widely used method in the world due to its simplicity and less time consumption for conclusion [1]. The Causality terms with their assessment criteria are shown in Table 11.1.

11.3.2.2 Naranjo's Method

This method of causality assessment is based on meticulously finalized questions with the scoring-based procedure, though it is time-consuming, it is very simple to remove most of the disagreement between the persons who are doing causality assessment [2]. The description of this method is presented in Figure 11.1.

11.3.2.3 UPDATED LOGISTIC METHOD

The Revised Logistic Method for Causality Assessment represents an enhanced approach to evaluating the likelihood of a causal relationship between a drug and an adverse event. This methodology refines the original Logistic Method introduced by Arimone et al. in 2006, subsequently updated by Theophile et al. in 2012 and 2013 to enhance certain criteria definitions and weights. A departure from traditional categorical labels, such as possible, probable or definite, the method employs a probabilistic approach by calculating the probability of drug causation as a numeric value ranging from 0 to 1.

TABLE 11.1

Causality Terms with Their Assessment Criteria

Causality Term	Assessment Criteria
Certain	1. Event or laboratory test abnormality, with a plausible time relationship to drug intake. 2. Cannot be explained by disease or other drugs. 3. Response to withdrawal plausible (pharmacologically, pathologically). 4. Event definitive pharmacologically or phenomenological (i.e., an objective and specific medical disorder or a recognized pharmacological phenomenon). 5. Re-challenge satisfactory, if necessary.
Probable/Likely	1. Event or laboratory test abnormality, with a reasonable time relationship to drug intake. 2. Unlikely to be attributed to disease or other drugs. 3. Response to withdrawal clinically reasonable. 4. Re-challenge not required.
Possible	1. Event or laboratory test abnormality, with a reasonable time relationship to drug intake. 2. Could also be explained by disease or other drugs. 3. Information on drug withdrawal may be lacking or unclear.
Unlikely	1. Event or laboratory test abnormality, with a time to drug intake that makes a relationship improbable (but not impossible). 2. Disease or other drugs provide plausible explanations.
Conditional/Unclassified	1. Event or laboratory test abnormality. 2. More data for proper assessment is needed or additional data under examination.
Unassessable/Unclassifiable	1. Report suggesting an adverse reaction. 2. Cannot be judged because information is insufficient or contradictory data cannot be supplemented or verified.

Integral to the Updated Logistic Method is the incorporation of expert consensus. Weightings are determined by a panel of experts who evaluate a substantial sample of drug-event pairs, providing a comprehensive reflection of the collective insights of seasoned pharmacovigilance professionals. The method systematically considers seven criteria deemed pertinent to causality assessment: time to onset, de-challenge, re-challenge, alternative causes, previous information, dose-response relationship and pharmacological plausibility.

This advanced method has undergone comparisons with alternative causality assessment techniques, including the Naranjo algorithm and the Liverpool algorithm. Notably, it exhibits superior agreement with consensual expert judgment and demonstrates reduced subjective variability. Additionally, the Updated Logistic Method is available in a computerized format, facilitating swift and straightforward application. This software tool guides users through the assessment process, automatically generating the probability of drug causation.

11.3.3 Role of Expert Judgment in Causality Assessment

Expert judgment has a significant role in the evaluation of the causal relationship between a drug treatment and the occurrence of an adverse event which can not be ignored [3].

11.4 BAYESIAN APPROACHES FOR CAUSALITY ASSESSMENT

The Bayesian approach for causality assessment is based on the probabilistic methods; this method transforms the prior estimate of probability to the posterior estimate of probability of drug-ADR pair combinations. Basically, these approaches work on the basis of Bayes Theorem of Probability,

To assess the adverse drug reaction, please answer the following questionnaire and give the pertinent score.	Yes	No	Do Not Know	Score
1. Are there previous *conclusive* reports on this reaction?	+1	0	0	____
2. Did the adverse event appear after the suspected drug was administered?	+2	-1	0	____
3. Did the adverse reaction improve when the drug was discontinued or a *specific* antagonist was administered?	+1	0	0	____
4. Did the adverse reactions appear when the drug was readministered?	+2	-1	0	____
5. Are there alternative causes (other than the drug) that could on their own have caused the reaction?	-1	+2	0	____
6. Did the reaction reappear when a placebo was given?	-1	+1	0	____
7. Was the drug detected in the blood (or other fluids) in concentrations known to be toxic?	+1	0	0	____
8. Was the reaction more severe when the dose was increased, or less severe when the dose was decreased?	+1	0	0	____
9. Did the patient have a similar reaction to the same or similar drugs in *any* previous exposure?	+1	0	0	____
10. Was the adverse event confirmed by any objective evidence?	+1	0	0	____
			Total Score	____

Total Score	ADR Probability Classification
9	Highly Probable
5-8	Probable
1-4	Possible
0	Doubtful

FIGURE 11.1 Naranjo Scale for causality assessment.

which can be understood by that it strengthens the prior drug-ADR pair combination outcome hypothesis, based on new evidences that were received recently. It updates our initial findings about causality based on new information from the drug-ADR pair combination.

Bayesian approaches for causality assessment have some advantages over other methods, such as expert judgment or algorithms. They are more logical, consistent, transparent and flexible. They can incorporate different types and sources of information, and they can be updated as new evidence becomes available. However, they also have some limitations, such as the difficulty of obtaining reliable and valid data for the prior odds and the likelihood ratios, the subjectivity and variability of the inputs and the complexity and computational burden of the calculations [4].

11.5   CASE STUDIES

Case Report 1

ML, a 65-year-old male patient was admitted to the hospital on 12.01.2016 with chief complaints of pain in the upper abdomen and nausea for the last five days. On physical examination, he had yellowish discolouration of the palm, conjunctiva and nail bed. His weight was 72 kg. He had few episodes of psychotic attacks, for which he was on Chlorpromazine therapy for the last four weeks. On enquiry, he told that he was taking Tab. Largactil (Chlorpromazine) 100 mg, four tablets at bedtime. He was also taking Tab. Diclofenac 50 mg twice a day (self-medication) for abdominal pain for three days before admission to the hospital. He was investigated on the day of admission for laboratory parameters which are as follows:

- Alkaline Phosphatase = 180U/L (normal range: 25–100U/L)
- ALT = 205U/L (normal range: 10–40 U/L)
- Total Bilirubin = 5.0 mg/Dl (normal range: 0.8–1.2 mg/dL)

- On admission, Chlorpromazine and Diclofenac therapy were stopped. After seven days of stopping the medications, the intensity of pain decreased. Also, he was re-investigated for the above parameters which are as follows:
  - Alkaline phosphatase = 110 IU/L
  - ALT = 98 U/L
  - Total Bilirubin = 1.8 mg/dL

- *Note*: Tab Chlorpromazine
- *Brand Name*: XXXX, Manufacturer: XXXX
- *Batch number*: LGL0881, Expiry date: Dec 2018

Description of Reaction or Problem

The patient has been taking Chlorpromazine since 12.12.2015. He developed pain in his abdomen and nausea on 07.01.2016. Examination revealed yellowish discolouration of conjunctiva, palm and nails. Pt. was admitted on 12.01.2016 and investigated. Liver function tests indicated raised serum bilirubin, ALT and alkaline phosphatase. Drugs were discontinued. On discontinuation of the drugs, the reaction subsided in one week.

Suspected Medication

- *Name (Brand and/or Generic)*: Tab. Chlorpromazine (XXXX)
- *Manufacturer (if known)*: XXXX
- *Batch No./Lot No.*: LGL0881
- *Exp. Date (if known)*: Dec. 2018
- *Dose used*: 400 mg
- *Route used*: Oral
- *Frequency (OD, BD etc.)*: OD
- Therapy dates

Date Started: 12.12.2015; Date Stopped: 12.01.2016

- *Indication*: Psychosis

11.5.1   CAUSALITY ASSESSMENT AS WHO CAUSALITY ASSESSMENT SCALE

Probable

Category	Time Relationship	Other Drugs/ Disease Ruled Out	De-Challenge	Re-Challenge
Certain	Yes	Yes	Yes	Yes
Probable	Yes	Yes	Yes	No
Possible	Yes	No	No	No
Unlikely	No	No	No	No

Case Report 2

Mr. SG, a 30-year-old male with 68 kg weight, was diagnosed with a case of bacterial meningitis. He was started empirically with Inj. Ceftriaxone 1 g IV BD and Inj. Vancomycin 500 mg IV QID on 12.01.2016. The first dose of Inj. Ceftriaxone was given at 8 am and Inj. Vancomycin was given at 9 am on 12.01.2016. After 10 minutes of the second drug administration, he started developing chills, rigours, fever, urticaria and intense flushing. He was treated with Inj. Pheniramine 25 mg IM, following which the reaction completely subsided. Inj. Ceftriaxone was continued. However, the next doses of Inj. Vancomycin scheduled on day 1 were not given. On day 2, the Inj. Vancomycin was re-introduced at 9 am to the patient. Similar symptoms developed again and quickly resolved after Inj. Pheniramine 25 mg IM.

- Inj. Vancomycin
  - *Brand Name*: XXXX
  - *Manufacturer*: XXXX
  - *Batch number*: KKIL098
  - *Expiry date*: Mar 2016

- Inj. Ceftriaxone
  - *Brand Name*: XXXX
  - *Manufacturer*: XXXX
  - *Batch number*: OPO659
  - *Expiry date*: Dec 2016

11.5.1.1   Causality Assessment as WHO Causality Assessment Scale

- *Certain*: Causality Assessment plays a significant role in preventable ADRs.

11.5.2   ADR REPORTING STRATEGIES RECEIVE A LOT OF ATTENTION

In the current scenario, the collection of data related to adverse drug reactions receive larger attention from the medical fraternities along with the Marketing Authorization Holders as a regulatory requirement. After the collection of ADRs, their causality plays an important role in preventive measures in terms of patient safety.

There are many studies that have been performed to find out the economic burden and ascertain whether the ADRs are preventable or not. At least one in ten admissions in hospitals of elderly patients is due to ADRs and which is preventable [5]. During causality assessment it can be found which medicine is responsible for adverse drug reactions in patients and preventive measures may be taken to avoid such events based on the clinical judgment of the clinician. Cases like this need to be reported to the regulatory authority if they require any actions from the regulation point of view [6]. These are allergic reactions, dose-related effects, idiosyncrasy, drug interactions and prescribing errors all fall under the category of preventable ADRs [7].

11.5.3 Prevention of ADRs Is Becoming Less Important

Generally, the prevention of ADR becomes less important while reporting adverse drug reactions, whereas if we focus more on the preventability of ADR, then it will be a significant outcome of the pharmacovigilance system in the healthcare facilities or at MAH's site.

11.5.4 Some ADRs Are Unavoidable, But Mostly Preventable

The adverse drug reaction is unavoidable because these are foreign bodies for any diseased individual, and when entered into the bloodstream, then the body's defence mechanism acts against them, which leads to an adverse drug reaction in the patient. Allergic reactions, dose-related effects, idiosyncrasy, drug interactions and prescribing errors – all fall in the preventable category of adverse drug reactions. The data related to such reactions further strengthen the healthcare system for patient safety.

11.5.5 Predictable and Preventable Adverse Drug Reactions

The clinical practice and laboratory-based methods are available to predict preventable ADRs, such as the identification of certain genes responsible for certain adverse drug reactions like ADRs like identification of certain genes responsible for certain adverse drug reactions like association of Drug Reaction with Eosinophilia and Systemic Symptoms (DRESS) Syndrome with HLA alleles.

A study was conducted by Swenn et al. (2023) in which they highlighted that the identification of a 12-gene pharmacogenetic panel which was found to significantly reduce the incidence of adverse drug reactions of specific drugs in the European population, Though the study was done on a small population, it could be performed on the large-scale population to make drug therapy safe for patients [8].

11.5.6 Identification and Understanding of Preventable ADRs

It requires a lot of clinical acumen and efforts to identify the preventable ADR. It should be the aim of all healthcare professionals in tandem to identify necessary actions for the prevention of ADRs.

11.5.7 Determination of ADR Preventability Should Be Part of the ADR Reporting and Reviewing Process

Currently. the pharmacovigilance teams across the world are focused on ADR reporting only. To identify preventable ADRs, the reviewing process must be done in a fast and timely manner, This will help in guiding programmes and system changes for reducing ADRs.

11.5.8 Set of Questions Developed by Schumock and Thornton (UIC, 1992) for This Purpose

- Was the drug causing ADR not appropriate?
- Was the dose, route and frequency of admn. inappropriate for pts. Age, Weight and Disease?
- Was TDM or other lab tests not performed?
- Was there H/O allergy or other reaction to the drug?
- Was a drug interaction involved?
- Was a toxic serum level documented?
- Was poor compliance involved in the reaction?
- An answer of YES to one or more of these questions would suggest that the ADR could have been prevented.

11.6   CURRENT CHALLENGES AND FUTURE PERSPECTIVES IN CAUSALITY ASSESSMENT OF ICSRs

- *Lack of Data in ICSRs*: ICSRs lack the data which is required for causality assessment like time to onset of event, Patient History and dose and indication of the suspected medicine. So, it became very challenging to assess the causality of the event.
- *Methodological Confusion*: There are many Causality Assessment methods available. It becomes difficult for the assessor to decide which method needs to be used.
- *Diagnosis Challenges*: Diagnosis of many diseases where we need clinical judgment for identification of events like vascular cognitive impairment (VCI) is very complex due to the variability in how it presents in patients and the lack of standardized diagnostic criteria.
- *Brain Injury Assessment*: Assessing the severity of vascular brain injury is often imprecise. This is because the relationship between the amount of vascular damage seen on brain imaging and the patient's cognitive function can vary greatly between individuals.
- *Clinical Practice*: These challenges highlight the need for more precise and reliable methods to assess Causality Assessment in clinical practice, ensuring the events have been analysed effectively.
- *Role of Artificial Intelligence in Causality Assessment*: Artificial intelligence-based tools may be used for causality assessment, but the involvement of expert judgment may contribute to the effectiveness of the outcome received through the use of artificial intelligence. It needs an integrated approach to use AI in the field of pharmacovigilance and specifically in the Causality Assessment.

11.7   CONCLUSION

Despite having all the causality assessment methods based on certain algorithms, the clinical judgment of the experts always prevails over all types of statistical methods of causality assessment. There are a number of factors such as confounding factors that may influence the causality assessment. The causality assessment methods will help you to reduce the data for analysis and make your work of data analysis a little simpler [9–11].

- ADRs and PVig remain a challenging clinical and scientific discipline.
- It continues to expand its domain due to the ever-increasing number of medicinal products.
- The benefits of drug therapy should always be greater than the risks.
- Risk can be minimized by ensuring good quality, safe, efficacious and rationally used medicines.
- Knowledge, education and risk management strategies are crucial for preventing/minimizing ADRs.

REFERENCES

1. Belhekar MN, Taur SR, Munshi RP. A study of agreement between the Naranjo algorithm and WHO-UMC criteria for causality assessment of adverse drug reactions. Indian J Pharmacol. 2014;46(1):117–120. https://doi.org/10.4103/0253-7613.125192
2. Naranjo CA, Busto U, Sellers EM, Sandor P, Ruiz I, Roberts EA, Janecek E, Domecq C, Greenblatt DJ. A method for estimating the probability of adverse drug reactions. Clin Pharmacol Ther. 1981;30(2):239–245. https://doi.org/10.1038/clpt.1981.154
3. Oscanoa TJ, Lizaraso F, Carvajal A. Hospital admissions due to adverse drug reactions in the elderly. A meta-analysis. Eur J Clin Pharmacol. 2017;73(6):759–770. https://doi.org/10.1007/s00228-017-2225-3
4. Schmiedl S, Rottenkolber M, Szymanski J, Drewelow B, Siegmund W, Hippius M, Farker K, Guenther IR, Hasford J, Thuermann PA, German Net of Regional Pharmacovigilance Centers (NRPC). Preventable ADRs leading to hospitalization – results of a long-term prospective safety study with 6,427 ADR cases focusing on elderly patients. Expert Opin Drug Saf. 2018;17(2):125–137. https://doi.org/10.1080/14740338.2018.1415322

5. Khalil H, Huang C. Adverse drug reactions in primary care: A scoping review. BMC Health Serv Res. 2020;20(1):5. doi: 10.1186/s12913-019-4651-7. PMID: 31902367; PMCID: PMC6943955.

6. Swen JJ, van der Wouden CH, Manson LE, Abdullah-Koolmees H, Blagec K, Blagus T, Böhringer S, Cambon-Thomsen A, Cecchin E, Cheung KC, Deneer VH. A 12-gene pharmacogenetic panel to prevent adverse drug reactions: an open-label, multicentre, controlled, cluster-randomised crossover implementation study. Lancet. 2023;401(10374):347–356.

7. Arimone Y, Bégaud B, Miremont-Salamé G, Fourrier-Réglat A, Molimard M, Moore N, Haramburu F. A new method for assessing drug causation provided agreement with experts' judgment. J Clin Epidemiol. 2006;59(3):308–314.

8. Théophile H, André M, Arimone Y, Haramburu F, Miremont-Salamé G, Bégaud B. An updated method improved the assessment of adverse drug reaction in routine pharmacovigilance. J Clin Epidemiol. 2012;65(10):1069–1077.

9. Théophile H, André M, Miremont-Salamé G, Arimone Y, Bégaud B. Comparison of three methods (an updated logistic probabilistic method, the Naranjo and Liverpool algorithms) for the evaluation of routine pharmacovigilance case reports using consensual expert judgement as reference. Drug Saf. 2013;36:1033–1044.

10. Behera SK, Das S, Xavier AS, Velupula S, Sandhiya S. Comparison of different methods for causality assessment of adverse drug reactions. Int J Clin Pharm. 2018;40:903–910.

11. Agbabiaka TB, Savović J, Ernst E. Methods for causality assessment of adverse drug reactions: a systematic review. Drug Saf. 2008;31:21–37.

12 Signal Analysis of Drug–Drug Interactions Using Spontaneous Reporting Methods

Rima Singh, Ruchika Sharma, Deepti Pandita, and Anoop Kumar

12.1 INTRODUCTION

A drug–drug interaction (DDI) arises when two or more drugs interact with each other in a way that affects their effectiveness or leads to adverse effects [1]. These interactions can occur in numerous ways, including pharmacokinetic interactions, metabolism, excretion, receptor interactions, antagonistic effects, combined toxicity, idiosyncratic interactions, food–drug interactions, and herbal and dietary supplement interactions. Some examples of DDIs include:

(1) *Grapefruit juice and statins*: Grapefruit juice can inhibit the enzyme responsible for breaking down certain statin medications, such as atorvastatin or simvastatin [2]. This can lead to higher levels of the statin in the bloodstream, increasing the risk of side effects like muscle pain and liver damage.

(2) *Warfarin and Antibiotics*: Some antibiotics, such as ciprofloxacin and erythromycin, can interfere with the metabolism of warfarin which is a blood thinner and can upsurge the risk of bleeding because warfarin's effects are enhanced [3].

(3) *Nonsteroidal anti-inflammatory drugs [NSAIDs] and Anticoagulants*: Taking NSAIDs like ibuprofen or aspirin with anticoagulants like warfarin or heparin can elevate the chances of bleeding due to their combined blood-thinning effects [4].

(4) *Digoxin and Diuretics*: Diuretics can lower potassium levels in the body, and when combined with digoxin, a medication used for heart conditions, it can lead to an increased risk of digoxin toxicity, which can affect the heart rhythm [5].

Adverse drug events (ADEs) triggered by DDIs are becoming added threats to our lives. Medicines may interact with each other when they are concurrently used. A survey conducted in Australia discovered that between 43% and 91% of individuals use two or more medicines and that such simultaneous use of medicines is more common among the elderly. Ultimately, polypharmacy has intensified the commonness of adverse DDIs [6].

12.2 DATA SOURCES

The spontaneous reporting system (SRS) has been used by many supervisory bodies and plays a significant role in post-market pharmaco surveillance. Detecting DDIs using the FDA Adverse Event Reporting System (FAERS) includes a systematic methodology that combines data mining techniques and pharmacovigilance principles (Figure 12.1). There are three types of data sources involved in adverse DDI detection investigations: (1) Free-text data, such as medical literature, documents, and social media; in this the studies extract DDI mentions from medical literature, documents, or even social media. (2) Medicine chemical or pharmaceutical property data stored in databases of domain knowledge, such as Drug Bank; in this the goal is to use medicine chemical

DOI: 10.1201/9781032629940-12

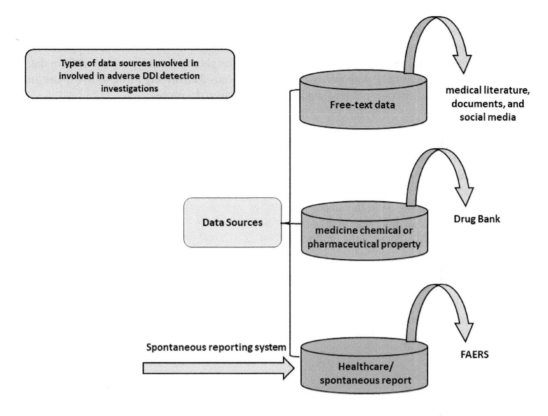

FIGURE 12.1   Three types of data sources involved in adverse DDI detection investigations.

or pharmaceutical data for the prediction of DDIs. (3) Healthcare or spontaneous report data, such as FAERS; in this, particularly those used spontaneous report data for the signal detection of DDIs, are most relevant to the detection of DDIs [7].

12.3   STEPS INVOLVED IN IDENTIFICATION OF SIGNALS OF DDI

Various steps are required to identify the signals of DDI, which are mentioned below.

12.3.1   DATA ACQUISITION

For data procurement, it is necessary to access the FAERS database, as it is widely accessible on the Food and Drug Administration (FDA) website which contains a huge quantity of spontaneous adverse event reports submitted by healthcare professionals, consumers, and manufacturers.

12.3.2   DATA PREPROCESSING

It is required to clean and preprocess the data to eliminate duplicates, variations, and inappropriate information. Standardization of drug names is important to ensure consistency in illustration. While preprocessing the data it is important to anonymize patient information to guard confidentiality.

12.3.3   SIGNAL DETECTION ALGORITHMS

After the data is processed it is necessary to use some algorithms to identify potential DDIs by various signal detection algorithms they include: (1) proportional reporting ratio (PRR) which measures the reporting rate of a particular DDI compared to all other drug pairs. (2) Reporting odds ratio (ROR) compares the odds of reporting a particular DDI in FAERS comparative to a reference group. (3) Bayesian data mining is a method of data analysis and modelling that influences Bayesian statistics and probabilistic reasoning to excerpt appreciated insights and make predictions from data. It combines the principles of Bayesian statistics with numerous data mining methods to explore and analyse large datasets, discover patterns, and make data-driven decisions.

12.3.4 Threshold Setting and Data Visualization

Threshold setting and data visualization play a key role in detecting DDIs. It is necessary to outline significance thresholds for the signal detection approaches to filter out spurious relations and emphasize relative DDIs. Typically, a combination of statistical significance and clinical relevance is considered when setting these thresholds. Thereafter, the generation of graphical representations is much more needed, such as heat maps or network diagrams, to visualize potential DDIs. These visualizations can help recognize the outlines and relationships between drugs and adverse events.

12.3.5 Expert Review

Expert review includes domain professionals, such as pharmacologists and toxicologists, to review and authenticate the detected signals. Experts can evaluate the clinical significance of probable DDIs and prioritize them based on severity and probability.

12.3.6 Signal Refinement

Furthermore, signal refinement of potential DDIs is also to be done by considering factors like the strength of the signal, the number of reports, and the biological plausibility of the interaction, and afterwards, evaluation of the potential mechanism of action for the DDI is to be done.

12.3.7 Reporting and Documentation

Reporting and documentation have equal importance to other aspects of the DDI method. It is necessary to document the detected DDIs, including details such as drug pairs, adverse events, signal strength, and supporting evidence, and make reports for regulatory authorities, healthcare professionals, and the public as needed.

12.3.8 Follow-Up Studies

Follow-up studies are necessary to investigate the identified DDIs more thoroughly. On the basis of necessity, follow-up studies, such as case-control studies or retrospective cohort studies, should be conducted, and collection of additional evidence to approve or disprove the associations.

12.3.9 Continuous Monitoring

The establishment of a system for ongoing monitoring of DDIs in FAERS data is equally important. It has a vital role in updating signal detection algorithms and thresholds as new data and making it available [8, 9].

12.4 IDENTIFICATION OF SIGNALS OF DDI USING THE FAERS DATABASE

The FAERS is a freely available database which is an extremely interactive web-based implement that allows for the querying of data in a user-approachable manner. The purpose of this tool is to develop an assessment of FAERS data for the wide-ranging community to explore for information

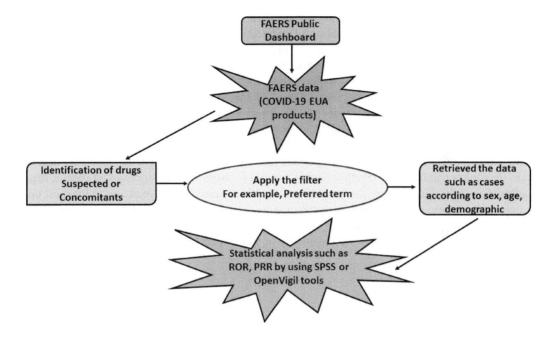

FIGURE 12.2   FAERS public dashboard.

connected to human adverse events reported to the FDA by the pharmaceutical industry, healthcare providers, and consumers [10].

For example, if DDI of any combination of drugs needs to be found in the case of COVID-19 reports using FAERS, first it is mandatory to open FAERS Public Dashboard which is an extremely cooperating web-based tool that lets operators to query FAERS data in a clear and approachable manner. For the COVID-19 pandemic, the FDA introduced this database for COVID-19 EUA products. In this dashboard, the effects of drugs are characterized as 'suspected drug' and 'concomitant drug,' and adverse events are determined with the 'preferred term' of the Medical Dictionary for Regulatory Activities (MedDRA). Afterwards, the combination of drugs is to be chosen as a suspected drug or a concomitant one according to the data provided by FAERS. Also, relatable information can be retrieved from the same data such as the age of the patient, sex, country many more [11]. Thereafter, statistical analysis can be done to analyse the exact values of cases including sex, age, and adverse events by using the Statistical Package for Social Sciences (SPSS) tool or by disproportionality analysis such as ROR, PRR, and Chi-square values from OpenVigil. The method is shown in Figure 12.2.

12.5   ADVANTAGES OF SRSs IN DDI DETECTION

SRSs, such as the FAERS, are of paramount importance in the study and monitoring of DDIs for several reasons: SRSs allow healthcare professionals, patients, and drug manufacturers to report adverse events and suspected DDIs that may not have been identified in clinical trials. This early detection is crucial for promptly recognizing and addressing potential safety concerns. SRSs collect real-world data, reflecting the diverse patient populations and medical settings in which drugs are used. This data provides a comprehensive view of DDIs that might not be apparent in controlled clinical trials, where patient characteristics and drug combinations are often limited. Clinical trials typically involve a smaller number of patients and focus on common drug interactions. It helps uncover rare or previously unknown DDIs that may only manifest in specific patient groups or with long-term use. DDIs can emerge after a drug has been approved and is on the market. SRSs serve as a critical component of post-marketing surveillance, allowing for the continuous monitoring of drug safety and the identification of previously undetected interactions. These systems employ data mining and statistical techniques to detect 'signals' of potential DDIs, helping regulatory agencies and healthcare professionals prioritize further investigation. This proactive approach can lead to timely interventions and safer drug use. Spontaneous reports often include information about the severity of adverse events associated with DDIs. This information helps healthcare providers and regulators understand the clinical impact of these interactions. Spontaneous reporting data can aid in identifying patient-specific factors, such as age, gender, and concomitant medical conditions, that increase the risk of DDIs. This information allows healthcare providers to tailor treatment plans to individual patient needs. Regulatory agencies like the FDA use spontaneous reporting data to make decisions regarding drug labelling, safety warnings, and potential restrictions on specific drug combinations. This helps ensure that patients and healthcare providers have access to the most up-to-date safety information. By facilitating the monitoring and evaluation of DDIs, SRSs contribute to overall public health and patient safety. Timely identification and management of DDIs reduce the risk of adverse events and improve the quality of patient care [12, 13].

12.6 CASE STUDIES

12.6.1 Nirmatrelvir/Ritonavir and Tacrolimus Interactions

Qin et al. [11] conducted a study of the DDI of nirmatrelvir/ritonavir and tacrolimus and did a potential risk disproportionality analysis of nephrotoxicity from COVID-19 reports using FAERS. From this study they concluded that a significant DDI between nirmatrelvir/ritonavir and tacrolimus and confirmed that their combination for COVID-19 treatment greatly increases the risk of acute kidney injury with 41.13%, serum creatinine increased by 14.18%, renal failure with 2.84%, and renal impairment with 2.84% [11].

12.6.2 Interaction among Antibodies

Zhao et al. conducted a study on the evaluation of adverse events of bamlanivimab, bamlanivimab/etesevimab used for COVID-19 based on the FAERS database and found that bamlanivimab, bamlanivimab/etesevimab is relatively safe, while the risk signals such as acute respiratory failure and infusion-related reaction require monitoring [14].

12.6.3 Tizanidine and CYP1A2 Inhibitors Interactions

Another study by Villa-Zapata et al. [15] on disproportionality analysis of DDIs of tizanidine and CYP1A2 inhibitors from the FAERS. In this study, data was extracted for disproportionality analysis using FAERS reports from 2004 quarter 1 through 2020 quarter 3 to calculate the ROR of reports mentioning tizanidine in a suspect or interacting role, or having any role, a CYP1A2 inhibitor, and the following adverse events: hypotension, bradycardia, syncope, shock, cardiorespiratory arrest, and fall or fracture. They found that the concurrent use of tizanidine with CYP1A2 inhibitors may lead to serious health consequences associated with low blood pressure such as falls and fractures [15].

12.6.4 Interactions between Therapies for COVID-19 and Concomitant Medications

Jeong et al. [16] conducted a study on detecting DDIs between therapies for COVID-19 and concomitant medications through the FAERS. They found that there were no significant results found in the Liverpool database and recommended additional validation studies for the pharmacokinetic properties, metabolic pathways, and pharmacodynamics properties of the drug [16].

TABLE 12.1

Signal Analysis of Drug–Drug Interactions Using Spontaneous Reporting Methods

Sl. No.	Database Used	Interactions	Interaction Effects	Type of Study	Parameters Checked	Ref.
1.	COVID-19 EUA FAERS Public Dashboard	Nirmatrelvir/Ritonavir and tacrolimus	Increase the risk of acute kidney injury, serum creatinine increased, renal failure, and renal impairment	Frequentist method	Reporting odds ratio (ROR)	[11]
2.	COVID-19 EUA FAERS Public Dashboard	Bamlanivimab and combination of bamlanivimab/ etesevimab	It may increase the risk of acute respiratory failure and infusion-related reaction	Frequentist method and Bayesian method	ROR, proportional reporting ratio (PRR), Empirical Bayes Geometric Mean (EBGM)	[14]
3.	FAERS	Tizanidine and CYP1A2 inhibitors	It may lead to serious health consequences associated with low blood pressure such as falls and fractures	Frequentist method	ROR	[15]
4.	FAERS	Interaction between therapies for COVID-19 and concomitant medications	No major issues were observed	Firth logistic regression, Monte Carlo simulations	relative excess risk due to interaction (RERI), odds ratio	[16]
5.	FAERS, PubMed, EMBASE	Drug–drug interaction with concomitant use of colchicine and clarithromycin	It can lead to pancytopenia, bone marrow failure and related life-threatening issues	Frequentist method	Drug Interaction Probability Scale (DIPS) rating	[17]

12.6.5 INTERACTIONS BETWEEN COLCHICINE AND CLARITHROMYCIN

Villa Zapata et al. [17] have done an experimental study on evidence of clinically meaningful DDI with concomitant use of colchicine and clarithromycin. They concluded that clinical indicators of colchicine-clarithromycin interaction may resemble other systemic diseases and may be life-threatening [17]. Case studies of DDIs are shown in Table 12.1.

12.7 CONCLUSION

The utilization of SRS for DDI detection and evaluation is a valuable approach that complements controlled clinical trials and contributes to a deeper understanding of real-world drug interactions. Further, research and ongoing collaboration are crucial to harness the full potential of SRS data for improving patient safety and optimizing medication management.

REFERENCES

1. Magro, Lara, Ugo Moretti, and Roberto Leone. "Epidemiology and characteristics of adverse drug reactions caused by drug–drug interactions." *Expert Opinion on Drug Safety* 11, no. 1 (2012): 83–94.
2. Azemawah, Veronica, Mohammad Reza Movahed, Patrick Centuori, Ryan Penaflor, Pascal L. Riel, Steven Situ, Mehrdad Shadmehr, and Mehrnoosh Hashemzadeh. "State of the art comprehensive review of individual statins, their differences, pharmacology, and clinical implications." *Cardiovascular Drugs and Therapy* 33 (2019): 625–639.
3. Holbrook, Anne M., Jennifer A. Pereira, Renee Labiris, Heather McDonald, James D. Douketis, Mark Crowther, and Philip S. Wells. "Systematic overview of warfarin and its drug and food interactions." *Archives of Internal Medicine* 165, no. 10 (2005): 1095–1106.
4. Abebe, Worku. "Review of herbal medications with the potential to cause bleeding: Dental implications, and risk prediction and prevention avenues." *EPMA Journal* 10 (2019): 51–64.
5. Wang, Meng-Ting, Chen-Yi Su, Agnes LF Chan, Pei-Wen Lian, Hsin-Bang Leu, and Yu-Juei Hsu. "Risk of digoxin intoxication in heart failure patients exposed to digoxin–diuretic interactions: A population-based study." *British Journal of Clinical Pharmacology* 70, no. 2 (2010): 258–267.
6. Lange, Beverly J., Bruce C. Bostrom, Joel M. Cherlow, Martha G. Sensel, Mei KL La, Wayne Rackoff, Nyla A. Heerema, Robert S. Wimmer, Michael E. Trigg, and Harland N. Sather. "Double-delayed intensification improves event-free survival for children with intermediate-risk acute lymphoblastic leukemia: A report from the Children's Cancer Group." *Blood, The Journal of the American Society of Hematology* 99, no. 3 (2002): 825–833.
7. Ibrahim, Heba, A. Abdo, Ahmed M. El Kerdawy, and A. Sharaf Eldin. "Signal detection in pharmacovigilance: A review of informatics-driven approaches for the discovery of drug–drug interaction signals in different data sources." *Artificial Intelligence in the Life Sciences* 1 (2021): 100005.
8. Vilar, Santiago, Carol Friedman, and George Hripcsak. "Detection of drug–drug interactions through data mining studies using clinical sources, scientific literature and social media." *Briefings in Bioinformatics* 19, no. 5 (2018): 863–877.
9. Iyer, Srinivasan V., Rave Harpaz, Paea LePendu, Anna Bauer-Mehren, and Nigam H. Shah. "Mining clinical text for signals of adverse drug–drug interactions." *Journal of the American Medical Informatics Association* 21, no. 2 (2014): 353–362.
10. Raschi, Emanuel, Elisabetta Poluzzi, Ariola Koci, Paolo Caraceni, and Fabrizio De Ponti. "Assessing liver injury associated with antimycotics: Concise literature review and clues from data mining of the FAERS database." *World Journal of Hepatology* 6, no. 8 (2014): 601.
11. Qin, Fuhong, Huiling Wang, Meng Li, Shengnan Zhuo, and Wei Liu. "Drug–drug interaction of Nirmatrelvir/ritonavir and tacrolimus: A potential risk disproportionality analysis of nephrotoxicity from COVID-19 reports in FAERS." *Expert Opinion on Drug Safety* 6, (2023): 1–7.
12. Gravel, Christopher A., Daniel Krewski, Donald R. Mattison, Franco Momoli, and Antonios Douros. "Concomitant use of statins and sodium-glucose co-transporter 2 inhibitors and the risk of myotoxicity reporting: A disproportionality analysis." *British Journal of Clinical Pharmacology* 89, no. 8 (2023): 2430–2445.
13. Kovačević, Milena, Sandra Vezmar Kovačević, Branislava Miljković, Slavica Radovanović, and Predrag Stevanović. "The prevalence and preventability of potentially relevant drug–drug interactions in patients admitted for cardiovascular diseases: A cross-sectional study." *International Journal of Clinical Practice* 71, no. 10 (2017): e13005.
14. Zhao, Yunfei, Huiling Wang, Qingsong Zhang, Yongxin Hu, Yulong Xu, and Wei Liu. "Evaluation of adverse events of bamlanivimab, bamlanivimab/etesevimab used for COVID-19 based on FAERS database." *Expert Opinion on Drug Safety* 22, no. 4 (2023): 331–338.
15. Villa-Zapata, Lorenzo, Ainhoa Gómez-Lumbreras, John Horn, Malinda S. Tan, Richard D. Boyce, and Daniel C. Malone. "A disproportionality analysis of drug–drug interactions of tizanidine and CYP1A2 inhibitors from the FDA adverse event reporting system (FAERS)." *Drug Safety* 45, no. 8 (2022): 863–871.
16. Jeong, Eugene, Scott D. Nelson, Yu Su, Bradley Malin, Lang Li, and You Chen. "Detecting drug–drug interactions between therapies for COVID-19 and concomitant medications through the FDA adverse event reporting system." *Frontiers in Pharmacology* 13 (2022): 938552.
17. Villa Zapata, Lorenzo, Philip D. Hansten, John R. Horn, Richard D. Boyce, Sheila Gephart, Vignesh Subbian, Andrew Romero, and Daniel C. Malone. "Evidence of clinically meaningful drug–drug interaction with concomitant use of colchicine and clarithromycin." *Drug Safety* 43 (2020): 661–668.

13 Use of Network Pharmacological Approaches in Signal Analysis

Vipin Bhati, Ruchika Sharma, Deepti Pandita, and Anoop Kumar

13.1 INTRODUCTION

Network pharmacology (NP) is a recently burgeoning and widely adopted methodology employed by numerous scientists during drug development. It integrates genomic technology, computational methods, and systems biology to elucidate the intricate connections among biological systems, pharmaceuticals, and medical conditions [1, 2]. It incorporates the concepts of polypharmacology [3], *in silico* pharmacology, and system pharmacology [4, 5]. In NP, we investigate and assess a network, which takes on a mesh-like structure consisting of nodes and edges. Here, nodes symbolize targets or proteins, while edges depict the connections or interactions between them [6].

The foundation of NP centres around the idea of simultaneously targeting multiple nodes within interconnected and complex molecular systems. These systems are intricately linked and operate synergistically, potentially leading to increased effectiveness and reduced side effect [7]. The traditional method of drug discovery relies on the concept of one drug targeting one specific gene associated with one disease. Nevertheless, it's now recognized that multiple genes and targets play roles in the development of any disease [8]. Hence, network pharmacological strategies can have a substantial impact on comprehending the involvement of numerous targets, genes, or pathways in the development of a specific disease. This understanding can aid in the creation of molecules that are both efficient and safe [9]. Many diseases encompass multiple pathways in their development; thus, interventions targeting a single specific target may not be highly effective. Network pharmacological methods can be pivotal in grasping disease development and in pinpointing potential medications, as they take into account numerous targets and networks involving various signalling molecules associated with diseases [10]. The initial and primary stage in NP is the recognition of essential disease-related pathways and signalling pathways. Subsequently, the network is examined and depicted through the utilization of diverse topological measurements, including nodes, edges, node degrees, shortest paths, and modules, among others. Considering these criteria, central nodes with a more significant impact on disease-linked pathways are identified [11]. This chapter furnishes comprehensive insights into network pharmacological methodologies and their specific application in signal analysis.

The WHO has defined "signal" as "Reported information on a possible causal relationship between an adverse event and a drug, the relationship being unknown or incompletely documented" [12].

13.2 NETWORK PHARMACOLOGICAL APPROACHES

NP generally involves a three-step process: first, identifying and predicting targets; second, constructing a network, which includes scrutinizing and verifying it; and third, performing pathway enrichment analysis and then validating the results obtained from this technique. A logical starting point is to construct an intricate physiological network using a substantial pre-existing database. Ultimately, determine the key nodes within the network and anticipate the vital biological processes by employing hierarchical clustering and analytical techniques. Finally, additional network assessment is required to validate the accuracy of the projected outcomes [13]. The primary objective of NP research is to identify common genes between drugs and disease-related targets, establish associations between them, and subsequently analyse and visualize their interconnected networks [14]. The process of selecting targets and predicting compounds relies on several databases, such as the Swiss database. Constructing a network of protein and gene interactions using resources like STRING and STITCH helps us to identify crucial nodes associated with the disease. Subsequently, we can anticipate the primary biological pathways by employing Kyoto Encyclopedia of Genes and Genomes (KEGG) pathway network analysis based on these identified nodes [15]. To validate the link between potent active components and their potential targets, additional network validation steps are carried out [16]. Each step involved in NP is described below.

13.2.1 TARGET IDENTIFICATION

The initial stage in NP involves identifying potential targets for desired compounds, as well as targets associated with the disease of interest's development. Various databases are accessible for locating potential compound and disease targets. Prominent databases for forecasting compound targets include the Swiss Target Prediction Database (http://www.swisstargetprediction.ch/) and ChEMBL. It's worth noting that the Swiss target predictor necessitates the use of SMILES notation for compound searches. Typically, researchers can obtain SMILES notation for compounds from various databases such as PubChem (https://pubchem.ncbi.nlm.nih.gov/), ChEMBL (https://www.ebi.ac.uk/chembl/), and others [17]. The subsequent phase involves choosing the species name according to the researcher's needs. Additionally, it's crucial to identify the targets associated with the specific disease. The commonly utilized databases for predicting disease-related targets include the Human Gene database (https://www.genecards.org/) and the DisGeNET database (https://www.disgenet.org/) [18].

13.2.2 COMMON TARGET IDENTIFICATION

After the identification of potential targets of drug and disease, we have to find common targets from the identified targets. Venn diagrams are the preferred tools for pinpointing shared targets of diseases and compounds. The primary goal of this step is to forecast genes associated with diseases and then discern the shared genes between diseases and compounds. There are number of tools by which we can make Venn diagram such as Venny 2.1 and InteractiVenn (http://www.interactivenn.net/) [7].

13.2.3 PROTEIN-PROTEIN INTERACTIONS

We know that targets within the human body do not function in isolation; they interact with one another. Hence, it is crucial to examine these interactions among targets. The widely favoured and frequently employed resource for assessing protein-protein interactions is the STRING database, accessible for free at https://string-db.org/. Moreover, the STRING database offers the capability to select a specific species of interest. The examination of protein-protein interaction networks involves the analysis of various topological characteristics, including nodes, edges, and degrees. Nodes can represent various entities such as genes, proteins, chemicals, and disease symptoms. A bridge node serves as a connector between two modules, whereas an edge represents a linkage between two nodes, which can signify a connection between two proteins, a chemical and a target, or the pathways associated with a disease and its targets [19].

13.2.4 ANALYSIS AND VISUALIZATION OF NETWORKS

Cytoscape stands out as the predominant software choice for network analysis and visualization. It offers several iterations, with the latest being Cytoscape version 3.10.1. The interactions between targets as found in the STRING database are typically examined and presented within Cytoscape for visualization and analysis. Network analysis tools and software are then employed to study and visualize the activities of the nodes within the network [20].

13.2.5 FUNCTIONAL ENRICHMENT ANALYSIS

Pathway analysis is conducted to enhance our understanding of chemicals, proteins, or enzymes and to explore the interconnected biological processes, as well as their interactions within the extensive signalling pathways within our body. For understanding the biological, molecular, and cellular processes associated with our gene of interest, the most effective tools are KEGG and GO (Gene Ontology) pathway analysis [21]. Additionally, it assists in uncovering how the compounds exert their therapeutic effects on the specific ailment. GO enrichment was predominantly employed to grasp the primary mechanism of the target, while KEGG pathway analysis was conducted to explore the distribution of the targets within the network [22].

13.2.6 VALIDATION AND CONFIRMATION OF THE RESULTS

It is crucial to validate the outcomes obtained from the earlier methods. The efficiency of expected molecular targets can be confirmed through various validation approaches. While in-vitro and in-vivo methods are generally considered the most reliable, they are also known to be labour-intensive and costly [23]. Hence, in response to the growing integration of technology into research and development, numerous computational methods have been devised to validate and bolster research findings. Among these methods, molecular docking stands out as a valuable tool for forecasting the most energetically favourable binding orientation of a ligand within a target's active site [24]. Typically, researchers aim to determine the potential mechanism of specific compounds in combating diseases by identifying potential compound targets. However, NP methods do not provide insights into the interactions between chosen compounds and their designated targets. To assess these interactions effectively, molecular docking emerges as the optimal approach. Various commercial and freely available software options exist to facilitate molecular docking experiments [25]. The Maestro tool from Schrödinger is a popular commercial software extensively employed, while the widely recognized freely available option is AutoDock Vina [26]. In various network pharmacological approaches, researchers have utilized gene expression microarray data as a method to corroborate their discoveries. Gene expression microarray analysis measures the activity of numerous genes concurrently, providing a comprehensive overview of cellular functions. These traits can be employed to distinguish cells undergoing active division or to illustrate how cells respond to a specific drug [27]. Among the most potent and versatile methods for examining gene expression patterns in diverse tissues or within a single tissue across different experimental conditions is high-density microarray analysis. After successfully completing the gene expression analysis, one can utilize real-time polymerase chain reaction (RT-PCR) to confirm the identified target genes exhibiting differential expression [28]. Another reliable approach for confirming the results is Western blotting, a frequently employed technique by researchers to validate their discoveries regarding interactions within target-pathway networks. The reliability and accuracy of Western blot analysis outcomes instil greater confidence among researchers, thereby promoting the exploration and advancement of new medications [29].

13.3 USE OF NETWORK PHARMACOLOGY IN SIGNAL ANALYSIS

Nowadays, NP is playing an important in the identification and analysis of novel signals. Recently, published reports have been demonstrated the role of NP in the analysis of mechanism involved in the development of the novel adverse drug reaction (ADR). Recently, Singh et al. [30] have found Raynaud's phenomenon as novel signal associated with Calcitonin gene-related peptide (CGRP) antagonists. They have extracted the data of CGRP antagonists by the help of FAERS database tool OpenVigil which is an open tool for data mining and analysis of pharmacovigilance data. After that statistical analysis was performed for the extracted data and the signal of Raynaud's phenomenon with CGRP antagonists was confirmed. To determine the mechanism behind the progression of the Raynaud's phenomenon, NP was performed. In which common genes of CGRP antagonists and Raynaud's phenomenon were identified by the help of a Venn diagram and these common genes were incorporated into a string database to check the gene-gene interaction. This gene-gene interaction data was downloaded in an Excel file. Lastly, these Excel files were imported into Cytoscape to construct a network between the genes and top ten genes identified by the help of the Cytohubba application of Cytoscape. Among these top ten genes, insulin-like growth factor 1-receptor (IGF1R) pathways were found in the progression of Raynaud's phenomenon, and on this basis they concluded that RP is recognized as a novel signal with all CGRP antagonists [30].

Bhati et al. [31] also performed NP approaches for confirmation of positive signal of progressive multifocal leukoencephalopathy (PML) with temozolomide. In this study OpenVigil tool was used to extract the individual case safety reports of temozolomide. The disproportionality analysis was performed to check either the signal is positive or not and by the help of disproportionality analysis they have found that the PML has a positive association with temozolomide. After that, network pharmacological approaches were used to identify the mechanism behind the temozolomide-induced PML. By the help of NP, they observed that human mitogen-activated protein kinase, 3-phosphoinositide-dependent protein kinase 1 protein, human mTOR complex protein, Phosphatidylinositol 4,5-bisphosphate 3-kinase protein, and glycogen synthase kinase-3 beta protein could be responsible for the progression of temozolomide-induced PML. The docking results have indicated that glycogen synthase kinase-3 beta and mitogen-activated protein kinase 1 have shown good interaction with temozolomide as compared to other identified targets, and on the basis of these results, they have concluded that the PML is identified as novel signal with temozolomide [31].

13.4 CURRENT PERSPECTIVES

Currently, various researchers are moving towards on the identification of the mechanism involved in the ADR. Network pharmacological methods can play an important role for researchers for the identification of the mechanism. Nowadays, researchers focus on the signal detection, but by the help of NP, they can be able to identify possible mechanism and pathways involved in the signal or ADR. These types of studies done by the help of various computational tools so the researchers should have some knowledge about computer and data science. Currently, various computational methods are being used in the field of signal detection. NP is also the one of the computational methods and some studies are being published by the help of NP.

13.5 CONCLUSION

Signal analysis involves the identification of novel ADRs associated with drugs. Therefore, network pharmacological approaches could play a significant role in the identification of possible mechanism behind the signal. However, these techniques should be used as per the standard protocol to get the good and authentic result.

REFERENCES

1. Chandran, U., Mehendale, N., Patil, S., Chaguturu, R., Patwardhan, B., 2017. Network Pharmacology. Innov. Approaches Drug Discov. Ethnopharmacol. Syst. Biol. Holist. Target. 25, 127–164. https://doi.org/10.1016/B978-0-12-801814-9.00005-2
2. Saima., Latha, S., Sharma, R., Kumar, A., 2024. Role of Network Pharmacology in Prediction of Mechanism of Neuroprotective Compounds. In Neuroprotection: Method and Protocols (pp. 159–179). New York, NY: Springer US.

3. Ekins, S., Mestres, J., Testa, B., 2007. In Silico Pharmacology for Drug Discovery: Methods for Virtual Ligand Screening and Profiling. Br. J. Pharmacol. 152, 9–20. https://doi.org/10.1038/sj.bjp.0707305

4. Potter, E., 2008. 基因的改变NIH Public Access. Bone 23, 1–7.

5. Arrell, D.K., Terzic, A., 2010. Network Systems Biology for Drug Discovery. Clin. Pharmacol. Ther. 88, 120–125. https://doi.org/10.1038/clpt.2010.91

6. Li, S., Zhang, B., 2013. Traditional Chinese Medicine Network Pharmacology: Theory, Methodology and Application. Chin. J. Nat. Med. 11, 110–120. https://doi.org/10.1016/S1875-5364(13)60037-0

7. Noor, F., Tahir ul Qamar, M., Ashfaq, U.A., Albutti, A., Alwashmi, A.S.S., Aljasir, M.A., 2022. Network Pharmacology Approach for Medicinal Plants: Review and Assessment. Pharmaceuticals 15, 572. https://doi.org/10.3390/ph15050572

8. Casas, A.I., Hassan, A.A., Larsen, S.J., Gomez-Rangel, V., Elbatreek, M., Kleikers, P.W.M., Guney, E., Egea, J., López, M.G., Baumbach, J., Schmidt, H.H.H.W., 2019. From Single Drug Targets to Synergistic Network Pharmacology in Ischemic Stroke. Proc. Natl. Acad. Sci. 116, 7129–7136. https://doi.org/10.1073/pnas.1820799116

9. Zhang, R., Zhu, X., Bai, H., Ning, K., 2019. Network Pharmacology Databases for Traditional Chinese Medicine: Review and Assessment. Front. Pharmacol. 10. https://doi.org/10.3389/fphar.2019.00123

10. Zhang, G., Li, Q., Chen, Q., Su, S., 2013. Network Pharmacology: A New Approach for Chinese Herbal Medicine Research. Evidence-Based Complement. Altern. Med. 2013, 1–9. https://doi.org/10.1155/2013/621423

11. Lee, W.-Y., Lee, C.-Y., Kim, Y.-S., Kim, C.-E., 2019. The Methodological Trends of Traditional Herbal Medicine Employing Network Pharmacology. Biomolecules 9, 362. https://doi.org/10.3390/biom9080362

12. Hauben, M., Aronson, J.K., 2009. Defining 'Signal' and Its Subtypes in Pharmacovigilance Based on a Systematic Review of Previous Definitions. Drug Saf. 32, 99–110. https://doi.org/10.2165/00002018-200932020-00003

13. Zhang, Y., Yuan, T., Li, Y., Wu, N., Dai, X., 2021a. Network Pharmacology Analysis of the Mechanisms of Compound Herba Sarcandrae (Fufang Zhongjiefeng) Aerosol in Chronic Pharyngitis Treatment. Drug Des. Devel. Ther. 15, 2783–2803. https://doi.org/10.2147/DDDT.S304708

14. Liang, X., Li, H., Li, S., 2014. A Novel Network Pharmacology Approach to Analyse Traditional Herbal Formulae: The Liu-Wei-Di-Huang Pill as a Case Study. Mol. BioSyst. 10, 1014–1022. https://doi.org/10.1039/C3MB70507B

15. Luo, T., Lu, Y., Yan, S., Xiao, X., Rong, X., Guo, J., 2020. Network Pharmacology in Research of Chinese Medicine Formula: Methodology, Application and Prospective. Chin. J. Integr. Med. 26, 72–80. https://doi.org/10.1007/s11655-019-3064-0

16. Dai, Y., Guo, M., Jiang, L., Gao, J., 2022. Network Pharmacology-Based Identification of miRNA Expression of *Astragalus membranaceus* in the Treatment of Diabetic Nephropathy. Medicine (Baltimore) 101, e28747. https://doi.org/10.1097/MD.0000000000028747

17. Mering, C. V., 2003. STRING: A Database of Predicted Functional Associations between Proteins. Nucleic Acids Res. 31, 258–261. https://doi.org/10.1093/nar/gkg034

18. Li, X., Tang, Q., Meng, F., Du, P., Chen, W., 2022. INPUT: An Intelligent Network Pharmacology Platform Unique for Traditional Chinese Medicine. Comput. Struct. Biotechnol. J. 20, 1345–1351. https://doi.org/10.1016/j.csbj.2022.03.006

19. Zhang, Y., Yuan, T., Li, Y., Wu, N., Dai, X., 2021b. Network Pharmacology Analysis of the Mechanisms of Compound Herba Sarcandrae (Fufang Zhongjiefeng) Aerosol in Chronic Pharyngitis Treatment. Drug Des. Devel. Ther. 15, 2783–2803. https://doi.org/10.2147/DDDT.S304708

20. Xie, R., Li, B., Jia, L., Li, Y., 2022. Identification of Core Genes and Pathways in Melanoma Metastasis via Bioinformatics Analysis. Int. J. Mol. Sci. 23, 794. https://doi.org/10.3390/ijms23020794

21. Aslam, S., Ahmad, S., Noor, F., Ashfaq, U.A., Shahid, F., Rehman, A., Tahir ul Qamar, M., Alatawi, E.A., Alshabrmi, F.M., Allemailem, K.S., 2021. Designing a Multi-Epitope Vaccine against Chlamydia Trachomatis by Employing Integrated Core Proteomics, Immuno-Informatics and In Silico Approaches. Biology (Basel) 10, 997. https://doi.org/10.3390/biology10100997

22. Tao, Q., Du, J., Li, X., Zeng, J., Tan, B., Xu, J., Lin, W., Chen, X., 2020. Network Pharmacology and Molecular Docking Analysis on Molecular Targets and Mechanisms of Huashi Baidu Formula in the Treatment of COVID-19. Drug Dev. Ind. Pharm. 46, 1345–1353. https://doi.org/10.1080/03639045.2020.1788070

23. Hsin, K.-Y., Ghosh, S., Kitano, H., 2013. Combining Machine Learning Systems and Multiple Docking Simulation Packages to Improve Docking Prediction Reliability for Network Pharmacology. PLoS One 8, e83922. https://doi.org/10.1371/journal.pone.0083922

24. Gupta, M., Sharma, R., Kumar, A., 2020. Docking Techniques in Toxicology: An Overview. Curr. Bioinform. 15, 600–610. https://doi.org/10.2174/1574893614666191003125540

25. Navyashree, V., Kant, K., Kumar, A., 2020. Natural Chemical Entities from Arisaema Genus Might Be a Promising Break-through against Japanese Encephalitis Virus Infection: A Molecular Docking and Dynamics Approach. J. Biomol. Struct. Dyn.. https://doi.org/10.1080/07391102.2020.1731603

26. Kant, K., Rawat, R., Bhati, V., Bhosale, S., Sharma, D., Banerjee, S., Kumar, A., 2020. Computational Identification of Natural Product Leads That Inhibit Mast Cell Chymase: An Exclusive Plausible Treatment for Japanese Encephalitis. J. Biomol. Struct. Dyn. https://doi.org/10.1080/07391102.2020.1726820

27. Song, S., Zhou, J., Li, Y., Liu, J., Li, J., Shu, P., 2022. Network Pharmacology and Experimental Verification Based Research into the Effect and Mechanism of Aucklandiae Radix–Amomi Fructus against Gastric Cancer. Sci. Rep. 12, 9401. https://doi.org/10.1038/s41598-022-13223-z

28. Guo, Q., Zheng, K., Fan, D., Zhao, Y., Li, L., Bian, Y., Qiu, X., Liu, X., Zhang, G., Ma, C., He, X., Lu, A., 2017. Wu-Tou Decoction in Rheumatoid Arthritis: Integrating Network Pharmacology and In Vivo Pharmacological Evaluation. Front. Pharmacol. 8. https://doi.org/10.3389/fphar.2017.00230

29. Dong, R., Huang, R., Shi, X., Xu, Z., Mang, J., 2021. Exploration of the Mechanism of Luteolin against Ischemic Stroke Based on Network Pharmacology, Molecular Docking and Experimental Verification. Bioengineered 12, 12274–12293. https://doi.org/10.1080/21655979.2021.2006966

30. Singh, R., Kumar, A., Lather, V., Sharma, R., Pandita, D., 2023. Identification of Novel Signal of Raynaud's Phenomenon with Calcitonin Gene-Related Peptide (CGRP) Antagonists Using Data Mining Algorithms and Network Pharmacological Approaches. Expert Opin. Drug Saf. 1–8. https://doi.org/10.1080/14740338.2023.2248877

31. Bhati, V., Kumar, A., Lather, V., Sharma, R., Pandita, D., 2023. Association of Temozolomide with Progressive Multifocal Leukoencephalopathy: A Disproportionality Analysis Integrated with Network Pharmacology. Expert Opin. Drug Saf. https://doi.org/10.1080/14740338.2023.2278682

14 Docking Studies in Signal Analysis

An Overview

Kumari Kala Shah, Ruchika Sharma, S Latha, and Anoop Kumar

14.1 INTRODUCTION

In docking studies, the binding affinity of small molecules (ligands) to specific receptors (proteins) is predicted through the application of mathematical, biological, and computer-based models. It helps to understand how two or more molecular structures fit together. In simple terms, it's like finding the best way to fit a key into a lock. In order to save the costs of long developmental processes and time indulging in the process, docking studies can be used to analyze the drug-protein binding or interaction in less span of time. Various computational algorithms and scoring functions are involved to predict the best fitting position and orientation of the ligand (drug) within the target protein's binding site (receptor). The scoring function calculates the free energy of binding or binding constant which is obtained by the Gibbs-Helmholtz equation [1, 2].

$$\Delta G = \Delta H - T\Delta S$$

where

    ΔG = Free energy of binding
    ΔH = Enthalpy
    T   = Temperature in Kelvin and
    ΔS = Entropy

The low binding energy of ligands with protein indicates good contact time with the receptor, hence, the high stability of ligand-protein complex.

$$\downarrow \text{Binding energy} = \uparrow \text{Stability of (L*P) complex}$$

Signal refers to the information that suggests a potential causal association between an adverse event and a drug, vaccine, or other intervention. This information is typically derived from a variety of sources, including adverse event reports, scientific literature, and findings from clinical and nonclinical studies [3].

For example, suppose a signal is detected, i.e. hepatotoxicity is induced by drug X. Molecular docking can be helpful in confirming this signal by predicting the strength and type of interaction between ligand (drug) and protein (receptor).

14.2 DOCKING PROTOCOL

14.2.1 Selection of Target

The targets are normally selected and downloaded from Protein Data Bank (PDB) ID which is available at https://www.rcsb.org/?ref=nav_home. The structure of the target should be selected

DOI: 10.1201/9781032629940-14

based on the following characteristics: The 3D structure of the protein should have been determined by X-ray diffraction, and the protein structure should not have mutation strain. Resolution is a representation of the diffraction pattern and the degree of detail in the electron density map. An electron density map with a high-resolution structure (resolution value of 1 Å or so) makes it possible to see every atom since the electrons are organized well. On the other hand, low-resolution structures (resolution values of 3 Å or above) simply show the protein chain's basic components. Therefore, protein structures with a resolution value of less than 2 Å are suggested to be selected.

14.2.2 PREPARATION OF TARGET

The selected targets are prepared by using a protein preparation module of suitable software. Normally, the structure is prepared by the addition of H^+ atoms, removal of water molecules, solvent, and extra chains. If the protein structure is not found in the PDB, homology modeling is used to construct the structure. After preparation, the protein model is validated using the Ramachandran plot or other techniques to increase prediction accuracy.

14.2.3 SELECTION AND PREPARATION OF LIGANDS

Based on the outcomes of the signal analysis, the ligand's structure can be drawn using ChemDraw software or can be taken from Chemistry Library, PubChem, or ZINC database and is prepared for docking by minimizing the energy using suitable software.

14.2.4 GRID GENERATION

The active sites on the target (protein) are generated using grid generation in the software.

14.2.5 SELECTION OF TYPE OF DOCKING STUDY

There are two main types of molecular docking: flexible docking and rigid docking. In the former approach, the ligand is flexible, while the target is considered a rigid molecule. The aim is to anticipate the ligand's binding conformation to the target, considering flexibility of the ligand. In the latter approach, both the target and the ligand are considered rigid molecules. The aim is to identify the ligand's binding conformation to the target, assuming that both molecules maintain their rigid structure during the docking process.

The type of docking that is selected depends on what is required for the researchers. Although rigid docking is extremely quick, it is not practical since it ignores the degree of conformational freedom that ligands possess. On the other hand, flexible docking, although it requires far more time and processing power, is extensively used because it permits conformational alterations. There are three distinct docking modes:

1. HTVS (high-throughput virtual screening)
2. SP (standard precision)
3. XP (extra precision)

Large libraries (7,000–9,000) of compounds are screened very quickly by HTVS with rough scoring functions. The top 20% of ligands are filtered by SP mode, which is more accurate and ten times slower than HTVS. The SP approach takes more time and is more thorough and accurate than HTVS. The best 50% of the ligands that were derived via SP are examined further with the more precise and highly computational methods. Ligands are shortlisted based on docking score, i.e. −7 and above from XP.

The following types of interactions such as pi-pi interactions, halogen bonds, aromatic bonds, salt bridges, hydrogen bonds, and pi-cations are examined which all contribute to maintaining the protein-ligand complexes' stability.

14.2.6 ANALYSIS OF RESULTS

Using the appropriate tools, the produced ligands are docked with the targets' active sites. Based on the docking score, the most stable ligand configuration in the target's active site is chosen. A dock score of −7 to −8 is considered an ideal score. Theoretically, the lower Gibbs free energy of a complex indicates that the protein-ligand complex is more stable [4–6].

14.3 PREDICTION OF THE POSSIBLE MECHANISM OF THE IDENTIFIED SIGNAL USING DOCKING STUDIES

Docking studies are useful for studying the interactions between particular drugs and potential targets which can involve adverse events. The best targets based on dock score can further be validated using molecular dynamics studies and experimental studies. Recent studies have used docking techniques in the prediction of possible mechanism of identified signals of drugs. Singh et al. [7] have used docking techniques to predict the possible mechanism of Raynaud's phenomenon (novel signal) with calcitonin gene-related peptide (CGRP) antagonists. Bhati et al [8] have used molecular docking techniques to predict possible mechanism of positive signal of progressive multifocal leukoencephalopathy (PML) with temozolomide.

14.4 CONCLUSION

Docking techniques could play an important role in the prediction of possible mechanism of identified novel signals. However, further molecular dynamics and experimental studies are needed to confirm the mechanism.

REFERENCES

1. Gupta M, Sharma R, Kumar A. Docking techniques in pharmacology: How much promising? Computat Biol Chem. 2018;76:210–217.
2. Gupta M, Sharma R, Kumar A. Docking techniques in toxicology: An overview. Curr Bioinform. 2020;15(6):600–610.
3. Hauben M., Zhou X., Quantitative methods in pharmacovigilance: Focus on signal detection. Drug Saf. 2003;26:159–186.
4. Van Den Driessche G, Fourches D. Adverse drug reactions triggered by the common HLA-B* 57: 01 variant: A molecular docking study. J Cheminformat. 2017;9(1):1–7.
5. LaBute MX, Zhang X, Lenderman J, Bennion BJ, Wong SE, Lightstone F.C. Adverse drug reaction prediction using scores produced by large-scale drug-protein target docking on high-performance computing machines. PLoS One. 2014;9(9):e106298.
6. Kant K, Lal UR, Kumar A, Ghosh M. A merged molecular docking, ADME-T and dynamics approaches towards the genus of Arisaema as herpes simplex virus type 1 and type 2 inhibitors. Comput Biol Chem. 2019;78:217–226.
7. Singh, R., Kumar, A., Lather, V., Sharma, R., Pandita, D. Identification of novel signal of Raynaud's phenomenon with Calcitonin Gene-Related Peptide(CGRP) antagonists using data mining algorithms and network pharmacological approaches. Expert Opin. Drug Saf. 2023; 1–8. https://doi.org/10.1080/14740338.2023.2248877
8. Bhati, V., Kumar, A., Lather, V., Sharma, R., Pandita, D. Association of temozolomide with progressive multifocal leukoencephalopathy: a disproportionality analysis integrated with network pharmacology. Expert Opin. Drug Saf. 2023. https://doi.org/10.1080/14740338.2023.2278682

15 Molecular Dynamics Simulations in Signal Validation

Debanjan Dey and Anoop Kumar

15.1 INTRODUCTION

Novel drug candidates are being identified through various phases of drug discovery, such as target identification, target validation, and lead identification, as well as lead optimization. Effectiveness, safety, compliance with clinical and commercial concerns, and, most importantly, "druggability" are some of the requirements for a good drug candidate. Increased confidence in the association between the target and disease is made possible by accurate target identification and validation [1]. Currently, data mining of already existing biological data helps in the identification of targets, which are further validated using various experimental techniques [2]. The lead compounds are identified against the validated targets using various computational techniques. Lead compounds are those that exhibit desired biological or pharmacological actions and could serve as the starting point for the creation of a new chemical entity (NCE) [3]. However, there has been a recent resurgence in the phenotypic drug discovery (PDD) approach. More precisely, PDD is defined as compounds that are discovered despite not knowing the specific molecular targets [4]. Comprehensive research of the pathophysiology of human diseases led to many decades of aimed target-based drug discovery. Yet, during the past ten years, we have observed an increase in the development and application of computational methodologies in the identification and testing of novel drug candidates [5].

These techniques are helpful in the reduction of time; one such important computational tool is molecular dynamics (MD). MD is defined as a computational simulation for predicting as well as analysing the movements of atoms and molecules physically.

Classical MD simulations are depending upon Newton's equation of motion, i.e. $\vec{f} = m * \vec{a}$, where $\vec{f}$ equals the force, m represents the mass of a particle, and $\vec{a}$ represents acceleration. The acceleration of a particle and atoms of mass is directly proportional to the acting force.

MD provide the output of dynamical and atomistic insights into intra- and intermolecular motions of biological molecules at timescales ranging generally from picoseconds to microseconds. A primary understanding of how biological macromolecules work needs knowledge of structure and dynamics. MD simulations can detect not only ample dynamical structural information of bio-macromolecules but also a wealth of energetic information about protein and ligand interactions simultaneously. This information is crucial for directing the drug discovery and design process, knowing the target's structure-function relationship and the fundamentals of protein-ligand interactions [6].

Drug targets can be DNA or RNA molecules as well as other proteins, such as receptors or enzymes. The information regarding the conformation of proteins is useful in the design of ligands, as most of the proteins are associated with a variety of conformational changes.

Regarding drug design, MD simulations can offer crucial information on the dynamic nature of the target. The dynamics and function of proteins, as well as their interactions with other proteins and tiny molecules, have all been studied and understood through the use of MD simulations. This is why MD simulation is getting a high acceptance rate in order to find novel compounds in the modern drug discovery arena. In recent years, it has been seen that the impact of MD simulations has increased exponentially in the field of molecular biology. Various difficulties arise when a molecular biologist attempts to comprehend the function of a protein or other biomolecule. An atomic-level structure is very beneficial and often yields significant insight into how the biomolecule functions.

DOI: 10.1201/9781032629940-15

However, because the atoms in a biomolecule are constantly in motion, the dynamics of the individual molecules affect both their internal structure and how well they work together. Gaining such information is very useful for researchers working in the fields like molecular biology. Now, we can get and capture the behaviour of proteins like receptors, enzymes, and other biological molecules in full atomic resolution even before doing the laboratory experiment using highly advanced computational tools [7]. The flow chart of computer-aided MD is presented in Figure 15.1.

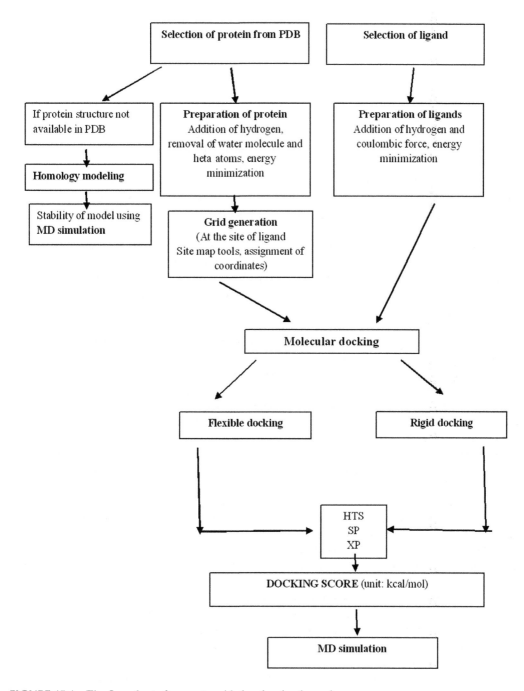

FIGURE 15.1   The flow chart of computer-aided molecular dynamics.

The process of actively looking for and identifying safety signals from a wide range of data sources is known as signal detection in pharmacovigilance. A safety signal is information about a new or well-documented adverse event that may be brought on by a medication and calls for additional research. Numerous sources, including spontaneous reports, clinical research, and scholarly literature, can be used to identify safety signals. In a nutshell, signals are essential for guaranteeing drug safety since they assist regulatory agencies and pharmaceutical firms in recognizing and addressing newly developing safety concerns related to drugs. By the application of MD, one can estimate the relation and validate drug-associated events.

15.2 MOLECULAR DYNAMICS

MD is defined as simulating the pattern-dependent characteristics of a given system of atoms. The basic concept of MD is very simple. MD is delivering all necessary details at the microscopic level, i.e., atomic positions and velocity. In addition, one can also estimate the strength released by every individual atom by all other atoms in a given system. That's why MD simulation is dependent on Newton's laws of motion to analyse the dimensional situation of every single atom with respect to time. More precisely, the result will show the three-dimensional visualization that elaborates the atomic-level configuration of the given system. These simulations play a significant role in scientific research in many fields, as information regarding the position and motion of every atom at every point with respect to time is not easy in the case of any experimental technique [7].

One of the important parameters which govern MD is called the "force field" [8]. In a more simple language, it is the force that is used in MD. It records the electrostatic interactions and other interatomic interactions between atoms and molecules during simulation [7]. The most commonly used force field in MD is Assisted Model Building with Energy Refinement (AMBER) [9] and Chemistry at Harvard Macromolecular Mechanics (CHARMM) [10]. Technology, computing hardware, and algorithms have been evolving day by day in order to pursue MD simulation efficiently. Supercomputers like graphical processing units (GPUs) made this possible in an excellent way to contribute to longer simulations [11]. These GPUs are bringing about simulations in a meaningful and accurate way, which contributes to biological research over many folds more than ever before [7]. After the MD simulation, one can get a specific idea regarding the overall compound's stability. Thus, MD is also called one of the best single mechanisms for basic computational in silico tools which record molecular as well as atomistic level fluctuations and give essential information regarding the stability of protein and ligand complexes. Although there are some common structure-based designs of a drug, approaches are available like molecular docking, side-by-side virtual screening, which can also predict some probable and shortlisted ligands with respect to desirable targets like proteins. However, MD can give information in a more detailed manner regarding ligand dynamical characteristics and stability against the desirable target-like proteins [12]. After doing MD, to get an idea of how tightly the ligand and protein molecules are interacting, one needs to find the binding free energy (ΔG). Predicted affinities for biomolecular complexes are provided via binding-free energy calculations based on MD simulations. A thorough description of the system, including its chemistry and the interactions between its parts, is the first step in these calculations. The system's simulations are then utilized to calculate thermodynamic data, including binding affinities. The ΔG of a system can be manipulated as an overall evaluation to calculate the stability of an experimented system. The equation is written as $\Delta G = \Delta H - T\Delta S$, where ΔH denotes the conversion in enthalpy of the system, T denotes the temperature in Kelvin, and ΔS denotes the conversion in the entropy of the system.

MMGBSA (Molecular mechanics generalized born surface area) [13] is defined as the binding affinity calculation between the ligand and target. The energy of optimized free receptors, free ligands, and complexes of the ligand and receptor are all calculated by MMGBSA. The following formula is used to compute the binding free energies (ΔG_{Bind}) of ligands at the protein's catalytic cavity, i.e. $\Delta G_{Bind} = G_{complex} - G_{protein} - G_{ligand}$, where $G_{complex}$, $G_{protein}$, and G_{ligand} are the complex, protein, and ligand-free energies, respectively.

The Hamiltonian portion of the associated Schrodinger's wave equation is used in density functional theory (DFT) to compute the ground state energy of systems utilizing various exchange correlation functionals. More precisely, DFT is a set of theoretical tools and computational models aiming to comprehend the structure and characteristics of atoms, molecules, and materials on atomic length scales [14].

15.3 AVAILABLE PACKAGES

Various packages are available which are either freely or commercially available to perform MD studies. A few of them, which are widely used, are discussed below.

15.3.1 AMBER (Assisted Model Building with Energy Refinement)

This was invented by Peter Kollman's group at the University of California, San Francisco [9]. Applications for applying the AMBER force fields to simulations of biomolecules are provided by the AMBER software suite. In AMBER, the parameter sets with names that begin with "ff" and have a two-digit year number, like "ff99," provide information regarding the parameters of peptides, proteins, and nucleic acids [15, 16]. Lipid14 serves as the main force field in the AMBER suit for lipids [17].

15.3.2 CHARMM (Chemistry at Harvard Macromolecular Mechanics)

In order to develop and maintain the CHARMM programme, Martin Karplus and his team at Harvard joined a global network of developers as part of the CHARMM Development Project. CHARMM19 [18], all-atom CHARMM22 [19], and its dihedral potential corrected from CHARMM22/CMAP, as well as subsequent versions CHARMM27 and CHARMM36 and other modifications like CHARMM36m and CHARMM36IDPSFF [20], are among the CHARMM force fields for proteins. CHARMM27 [21] is widely utilized for lipids, DNA, and RNA.

15.3.3 Desmond

This was developed at D.E. Shaw Research to conduct high-momentum MD of biological molecules using conventional computational approaches [22–25]. This is freely available. Desmond offers support for the usual MD algorithms, which are quick and precise. Using particle mesh Ewald-based techniques, one can compute long-range electrostatic energy and forces [26, 27]. The code can be run on a single computer or platform with many processors [28], using unique parallel algorithms [29] and side-by-side numerical methods [30] to achieve excellent performance. This package presently supports various series of CHARMM, Amber, and Freforce fields as well as a variety of water models.

15.3.4 GROMACS

This software package was invented at the Department of Biophysical Chemistry, lab of the University of Groningen [31–33]. This is a freely available package to run MD. It is programmed using a command line. Through various advanced and upgraded parallelization techniques, the newest version 5 achieves new performance heights. In the Folding@Home distributed computing project, GROMACS has been utilized for a long time [34] and is widely employed for meta-dynamics along with plumbing [35].

15.3.5 MBN Explorer (MesoBioNano Explorer)

This was developed by the MBN Research Center. It was developed as a traditional MD code in the year 2000 to simulate the interaction of many-body systems. It addresses the simulation of atomic clusters, nanoparticles as well as biomolecules, and nano-systems [36].

15.3.6   ORAC

In the year between 1989 and 1990, this was invented originally by Massimo Marchi while he was staying at International Business Machines (IBM), Kingston, USA [37]. It can accelerate complex simulation at the atomistic level.

15.3.7   SHARC (Surface Hopping Including Arbitrary Couplings)

It was created by the SHARC development team working under the direction of Prof. Leticia Gonzalez at the Institute of Theoretical Chemistry, University of Vienna, Austria. In October 2014, the SHARC software package became accessible to everyone. It is emphasized to record the dynamical behaviour of excited-state molecules [38]. This package is freely available for academic use, and it is declared under a General Public License.

15.3.8   Nanoscale Molecular Dynamics (NAMD, Formerly Not Another Molecular Dynamics Programme)

The theoretical and computational biophysics (TCB) group and the Parallel Programming Laboratory (PPL) at the University of Illinois at Urbana–Champaign worked together to develop it. It is a computer software for simulating MD that was created using the Charm++ parallel programming model [39].

15.3.9   OpenMM

OpenMM is a library for doing MD simulations on a range of hardware architectures. It was written by Peter Eastman in the Vijay S. Pande lab at Stanford University. It was initially published in January 2010. Using CUDA and OpenCL, OpenMM simulates protein dynamics on GPUs [40].

15.3.10   Tinker

Tinker is a collection of MD simulation software programmes, formerly abbreviated as TINKER. The codes offer a comprehensive and all-encompassing set of tools for MD and mechanics, together with certain unique characteristics for biomolecules. Windows, macOS, Linux, and Unix are all supported by Tinker [41].

15.3.11   XMD

A traditional MD programme called XMD was created to simulate issues in the field of materials research. The code was developed by Jon Rifkin of the University of Connecticut and is being distributed under GNU General Public License.

15.3.12   Pydlpoly

Pydlpoly is a MD simulation tool with Python language interface. Rochus Schmid, a German student at Ruhr University Bochum, was the author of Pydlpoly.

15.4   STEPS INVOLVED IN MOLECULAR DYNAMICS

Various steps are needed to undertake MD studies. These steps are somewhat similar across all packages. As GROMACS is the most widely used package, therefore the steps involved in GROMACS are summarized below.

15.4.1 PROTEIN AND LIGAND PREPARATION

The first step to running MD is the preparation of proteins and ligands. The structure of the protein can be downloaded from PDB (Protein Data Bank) and prepared by removing heta atoms, and addition of hydrogen atoms. Ligand preparation can be done by the addition of hydrogen atoms and coulombic force. During ligand preparation, it is additionally needed to expose the force fields to each ligand. Following that, a minimized ligand is ready for further simulation.

15.4.2 CHARACTERIZE THE BOX

It is necessary to specify a box, such as a dodecahedron box, octahedron, or cubic, with the right dimension for the protein in order to construct a system.

15.4.3 SOLVATING PROTEIN

The addition of solvate is done to keep track of how many water molecules have added to the system, which it then writes to the system's topology to reflect the changes that have been made.

15.4.4 ADDITION OF IONS

The ions are added to balance the overall charge of the system. The sodium ions are added in the negative charge system, whereas the chloride ions are added in the positive charge system.

15.4.5 ENERGY MINIMIZATION

After the addition of ions, it is necessary to relax the structure by executing an energy minimization (EM) method in order to eliminate any steric conflicts or odd geometry that could artificially increase the system's energy.

15.4.6 EQUILIBRATION

It is now important to equilibrate the solvent around the solute (usually proteins). Equilibration is carried out first under the guidance of an NVT (constant particle number, volume, and temperature) ensemble, and then under the guidance of an NPT (constant particle number, pressure, and temperature) ensemble.

15.4.7 FINAL MD RUN

Now one can let go of the posture restrictions because equilibration has been accomplished. Finally, the system is being prepared to run a production MD simulation.

Thereafter, it can be concluded that a trajectory (xtc file) is being generated, which specifies the system's atomic motion, after finishing the steps of running the process [42]. The various steps involved in MD using GROMACS are mentioned in Figure 15.2.

15.5 MOLECULAR DYNAMICS SIMULATION PARAMETERS

MD simulation parameter answers various research questions, like what intriguing molecular properties can be observed, and has the simulation sufficiently converged.

15.5.1 ROOT MEAN SQUARE DEVIATION (RMSD)

The average distance between a collection of atoms is measured by the root mean square deviation (RMSD), a commonly used metric of the structural distance among coordinate sets. It represents

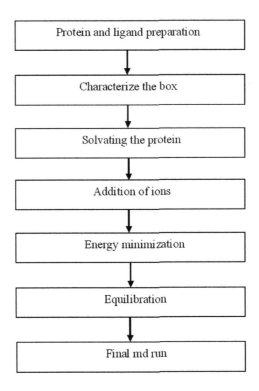

FIGURE 15.2 The various steps involved in MD.

the degree to which the protein structure has altered over the simulation period between different time points of the trajectory.

15.5.2 ROOT MEAN SQUARE FLUCTUATION (RMSF)

The deviation from a reference position over time is represented by the root mean square fluctuation (RMSF). It is used to determine if any fluctuation of amino acids happened in the protein during the simulation.

15.5.3 RADIUS OF GYRATION (RoG)

A protein's compactness can be assessed using the radius of gyration (RoG). A protein tends to be difficult to fold when it is exceedingly compact. Typically, RoG is shown in a protein-ligand combination, followed by MD. This clarifies the complex's stability in addition to the RMSD.

15.5.4 POTENTIAL ENERGY (PE)

To differentiate between conformations and configurations of the same molecular system, potential energy (PE) can be used. It is also one of the important parameters to be considered.

15.5.5 PRINCIPAL COMPONENT ANALYSIS (PCA)

Principal component analysis (PCA) can be used to project the trajectory in small dimensions in terms of main components, which helps to understand the dynamical behaviour of biomolecules [42, 43].

15.5.6 Dynamic Cross Co-Relation Matrix (DCCM)

A common technique for examining the trajectories of MD simulations is dynamic cross-correlation analysis [44].

15.6 APPLICATION OF MOLECULAR DYNAMICS IN THE PREDICTION OF POSSIBLE MECHANISMS OF IDENTIFIED SIGNALS

Various researchers across the globe have identified and reported various types of unidentified signals. Some of them are explained below. Sharma et al. (2023) in their study have concluded that abacavir, ganciclovir, lamivudine, lopinavir, nevirapine, ribavirin, ritonavir, and zidovudine have a positive DRESS signal [45]. Singh et al. have concluded that CGRP antagonists are linked to Raynaud's phenomenon (RP) using the FAERS database and data mining techniques [46]. Nango et al. have demonstrated that the over-representation of signals for acute pancreatitis may be detected for prednisolone, methylprednisolone, and some others in the FAERS database [47]. Wu et al. have demonstrated that tamoxifen side effects affect the neurological system, particularly in high-risk Alzheimer's patients [48].

MD can play a significant role in the validation of these drugs as well as in getting to know the connection between suspect drugs with targets. Some of the MD-based experimental findings are compiled in Table 15.1 [49–68].

TABLE 15.1

Some Studies Used by Molecular Dynamics (MD) Simulations in the Identification of Drug-Target Mechanism Understanding

References	MD Experimental Finding	Software Used	MD Running Time (ns)	Possible Therapeutic Indication
Srivastava [49]	Cinnamic acid, thymoquinone, and andrographolide (Kalmegh) were screened out, showing strong binding affinity to the active site of ACE2 receptor	Schrödinger	50	COVID-19
Arif [50]	Peptides (LIVA, TSEP, EKAI, LKHA, EALF, VAEK, DFGAS, and EPGGGG) were found to have best binding affinity with IR and SGLT1	AMBER	50	Diabetes
Kushwaha [51]	*Azadirachta indica* and *Aloe vera* phytochemicals interacted with two of the critical amino acids (Gln493 and Ser494), which are critical for viral spike protein	GROMACS	20	COVID-19
Arifuzzaman [52]	Podocarpusflavone A (PF), 7-*O*-methoxyquercitrin (MQ), proanthocyanidin (PA), and chimarrhinin (CM) found potential inhibitors of main protease	AMBER	100	COVID-19
Kant [53]	C00008437-TIPDB, C00014417-MPD3, and 8141903-MTDP are the ligands found potential inhibitors of chymase	GROMACS	100	Japanese encephalitis
Agarwal [54]	Zn03, Zn05-NH2, and Zn35 are the molecules with high affinity within the active site of the EGFR mutant protein	Schrödinger	100	Resistant lung cancer

(Continued)

TABLE 15.1 *(Continued)*

Some Studies Used by Molecular Dynamics (MD) Simulations in the Identification of Drug-Target Mechanism Understanding

References	MD Experimental Finding	Software Used	MD Running Time (ns)	Possible Therapeutic Indication
Manandhar [55]	Baicalin is a potent molecule with strong and stable binding with AChE protein 2. With DKK1 protein, chebulic acid has shown stable interactions in MD simulations with important interactions 3. For LRP6 protein, mangiferin has shown the stable RMSD and interactions with the key residues during MD simulation	Schrödinger	50	Alzheimer's disease
Schames [56]	5CITEP found a potential inhibitor of HIV-1 integrase	AMBER	2	HIV
Rafi [57]	CID44271905 (Lopinavir analogue) showed most binding energy found more potently targeting the main protease of SARS COV-2	GROMACS	50	COVID-19
Keretsu [58]	Saquinavir found more potent against 3-cymotrypsin-like protease receptor	GROMACS	50	COVID-19
Priya [59]	Toddanol, toddanone, and toddalenone showed the most binding affinity with HIV-1-RT	GROMACS	20	HIV
Dong [60]	Mitoxantrone and chlorhexidine found most effective against targeting human O-GlcNAcase	AMBER	30	Diabetes, cancer, Alzheimer's
Al-Obaidi [61]	ZINC000364721779 and ZINC000635903250 compounds have shown the highest binding affinities towards human GABA-Amino transferase	NAMD	50	Neurological disease
Jairajpuri [62]	ZINC02123811 can act as a potential lead in drug development against SARS-CoV-2 infection	GROMACS	100	COVID-19
Metwaly [63]	(Z)-6-(3-hydroxy-4-methoxystyryl)-4-methoxy-2H-pyran-2-one found potential inhibitor against SARS COV-2 helicase	NAMD	100	COVID-19
Mohamed [64]	mol1_069 and mol1_092 were experimented with as potential anti-EBOV inhibitors	NAMD	100	Ebola
Bepari [65]	S51765 from MyriaScreen Diversity Library found potential against SARS-CoV-2 3CL-PRO	GROMACS	100	COVID-19
Sharma [66]	Peimisine, gmelanone, and Epsilon-viniferin were identified, which showed promising high affinities against Mpro of SARS-CoV-2	GROMACS	100	COVID-19
Verma [67]	Their study confirmed the stability of the compound with a coagulin H in the trajectory of the protein Mpro of SARS-CoV-2	Schrödinger	50	COVID-19
Edache [68]	A series of salicylidene acylhydrazide derivatives were experimented with and found potential against *Chlamydia trachomatis* inhibitors	NAMD	1	*Chlamydia trachomatis*

15.7 CURRENT CHALLENGES AND FUTURE PERSPECTIVES

There is no doubt about the advantages of MD simulations in drug discovery and development; however, there are also many challenges. One of the biggest challenges is the creation of an environment exactly like that of the human body using computational tools. Therefore, the identified drug candidates can also fail when tested under *in vitro* studies. Another biggest challenge regarding MD studies is the generation of large amounts of data. Therefore, a computer system with high storage capacity is required to perform MD studies. The selection of the correct force field is also one of the challenges in MD studies, as it can generate much systematic error. Therefore, further refinement of the force field is required to run MD studies. The approximate time required for the completion of a single MD's simulation requires several hours to even several days, depending upon the molecular structure of protein and ligand. It is a time-consuming process. Sometimes, the MD of particular protein-ligand systems are shut-out in the middle of the simulations because of several factors like equilibration issues and force field issues. Feig et al. (2018) have mentioned the challenges and opportunities in connecting simulations with experiments. Despite all the challenges, this tool is very helpful for gaining more detailed information on the atomistic, dynamical, as well as conformational understanding of biological molecules, which can be revealed using a high-performance computer system. MD enables the assessment in the lead discovery and lead optimization phases, assisting in the selection of the most promising candidate compounds for further research. A number of MD-based drug discovery research papers are also increasing day by day. It is advantageous for pharmaceutical corporations because it can cut down on the time and money needed to develop fresh drug candidates. Patients awaiting the development of a medication for their specific illness, particularly those with uncommon disorders for which there is no known treatment, will also benefit from this. Overall, the future of MD-based drug discovery may contribute to a reduction in the burden of disease worldwide by being more efficient for more successful drug discovery processes. Computer power is acquiring strength day by day with civilization, which will be easy the MD simulations in the future.

15.8 CONCLUSION

MD simulations play a significant role in the identification as well as validation of the prediction of possible mechanisms of identified signals with suspect drugs.

REFERENCES

1. Hughes, J.; Rees, S.; Kalindjian, S.; Philpott, K. (2011) Principles of Early Drug Discovery. British Journal of Pharmacology, 162 (6), 1239–1249. https://doi.org/10.1111/j.1476-5381.2010.01127.x
2. Yang, Y.; Adelstein, S. J.; Kassis, A. I. (2009) Target Discovery from Data Mining Approaches. Drug Discovery Today, 14 (3–4), 147–154.
3. Modern Drug Discovery—Lead Identification https://pharmafactz.com/modern-drug-discovery-lead-identification Accessed 14 February 2023.
4. Lead Discovery|PerkinElmer https://www.perkinelmer.com/uk/category/lead-discovery Accessed 14 February 2023.
5. Ekins, S.; Mestres, J.; Testa, B. (2007) In Silico pharmacology for Drug Discovery: Methods for Virtual Ligand Screening and Profiling. British Journal of Pharmacology, 152 (1), 9–20.
6. Liu, X.; Shi, D.; Zhou, S.; Liu, H.; Liu, H.; Yao, X. (2017) Molecular Dynamics Simulations and Novel Drug Discovery. Expert Opinion on Drug Discovery, 13 (1), 23–37. https://doi.org/10.1080/17460441.2018.1403419
7. Hollingsworth, S. A.; Dror, R. O. (2018) Molecular Dynamics Simulation for All. Neuron, 99 (6), 1129–1143.
8. Durrant, J. D.; McCammon, J. A. (2011) Molecular Dynamics Simulations and Drug Discovery. BMC Biology, 9 (1).

9. Cornell, W. D.; Cieplak, P.; Bayly, C. I.; Gould, I. R.; Merz, K. M.; Ferguson, D. M.; Spellmeyer, D. C.; Fox, T.; Caldwell, J. W.; Kollman, P. A. A. (1995) Second Generation Force Field for the Simulation of Proteins, Nucleic Acids, and Organic Molecules. Journal of the American Chemical Society, 117 (19), 5179–5197. https://doi.org/10.1021/ja00124a002

10. Brooks, B. R.; Bruccoleri, R. E.; Olafson, B. D.; States, D. J.; Swaminathan, S.; Karplus, M. (1983) CHARMM: A Program for Macromolecular Energy, Minimization, and Dynamics Calculations. Journal of Computational Chemistry, 4 (2), 187–217. https://doi.org/10.1002/jcc.540040211

11. Shaw, D. E.; Deneroff, M. M.; Dror, R. O.; Kuskin, J. S.; Larson, R. H.; Salmon, J. K.; Young, C.; Batson, B.; Bowers, K. J.; Chao, J. C.; Eastwood, M. P.; Gagliardo, J.; Grossman, J. P.; Ho, C. R.; Ierardi, D. J.; Kolossváry, I.; Klepeis, J. L.; Layman, T.; McLeavey, C.; Moraes, M; Anton, A. (2008) A Special-Purpose Machine for Molecular Dynamics Simulation. Communications of the ACM, 51 (7), 91–97. https://doi.org/10.1145/1364782.1364802

12. Alturki, N. A.; Mashraqi, M. M.; Alzamami, A.; Alghamdi, Y. S.; Alharthi, A. A.; Asiri, S. A.; Ahmad, S.; Alshamrani, S. (2022) In-Silico Screening and Molecular Dynamics Simulation of Drug Bank Experimental Compounds Against SARS-CoV-2. Molecules, 27 (14), 4391. https://doi.org/10.3390/molecules27144391

13. Genheden, S.; Ryde, U. (2015) The MM/PBSA and MM/GBSA Methods to Estimate Ligand-Binding Affinities. Expert Opinion on Drug Discovery, 10 (5), 449–461. https://doi.org/10.1517/17460441.2015.1032936

14. Sharma, P. (2020) Density Functional Theory in Biology. Annals of Chemical Science Research, 2 (1). https://doi.org/10.31031/acsr.2020.02.000530

15. Maier, J. A.; Martinez, C.; Kasavajhala, K.; Wickstrom, L.; Hauser, K. E.; Simmerling, C. (2015) Ff14SB: Improving the Accuracy of Protein Side Chain and Backbone Parameters from Ff99SB. Journal of Chemical Theory and Computation, 11 (8), 3696–3713. https://doi.org/10.1021/acs.jctc.5b00255

16. The Amber Force Fields http://ambermd.org/AmberModels.php Accessed 3 March 2023.

17. Dickson, C. J.; Madej, B. D.; Skjevik, Å. A.; Betz, R. M.; Teigen, K.; Gould, I. R.; Walker, R. C. (2014) Lipid14: The Amber Lipid Force Field. Journal of Chemical Theory and Computation, 10 (2), 865–879. https://doi.org/10.1021/ct4010307

18. Reiher, W.H. III (1985) Theoretical studies of hydrogen bonding. PhD Thesis, Harvard University: Cambridge.

19. MacKerell, A. D.; Bashford, D.; Bellott, M.; Dunbrack, R. L.; Evanseck, J. D.; Field, M. J.; Fischer, S.; Gao, J.; Guo, H.; Ha, S.; Joseph-McCarthy, D.; Kuchnir, L.; Kuczera, K.; Lau, F. T. K.; Mattos, C.; Michnick, S.; Ngo, T.; Nguyen, D. T.; Prodhom, B.; Reiher, W. E. (1998) All-Atom Empirical Potential for Molecular Modeling and Dynamics Studies of Proteins. The Journal of Physical Chemistry B, 102 (18), 3586–3616. https://doi.org/10.1021/jp973084f

20. Mackerell, A. D.; Feig, M.; Brooks, C. L. (2004) Extending the Treatment of Backbone Energetics in Protein Force Fields: Limitations of Gas-Phase Quantum Mechanics in Reproducing Protein Conformational Distributions in Molecular Dynamics Simulations. Journal of Computational Chemistry, 25 (11), 1400–1415. https://doi.org/10.1002/jcc.20065

21. MacKerell, A. D.; Banavali, N.; Foloppe, N. (2000) Development and Current Status of the CHARMM Force Field for Nucleic Acids. Biopolymers, 56 (4), 257–265. https://doi.org/10.1002/1097-0282(2000)56:4<257::AID-BIP10029>3.0.CO;2-W

22. Bowers, K. J.; Chow, D. E.; Xu, H.; Dror, R. O.; Eastwood, M. P.; Gregersen, B. A.; Klepeis, J. L.; Kolossvary, I.; Moraes, M. A.; Sacerdoti, F. D.; Salmon, J. K.; Shan, Y.; Shaw, D. E. Scalable Algorithms for Molecular Dynamics Simulations on Commodity Clusters https://ieeexplore.ieee.org/document/4090217 Accessed 15 February 2023.

23. Jensen, M. O.; Borhani, D. W.; Lindorff-Larsen, K.; Maragakis, P.; Jogini, V.; Eastwood, M. P.; Dror, R. O.; Shaw, D. E. (2010) "Principles of Conduction and Hydrophobic Gating in K+ Channels. Proceedings of the National Academy of Sciences, 107 (13), 5833–5838. https://doi.org/10.1073/pnas.0911691107

24. Dror, R. O.; Arlow, D. H.; Borhani, D. W.; Jensen, M. O.; Piana, S.; Shaw, D. E. (2009) "Identification of Two Distinct Inactive Conformations of the 2-Adrenergic Receptor Reconciles Structural and Biochemical Observations. Proceedings of the National Academy of Sciences, 106 (12), 4689–4694. https://doi.org/10.1073/pnas.0811065106.

25. Shan, Y.; Seeliger, M. A.; Eastwood, M. P.; Frank, F.; Xu, H.; Jensen, M. O.; Dror, R. O.; Kuriyan, J.; Shaw, D. E. (2009) "A Conserved Protonation-Dependent Switch Controls Drug Binding in the Abl Kinase. Proceedings of the National Academy of Sciences, 106 (1), 139–144. https://doi.org/10.1073/pnas.0811223106.

26. Shan, Y.; Klepeis, J. L.; Eastwood, M. P.; Dror, R. O.; Shaw, D. E. (2005) Gaussian Split Ewald: A Fast Ewald Mesh Method for Molecular Simulation. The Journal of Chemical Physics, 122 (5), 054101. https://doi.org/10.1063/1.1839571.

27. Bowers, K. J.; Lippert, R. A.; Dror, R. O.; Shaw, D. E. (2010) Improved Twiddle Access for Fast Fourier Transforms. IEEE Transactions on Signal Processing, 58 (3), 1122–1130. https://doi.org/10.1109/tsp.2009.2035984.

28. D. E. Shaw Research: Resources https://www.deshawresearch.com/resources.html Accessed 3 March 2023.

29. Bowers, K. J.; Dror, R. O.; Shaw, D. E. (2006) The Midpoint Method for Parallelization of Particle Simulations. The Journal of Chemical Physics, 124 (18), 184109. https://doi.org/10.1063/1.2191489.

30. Lippert, R. A.; Bowers, K. J.; Dror, R. O.; Eastwood, M. P.; Gregersen, B. A.; Klepeis, J. L.; Kolossvary, I.; Shaw, D. E. (2007) A Common, Avoidable Source of Error in Molecular Dynamics Integrators. The Journal of Chemical Physics, 126 (4), 046101. https://doi.org/10.1063/1.2431176.

31. About GROMACS—GROMACS webpage https://www.gromacs.org/about.html# Accessed 3 March 2023.

32. Van Der Spoel, D.; Lindahl, E.; Hess, B.; Groenhof, G.; Mark, A. E.; Berendsen, H. J. C. (2005) GROMACS: Fast, Flexible, and Free. Journal of Computational Chemistry, 26 (16), 1701–1718. https://doi.org/10.1002/jcc.20291.

33. Hess, B.; Kutzner, C.; van der Spoel, D.; Lindahl, E. (2008) GROMACS 4: Algorithms for Highly Efficient, Load-Balanced, and Scalable Molecular Simulation. Journal of Chemical Theory and Computation, 4 (3), 435–447. https://doi.org/10.1021/ct700301q

34. Shirts, M.; Pande, V. S. (2000) COMPUTING: Screen Savers of the World Unite! Science (New York, N.Y.), 290 (5498), 1903–1904. https://doi.org/10.1126/science.290.5498.1903

35. Tribello, G. A.; Bonomi, M.; Branduardi, D.; Camilloni, C.; Bussi, G. (2014) PLUMED 2: New Feathers for an Old Bird. Computer Physics Communications, 185 (2), 604–613. https://doi.org/10.1016/j.cpc.2013.09.018

36. MBN Explorer | MBN Research Center http://www.mbnresearch.com/get-mbn-explorer-software Accessed 3 March 2023.

37. Procacci, P.; Darden, T.A.; Paci, E.; Marchi, M. (1997) "ORAC: A Molecular Dynamics Program to Simulate Complex Molecular Systems With Realistic Electrostatic Interactions. Journal of Computational Chemistry, 18 (15), 1848–1862. https://doi.org/10.1002/(SICI)1096-987X(19971130)18:15<1848::AID-JCC2>3.0.CO;2-O

38. JISCMail - MOLECULAR-DYNAMICS-NEWS Archives https://www.jiscmail.ac.uk/cgi-bin/wa-jisc.exe?A2=ind1410&L=MOLECULAR-DYNAMICS-NEWS&P=R2529 Accessed 3 March 2023.

39. Melo, M. C. R.; Bernardi, R. C.; Rudack, T.; Scheurer, M.; Riplinger, C.; Phillips, J. C.; Maia, J. D. C.; Rocha, G. B.; Ribeiro, J. V.; Stone, J. E.; Neese, F.; Schulten, K.; Luthey-Schulten, Z. (2018) NAMD Goes Quantum: An Integrative Suite for Hybrid Simulations. Nature Methods, 15 (5), 351–354.

40. Release OpenMM 8.0.0 · openmm/openmm. GitHub. https://github.com/openmm/openmm/releases/tag/8.0.0 Accessed 27 June 2023.

41. Lagardère, L.; Jolly, L.-H.; Lipparini, Filippo; Aviat, Félix; Stamm, B.; Jing, Z.; Harger, M.; Torabifard, H.; Cisneros, G. A.; Schnieders, M. J.; Gresh, N.; Maday, Y.; Ren, P.; Ponder, J. W.; Piquemal, J.-P. (2018) Tinker-HP: A Massively Parallel Molecular Dynamics Package for Multiscale Simulations of Large Complex Systems With Advanced Point Dipole Polarizable Force Fields. Chemical Science, 9 (4), 956–972. https://doi.org/10.1039/c7sc04531j.

42. Bray, S., Running molecular dynamics simulations using GROMACS (Galaxy ssTraining Materials). https://training.galaxyproject.org/training-material/topics/computational-chemistry/tutorials/md-simulation-gromacs/tutorial.html Online; Accessed 3 March 2023.

43. Barnett, C.; Senapathi, T.; Bray, S.; Goué, N., Analysis of molecular dynamics simulations (Galaxy Training Materials). https://training.galaxyproject.org/training-material/topics/computational-chemistry/tutorials/analysis-md-simulations/tutorial.html Online; Accessed 3 March 2023.

44. Kasahara, K.; Fukuda, I.; Nakamura, H. (2014) A Novel Approach of Dynamic Cross Correlation Analysis on Molecular Dynamics Simulations and Its Application to Ets1 Dimer–DNA Complex. PLoS ONE, 9 (11), e112419. https://doi.org/10.1371/journal.pone.0112419.

45. Sharma, A.; Roy, S.; Sharma, R.; Kumar, A. (2023). Association of Antiviral Drugs and Their Possible Mechanisms With DRESS Syndrome Using Data Mining Algorithms. Journal of Medical Virology, 95(3). https://doi.org/10.1002/jmv.28671

46. Singh, R.; Kumar, A.; Lather, V.; Sharma, R.; Pandita, D. (2023). Identification of Novel Signal of Raynaud's Phenomenon With Calcitonin Gene-Related Peptide (CGRP) Antagonists Using Data Mining Algorithms and Network Pharmacological Approaches. Expert Opinion on Drug Safety. https://doi.org/10.1080/14740338.2023.2248877

47. Nango, D.; Hirose, Y.; Goto, M.; Echizen, H. (2019). Analysis of the Association of Administration of Various Glucocorticoids With Development of Acute Pancreatitis Using US Food and Drug Administration Adverse Event Reporting System (FAERS). Journal of Pharmaceutical Health Care and Sciences, 5(1). https://doi.org/10.1186/s40780-019-0134-6

48. Wu, J.; Tang, B.; Xiao, X.; Wu, W. (2023). Signals of Adverse Drug Events of Tamoxifen: An Analysis of FDA Adverse Event Reporting System Data from 2015 to 2020. Authorea (Authorea). https://doi.org/10.22541/au.168421360.05647587/v1

49. Srivastava, N.; Garg, P.; Srivastava, P.; Seth, P. K. (2021) A Molecular Dynamics Simulation Study of the ACE2 Receptor With Screened Natural Inhibitors to Identify Novel Drug Candidate Against COVID-19. PeerJ, 9, e11171. https://doi.org/10.7717/peerj.11171.

50. Arif, R.; Ahmad, S.; Mustafa, G.; Mahrosh, H. S.; Ali, M.; Tahir ul Qamar, M.; Dar, H. R. (2021) Molecular Docking and Simulation Studies of Antidiabetic Agents Devised from Hypoglycemic Polypeptide-P of Momordica Charantia. BioMed Research International, e5561129. https://doi.org/10.1155/2021/5561129.

51. Kushwaha, P. P.; Singh, A. K.; Bansal, T.; Yadav, A.; Prajapati, K. S.; Shuaib, M.; Kumar, S. (2021) Identification of Natural Inhibitors Against SARS-CoV-2 Drugable Targets Using Molecular Docking, Molecular Dynamics Simulation, and MM-PBSA Approach. Frontiers in Cellular and Infection Microbiology, 11. https://doi.org/10.3389/fcimb.2021.730288.

52. Arifuzzaman; Mohammadi, M.; Rupa, F. H.; Khan, M. F.; Rashid, R. B.; Rashid, M. A. (2022) Identification of Natural Compounds With Anti-SARS-CoV-2 Activity Using Machine Learning, Molecular Docking and Molecular Dynamics Simulation Studies. Dhaka University Journal of Pharmaceutical Sciences, 21 (1), 1–13.

53. Kant, K.; Rawat, R.; Bhati, V.; Bhosale, S.; Sharma, D.; Banerjee, S.; Kumar, A. (2020) Computational Identification of Natural Product Leads That Inhibit Mast Cell Chymase: An Exclusive Plausible Treatment for Japanese Encephalitis. Journal of Biomolecular Structure and Dynamics, 39 (4), 1203–1212. https://doi.org/10.1080/07391102.2020.1726820.

54. Agarwal, S. M.; Nandekar, P.; Saini, R. (2022) Computational Identification of Natural Product Inhibitors Against EGFR Double Mutant (T790M/L858R) by Integrating ADMET, Machine Learning, Molecular Docking and a Dynamics Approach. RSC Advances, 12 (26), 16779–16789. https://doi.org/10.1039/d2ra00373b.

55. Manandhar, S.; Sankhe, R.; Priya, K.; Hari, G.; Kumar, B.; Mehta, H.; Nayak, C. H.; Pai, U. Y. (2022) Molecular Dynamics and Structure-Based Virtual Screening and Identification of Natural Compounds as Wnt Signaling Modulators: Possible Therapeutics for Alzheimer's Disease. Molecular Diversity, 26 (5), 2793–2811.

56. Schames, J. R.; Henchman, R. H.; Siegel, J. S.; Sotriffer, C. A.; Ni, H.; McCammon, J. A. (2004) Discovery of a Novel Binding Trench in HIV Integrase. Journal of Medicinal Chemistry, 47 (8), 1879–1881. https://doi.org/10.1021/jm0341913.

57. Rafi, Md. O.; Bhattacharje, G.; Al-Khafaji, K.; Taskin-Tok, T.; Alfasane, Md. A.; Das, A. K.; Parvez, Md. A. K.; Rahman, Md. S. (2020) Combination of QSAR, Molecular Docking, Molecular Dynamic Simulation and MM-PBSA: Analogues of Lopinavir and Favipiravir as Potential Drug Candidates Against COVID-19. Journal of Biomolecular Structure and Dynamics, 1–20. https://doi.org/10.1080/07391102.2020.1850355.

58. Keretsu, S.; Bhujbal, S. P.; Cho, S. J. (2020) Rational Approach Toward COVID-19 Main Protease Inhibitors via Molecular Docking, Molecular Dynamics Simulation and Free Energy Calculation. Scientific Reports, 10 (1). https://doi.org/10.1038/s41598-020-74468-0.

59. Priya, R.; Sumitha, R.; Doss, C. G. P.; Rajasekaran, C.; Babu, S.; Seenivasan, R.; Siva, R. (2015) Molecular Docking and Molecular Dynamics to Identify a Novel Human Immunodeficiency Virus Inhibitor from Alkaloids of Toddalia Asiatica. Pharmacognosy Magazine, 11 (Suppl 3), S414–S422. https://doi.org/10.4103/0973-1296.168947.

60. Dong, L.; Shen, S.; Chen, W.; Xu, D.; Yang, Q.; Lu, H.; Zhang, J. (2019) Discovery of Novel Inhibitors Targeting Human O-GlcNAcase: Docking-Based Virtual Screening, Biological Evaluation, Structural Modification, and Molecular Dynamics Simulation. Journal of Chemical Information and Modeling, 59 (10), 4374–4382. https://doi.org/10.1021/acs.jcim.9b00479.

61. Al-Obaidi, A.; Elmezayen, A. D.; Yelekçi, K. (2020) Homology Modeling of Human GABA-at and Devise Some Novel and Potent Inhibitors via Computer-Aided Drug Design Techniques. Journal of Biomolecular Structure and Dynamics, 39 (11), 4100–4110. https://doi.org/10.1080/07391102.2020.1774417.

62. Jairajpuri, D. S.; Hussain, A.; Nasreen, K.; Mohammad, T.; Anjum, F.; Tabish Rehman, Md.; Mustafa Hasan, G.; Alajmi, M. F.; Imtaiyaz Hassan, Md. (2021) Identification of Natural Compounds as

Potent Inhibitors of SARS-CoV-2 Main Protease Using Combined Docking and Molecular Dynamics Simulations. Saudi Journal of Biological Sciences, 28 (4), 2423–2431. https://doi.org/10.1016/j. sjbs.2021.01.040.

63. Metwaly, A. M.; Elwan, A.; El-Attar, A.-A. M. M.; Al-Rashood, S. T.; Eissa, I. H. (2022) Structure-Based Virtual Screening, Docking, ADMET, Molecular Dynamics, and MM-PBSA Calculations for the Discovery of Potential Natural SARS-CoV-2 Helicase Inhibitors from the Traditional Chinese Medicine. Journal of Chemistry, 1–23. https://doi.org/10.1155/2022/7270094.

64. Mohamed, E. A. R.; Abdelwahab, S. F.; Alqaisi, A. M.; Nasr, A. M. S.; Hassan, H. A. (2022) Identification of Promising Anti-EBOV Inhibitors: De Novo Drug Design, Molecular Docking and Molecular Dynamics Studies. Royal Society Open Science, 9 (9). https://doi.org/10.1098/rsos.220369.

65. Bepari, A. K.; Reza, H. M. (2021) Identification of a Novel Inhibitor of SARS-CoV-2 3CL-Pro Through Virtual Screening and Molecular Dynamics Simulation. PeerJ, 9, e11261. https://doi.org/10.7717/peerj.11261.

66. Sharma, P.; Joshi, T.; Mathpal, S.; Joshi, T.; Pundir, H.; Chandra, S.; Tamta, S. (2020) Identification of Natural Inhibitors Against Mpro of SARS-CoV-2 by Molecular Docking, Molecular Dynamics Simulation, and MM/PBSA Methods. Journal of Biomolecular Structure and Dynamics, 1–12. https://doi.org/10.1080/07391102.2020.1842806.

67. Verma, S.; Patel, C. N.; Chandra, M. (2021) Identification of Novel Inhibitors of SARS-CoV -2 Main Protease (M Pro) from *Withania* sp. by Molecular Docking and Molecular Dynamics Simulation. Journal of Computational Chemistry, 42 (26), 1861–1872.

68. Edache, E. I.; Uzairu, A.; Shallangwa, G. A.; Mamza, P. A. (2021) Virtual Screening, Pharmacokinetics, and Molecular Dynamics Simulations Studies to Identify Potent Approved Drugs for Chlamydia Trachomatis Treatment. Future Journal of Pharmaceutical Sciences, 7 (1). https://doi.org/10.1186/s43094-021-00367-4.

16 Identification and Validation of Novel Signals of Drugs Using the FAERS Database

A Case Study

Sweta Roy, Vipin Bhati, Ruchika Sharma, Deepti Pandita, and Anoop Kumar

16.1 INTRODUCTION

According to the WHO, a safety signal is defined as "reported information on a possible causal relationship between an adverse event (AE) and a drug, of which the relationship is unknown or incompletely documented previously" [1]. The concept of "signal detection" is primarily applied to approved medications, especially in cases where a safety issue is emerging that was not previously identified during the premarketing stages or when a new pattern of a known adverse drug reaction (ADR) is observed, taking into account factors such as severity, high-risk sub-populations, and frequency [2]. Signal detection can take two main forms: one is "qualitative," involving a detailed evaluation of individual case safety reports (ICSRs), while the other is "quantitative," employing artificial intelligence (AI) tools and data mining techniques on real-world data sources such as clinical trials registries, electronic health records (eHR), biomedical data and literature and different pharmacological databases. Signals identified through quantitative or qualitative methods require additional validation and confirmation through clinical assessment [3].

The FDA Adverse Event Reporting System (FAERS) is a repository that houses data related to reports of adverse events and medication errors sent to the FDA. The main objective of creating this database is to support the monitoring of drugs and therapeutic biological products [4]. The FAERS database follows the international safety reporting guidelines established by the International Conference on Harmonisation (ICH E2B). It encompasses reports of adverse events, medication errors, and product quality complaints that have led to adverse events and were submitted to the FDA. The FAERS Public Dashboard aims to increase public accessibility to FAERS data, allowing individuals to search for information regarding human adverse events reported to the FDA, which comes from sources such as the pharmaceutical industry, healthcare professionals and consumers [5]. The FAERS Public Dashboard is an exceptionally user-friendly web-based tool that enables easy access to FAERS data. It's crucial to emphasize that the presence of reports in FAERS related to a specific drug or biologic doesn't necessarily imply that the drug or biologic was the cause of the adverse event. FAERS data, on their own, should not be considered a determinant of the safety profile of the drug or biologic [6].

16.2 CASE STUDIES

16.2.1 Case Study 1: Identification of HMG-CoA Associated Myasthenia Gravis

Myasthenia gravis: Myasthenia gravis (MG) is an autoimmune neuromuscular disorder in which the antibodies attack the post-synaptic membrane components, which leads to improper

DOI: 10.1201/9781032629940-16

neuromuscular transmission and causes weakness and fatigue of muscles, especially skeletal muscles [7]. Most of the cases of MG are associated with the formation of antibodies against acetylcholine receptor (AChR), muscle-specific kinase (MuSK), or the protein associated with MuSK such as agrin and low-density lipoprotein receptor-related protein 4 (LRP4). MuSK is an integral membrane protein found in the post-synaptic region, which plays a key role in the formation and development of the neuromuscular junction (NMJ) in combination with agrin and LRP4 [8].

HMG-CoA: Statins belong to the anti-hyperlipidemic class of drugs, which are widely prescribed or recommended to lower the blood cholesterol level. Currently, different statins are prescribed, such as atorvastatin, cerivastatin, fluvastatin, lovastatin, pitavastatin, pravastatin, rosuvastatin and simvastatin. Statins bind competitively with HMG-CoA reductase, which is an important rate-limiting enzyme in the mevalonic acid pathway. Blocking of the active sites prevents the conversion of HMG-CoA to mevalonic acid, and it reduces the synthesis of cholesterol in the liver [9].

16.2.1.1   Materials and Methods

16.2.1.1.1   Databases

In this study, we have used the FAERS database, which contains millions of ICSRs that consumers, healthcare professionals and drug manufacturing companies submit. All reported cases of MG linked to HMG-CoA inhibitors were extracted using OpenVigil2.1-MedDRA-v2 (2004Q1-2022Q3) database from the time duration of January 1, 2004, to September 30, 2022. In the section on the name of the drug, HMG-CoA inhibitors were used and "Myasthenia gravis" as the preferred term was used in the vent section of OpenVigil2.1.

16.2.1.1.2   Subgroup Analysis

The secondary analysis was conducted by categorizing data into subgroups. The categorization was carried out based on gender (male and female) and age, such as children, adolescents, adults and the elderly (0–11 years, 12–17 years, 18–64 years and 65+ years). Cases were additionally gathered based on the reported country of origin. Subgroup analysis was done using the different filters of OpenVigil2.1-MedDRA-v2 tool, such as age; we can select the different age groups by using the age filter. The same was done for the other subgroup analysis criteria like sex and reporter country region.

16.2.1.1.3   Sensitivity Analysis

Sensitivity analysis was performed to determine the effect of the concomitant medication on the results. For performing sensitivity analysis, we downloaded the raw data from OpenVigil2.1-MedDRA-v2 tool. Then we checked each concomitant drug for their listedness and whether they have MG as an adverse event or not by using the summary of product characteristic (SmPC) of the product. The cases of MG caused by concomitant drugs were excluded using the filter "AND NOT" in OpenVigil2.1-MedDRA-v2 tool. After excluding concomitant drug cases, we found the cases of MG caused by the drug of interest only.

16.2.1.1.4   Statistical Analysis

Disproportionality analysis was conducted to identify potential associations between HMG-CoA inhibitors and the MG. Data mining algorithms such as reporting odds ratio (ROR) and proportional reporting ratio (PRR), along with Chi-square with Yates correction, were used to perform the signal analysis. The statistical analysis relies on a contingency table with two rows and two columns, and all Data Mining Alogrithms (DMAs), i.e., PRR, ROR and Chi-squared values are calculated using a formula. PRR with associated Chi-square value is the important DMA which is used to measure the disproportionality and predefined threshold for statistical significance set as PRR of ≥ 2.0 along with Chi-squared test statistic of ≥ 4.0 and the total number of cases occurrence should be more than 3 ($n \geq 3$).

16.2.1.2 Results

A total number (2310) of cases of MG were reported. Out of these, 254 were found to be associated with HMG-CoA inhibitors.

16.2.1.2.1 Cases of Myasthenia Gravis with HMG-CoA Inhibitors

The number of cases of MG found with HMG-CoA inhibitors was 254, as compiled in Table 16.1. The highest number of cases of MG was found with simvastatin (78 cases) followed by atorvastatin (74 cases), rosuvastatin (58 cases), pravastatin (28 cases), fluvastatin (8 cases), lovastatin (6 cases) and pitavastatin (2 cases) as compiled in Table 16.1.

Association of Myasthenia Gravis with HMG-CoA Inhibitors

Among all the HMG-CoA inhibitors, fluvastatin, pravastatin, rosuvastatin and simvastatin were found to be associated with MG as indicated by values of PRR with Chi-square and ROR as compiled in Table 16.1.

On the basis of PRR with Chi-square and ROR values, it was found that fluvastatin has shown strong association with MG with [PRR 10.455 (5.227; 20.916)] Chi-square value 59.085 and ROR value 10.477 (5.229; 20.992) followed by pitavastatin and rosuvastatin shown in Table 16.1.

16.2.1.2.2 Subgroup Analysis

Subgroup analysis was performed with HMG-CoA inhibitor drugs to know the association of MG in males or females and different age groups as well as on the basis of reporter country. The results of the subgroup analysis are discussed below.

16.2.1.2.2.1 Number of Cases Based on Gender

The number of cases of MG with HMG-CoA inhibitors was also categorized based on gender (Table 16.2). The highest number of cases of MG in males was observed with simvastatin (59) followed by atorvastatin (47), rosuvastatin (33) and pravastatin (21), whereas, in females, the highest number of cases was observed with atorvastatin (19), rosuvastatin (14) and simvastatin (14).

TABLE 16.1

Total Number of Myasthenia Gravis Cases Associated with HMG-CoA Reductase Inhibitors (Statins)

Drug	N	Chi-Squared with Yates' Correction	Proportional Reporting Ratio (PRR) and 95% Confidence Interval	Reporting Odds Ratio (ROR) and 95% Confidence Interval
Atorvastatin	74	41.742	2.128 (1.688; 2.683)	2.129 (1.688; 2.683)
Fluvastatin	8	59.085	10.455 (5.227; 20.916)	10.477 (5.229; 20.992)
Lovastatin	6	3.345	2.338 (1.05; 5.209)	2.339 (1.049; 5.213)
Pitavastatin	2	3.007	4.985 (1.247; 19.929)	4.989 (1.246; 19.977)
Pravastatin	28	50.66	3.651 (2.516; 5.299)	3.653 (2.516; 5.304)
Rosuvastatin	58	126.306	4.038 (3.112; 5.24)	4.041 (3.113; 5.245)
Simvastatin	78	95.459	2.953 (2.356; 3.701)	2.954 (2.357; 3.703)

TABLE 16.2

Subgroup Analysis on the Basis of Gender and Age Group

Drug Class	Drug	Characteristics	Myasthenia Gravis		Chi-Squared with Yates' Correction	Proportional Reporting Ratio (PRR) and 95% Confidence Interval	Reporting Odds Ratio (ROR) and 95% Confidence Interval
			Number of Cases	Total Cases			
3-Hydroxy 3-methylglutaryl coenzyme A (HMG-CoA) reductase inhibitors	Atorvastatin	Male	47	967	35.467	2.408 (1.796; 3.228)	2.409 (1.797; 3.23)
		Female	19	1,034	1.901	1.418 (0.901; 2.232)	1.418 (0.901; 2.233)
		0–11	0	23	16.041	0.0	0.0
		12–17	0	20	16.073	0.0	0.0
		18–64	27	665	31.644	2.955 (2.011; 4.343)	2.957 (2.011; 4.346)
		65 years and above	32	749	1.905	1.306 (0.917; 1.861)	1.306 (0.917; 1.861)
	Fluvastatin	Male	5	967	38.995	11.794 (4.905; 28.363)	11.828 (4.905; 28.522)
		Female	2	1,034	4.677	6.562 (1.64; 26.25)	6.568 (1.639; 26.321)
		0–11	0	23	173.812	0.0	0.0
		12–17	0	20	322.896	0.0	0.0
		18–64	6	665	112.094	24.566 (11.018; 54.773)	24.684 (11.027; 55.255)
		65 years and above	2	749	1.659	3.66 (0.914; 14.646)	3.663 (0.914; 14.69)
	Lovastatin	Male	6	967	12.982	4.509 (2.022; 10.057)	4.513 (2.022; 10.076)
		Female	0	1,034	0.36	0.0	0.0
		0–11	0	23	173.812	0.0	0.0
		12–17	0	20	46.647	0.0	0.0
		18–64	5	665	21.652	7.443 (3.09; 17.931)	7.453 (3.09; 17.978)
		65 years and above	0	749	0.845	0.0	0.0
	Pitavastatin	Male	2	967	8.564	10.119 (2.532; 40.442)	10.143 (2.528; 40.692)
		Female	0	1,034	0.715	0.0	0.0
		0–11	0	23	960.408	0.0	0.0
		12–17	0	20	0.0	0.0	0.0
		18–64	0	665	1.367	0.0	0.0
		65 years and above	2	749	4.888	6.76 (1.691; 27.031)	6.775 (1.688; 27.192)

(Continued)

TABLE 16.2 *(Continued)*

Subgroup Analysis on the Basis of Gender and Age Group

Drug Class	Drug	Characteristics	Myasthenia Gravis		Chi-Squared with Yates' Correction	Proportional Reporting Ratio (PRR) and 95% Confidence Interval	Reporting Odds Ratio (ROR) and 95% Confidence Interval
			Number of Cases	Total Cases			
	Pravastatin	Male	21	967	67.912	5.32 (3.454; 8.196)	5.326 (3.455 ; 8.21)
		Female	3	1,034	0.008	0.898 (0.289; 2.788)	0.898 (0.289 ; 2.788)
		0–11	0	23	42.716	0.0	0.0
		12–17	0	20	49.624	0.0	0.0
		18–64	14	665	66.779	7.164 (4.221; 12.16)	7.173 (4.223 ; 12.185)
		65 years and above	9	749	1.582	1.63 (0.845; 3.144)	1.63 (0.845; 3.146)
	Rosuvastatin	Male	33	967	78.265	4.324 (3.056; 6.118)	4.328 (3.058; 6.125)
		Female	14	1,034	8.967	2.306 (1.361; 3.908)	2.307 (1.361 ; 3.91)
		0–11	0	23	31.608	0.0	0.0
		12–17	0	20	31.408	0.0	0.0
		18–64	18	665	37.688	4.075 (2.552; 6.508)	4.078 (2.552 ; 6.515)
		65 years and above	28	749	34.8	3.034 (2.08; 4.424)	3.036 (2.081 ; 4.429)
	Simvastatin	Male	59	967	115.903	3.883 (2.985; 5.053)	3.886 (2.986 ; 5.058)
		Female	14	1,034	0.896	1.344 (0.793; 2.277)	1.344 (0.793; 2.277)
		0–11	0	23	15.183	0.0	0.0
		12–17	0	20	22.492	0.0	0.0
		18–64	40	665	146.682	5.852 (4.251; 8.054)	5.857 (4.254 ; 8.065)
		65 years and above	26	749	1.743	1.33 (0.9; 1.967)	.33 (0.899 ; 1.967)

Association of HMG-CoA Inhibitors with Myasthenia Gravis According to Gender
The PRR, along with the Chi-square value and ROR, was calculated for all HMG-CoA inhibitors drugs as compiled in Table 16.2. Among all drugs, fluvastatin and pravastatin have shown significant association with MG in males, whereas, in females, no significant association was found in MG and HMG-CoA inhibitors maculopathy as indicated by their ROR, PRR with Chi-square values as compiled in Table 16.2.

16.2.1.2.2.2   Number of Cases in Different Age Groups   The number of cases of MG with all HMG-CoA inhibitor drugs was also categorized based on different age groups (Table 16.2). The highest number of cases of MG in the age group 18–64 years was observed with simvastatin (40) followed by atorvastatin (27) and rosuvastatin (18), whereas, in the age group > 65 years, the highest number of cases was observed with atorvastatin (32) followed by rosuvastatin (28) and simvastatin (28).

Association of HMG-CoA Inhibitors with Myasthenia Gravis Different Age Groups
Among all drugs, fluvastatin has shown a significant association with MG in the 18–64 years age group, whereas in the > 65 years age group, rosuvastatin has shown a significant association. Apart from this no age group shows a significant association as indicated by its ROR, PRR with Chi-square values as compiled in Table 16.2.

16.2.1.2.3   Subgroup Analysis on the Basis of Reporter Country
Subgroup analysis was also done on the basis of the reporter's country. Results are shown in Table 16.3.

TABLE 16.3
Subgroup Analysis on the Basis of Reporter Country

Drug Class	Drug	Reporter Country	Myasthenia Gravis	
			Number of Cases	Total Cases
3-Hydroxy 3-methylglutaryl coenzyme A (HMG-CoA) reductase inhibitors	Atorvastatin	Africa	0	6
		Asia	8	437
		Europe	28	496
		North America	28	1,152
		Oceania	3	33
		South America	0	46
	Fluvastatin	Africa	0	6
		Asia	3	437
		Europe	1	496
		North America	3	1,152
		Oceania	0	33
		South America	0	46
	Lovastatin	Africa	0	6
		Asia	0	437
		Europe	0	496
		North America	5	1,152
		Oceania	0	33
		South America	0	46

(Continued)

TABLE 16.3 *(Continued)*

Subgroup Analysis on the Basis of Reporter Country

Drug Class	Drug	Reporter Country	Myasthenia Gravis	
			Number of Cases	Total Cases
	Pitavastatin	Africa	0	6
		Asia	2	437
		Europe	0	496
		North America	0	1,152
		Oceania	0	33
		South America	0	46
	Pravastatin	Africa	0	6
		Asia	3	437
		Europe	9	496
		North America	13	1,152
		Oceania	1	33
		South America	0	46
	Rosuvastatin	Africa	0	6
		Asia	5	437
		Europe	18	496
		North America	27	1,152
		Oceania	2	33
		South America	2	46
	Simvastatin	Africa	0	6
		Asia	7	437
		Europe	38	496
		North America	29	1,152
		Oceania	0	33
		South America	0	46

16.2.1.2.4   Sensitivity Analysis

Sensitivity analysis is a key part of the disproportionality analysis. Sensitivity analysis was done using OpenVigil2.1-MedDRA-v2 tool, and out of seven drugs of interest (atorvastatin, fluvastatin, lovastatin, pitavastatin, pravastatin, rosuvastatin and simvastatin), we have found that cases decreased from 74 to 73 for atorvastatin, followed by rosuvastatin (58–57) and simvastatin (78–73). After sensitivity analysis of atorvastatin, rosuvastatin and simvastatin, positive signal was found with rosuvastatin and simvastatin with MG indicated by PRR, Chi-square with Yates correction, and ROR values as shown in Tables 16.4 and 16.5.

TABLE 16.4

Number of Cases after Removal of Cases of Concomitant Drugs

Drug, Total Cases	Without Drug, Cases	Final Cases
Atorvastatin, 74	Valium, 73	73
Fluvastatin, 8	–	8
Lovastatin, 6	–	6
Pitavastatin, 2	–	2
Pravastatin, 28	–	28
Rosuvastatin, 58	Levofloxacin, 57	57
Simvastatin, 78	Valium, 73	73

TABLE 16.5
Sensitivity Analysis Results

Drug of Interest	Concomitant	Chi-Squared with Yates' Correction	Proportional Reporting Ratio (PRR) and 95% Confidence Interval	Reporting Odds Ratio (ROR) and 95% Confidence Interval
Atorvastatin	Without Valium	41.099	2.127 (1.684; 2.685)	2.127 (1.685; 2.686)
Fluvastatin	–	–	–	–
Lovastatin	–	–	–	–
Pitavastatin	–	–	–	–
Pravastatin				
Rosuvastatin	Without Levofloxacin	122.557	4.007 (3.081; 5.212)	4.01 (3.083; 5.216)
Simvastatin	Without Valium	98.226	3.014 (2.402; 3.782)	3.015 (2.402; 3.784)

16.2.2 CASE STUDY 2: ASSOCIATION OF PROPIONIC ACID DERIVATIVE WITH DRUG REACTION WITH EOSINOPHILIA AND SYSTEMIC SYMPTOMS: A DISPROPORTIONALITY ANALYSIS

***Propionic acid derivative*:** Naproxen, flurbiprofen and ketoprofen are propionic acid derivatives which are nonsteroidal anti-inflammatory drugs (NSAIDs). These drugs work by inhibiting cyclo-oxygenase (COX) enzymes, which are responsible for the production of prostaglandins. COX-1: A COX enzyme is responsible for the synthesis of prostaglandins that protect the stomach and kidneys. COX-2: A COX enzyme is involved in the production of prostaglandins that promote inflammation. Prostaglandins: A group of lipid molecules that play a role in a variety of bodily functions, including pain, inflammation, and blood clotting. Edema: The accumulation of fluid in tissues, which can cause swelling. Hypersensitivity reaction: An allergic reaction that can cause a variety of symptoms, including hives, rash and difficulty breathing. Naproxen, flurbiprofen and ketoprofen are propionic acid derivatives; all have analgesic, antipyretic and anti-inflammatory properties. They are generally well-tolerated, but ADRs can include headache, dizziness, gastrointestinal upset, somnolence, abdominal discomfort, diarrhea, nausea, and hypersensitivity reactions to edema [10].

The term Drug Reaction with Eosinophilia and Systemic Symptoms (DRESS) syndrome was coined by Bocquet and later, due to the variety of skin eruptions, the R was changed from RASH to REACTION. Over the past 80 years, the terms for the condition have changed from drug-induced pseudolymphoma to drug-induced delayed multiorgan hypersensitivity syndrome to drug-induced delayed hypersensitivity syndrome (DIHS). DRESS syndrome is categorized as severe cutaneous adverse reactions (SCARs) [2].

DRESS syndrome often manifests within two months after the offending drug's consumption, most frequently two to six weeks following the first usage. On the other hand, re-exposure may cause symptoms to manifest more quickly and worsen. The most typical symptom is a morbilliform cutaneous eruption accompanied by lymphadenopathy and fever. It is widely recognized that DRESS can occur due to immunological reactions that have been observed in less number of patients characterized by eosinophilia which is responsible for influencing the lymphocytic response. It can be possible that a delayed mediated immune response occurs in DRESS syndrome that is quite plausible because it needs to be sensitized and is repeatable through skin tests. In the DRESS syndrome, levels of several pro-inflammatory cytokines are increased, including tumor necrosis factor and interleukin-6 [11]. Several organ systems that could be adversely affected include gastrointestinal, hematologic, hepatic, renal, pulmonary, cardiac, neurologic, and endocrine systems. Overall, 10% of those who get this syndrome pass away; fulminant hepatitis with hepatic necrosis is the most common type [12]. It has been estimated that between 1 in 1,000 and 1 in 10,000 members of the population are at risk of drug exposure. Reactivation of the Human Herpes Virus (HHV) and the corresponding immune response in the host are associated with the systemic manifestations of

DRESS. As approximately 60%–80% of DRESS syndrome patients have HHV-6 detected in their blood, HHV-6 reactivation has been included in the diagnostic criteria for DRESS syndrome established by Japanese experts.

Additionally, DRESS syndrome can also be aggregated due to the reactivation of Epstein-Barr virus (EBV), HHV-7 and cytomegalovirus (CMV). There were two theories that have been proposed to explain the role of the virus. Firstly, cytokine storm was observed due to the immunological reaction against the medication; secondly, viral reactivation is involved in the majority of DRESS syndrome symptoms [13]. The role of virus activation in the pathophysiology of DRESS remains unclear despite the potential diagnostic utility of detecting HHV reactivation in DRESS.

16.2.2.1 Methods

16.2.2.1.1 Database

Database U.S. FAERS was used to collect the data on adverse effects of DRESS syndrome linked with naproxen, flurbiprofen and ketoprofen. The data was extracted from the OpenVigil2.1-MedDRA-v24 (2004Q1-2022Q3). We have used naproxen, flurbiprofen and ketoprofen as drug names and "Drug reaction with eosinophilia and systemic symptoms" as an adverse effect preferred term in "Search Window" of Openvigil 2.1. Statistical analysis of the alliance connecting naproxen, flurbiprofen, ketoprofen and ADRs was determined by using DMA such as ROR with a 95% confidence interval and PRR with associated Chi-square value. A PRR ≥ 2 with linked Chi-square value ≥ 4, ROR ≥ 2, and count of co-occurrence ≥ 3 is considered a possible signal. Listedness was performed for each drug by using SmPC. The analysis was done with the help of OpenVigil2.1-MedDRA.

16.2.2.1.2 Subgroup Analysis

These were performed on the basis of gender, which is categorized as male and female, age groups according to WHO child (0–11 years), adolescents (11–17 years), adult (18–64 years) and elder (65 years above) and reported continents.

16.2.2.1.3 Sensitivity Analysis

Sensitivity analyses for flurbiprofen were performed to investigate the effects of concomitant drugs (acetaminophen and ibuprofen), which outcomes in DRESS. Sensitivity analysis was performed by removing the drug, which is already known for DRESS syndrome. For ketoprofen, sensitivity analysis was also performed to inspect the effects of other drugs (esomeprazole, furosemide, spironolactone, acetaminophen and ibuprofen), which outcomes in DRESS syndrome.

16.2.2.2 Results

A total of 13,888 instances of DRESS were reported. A total of 112 for naproxen, 32 flurbiprofen and 30 ketoprofen out of 13,888 were discovered as shown in Tables 16.6 and 16.7.

The PRR and ROR, along with Chi-square with Yates' correction, reveal that they are not associated with DRESS syndrome and naproxen, flurbiprofen and ketoprofen as shown in Table 16.8.

TABLE 16.6

Cases of DRESS Syndrome with Naproxen and Flurbiprofen

Groups	Naproxen	Other Drugs	Total	Flurbiprofen	Other Drugs	Total
DRESS Syndrome	115	13,803	13,918	32	13,886	13,918
Other events	77,567	10,408,824	10,486,391	813	10,485,578	10,486,391
Total	77,682	10,422,627	10,500,309	845	10,499,464	10,500,309

TABLE 16.7

Cases of DRESS Syndrome with Ketoprofen

Ketoprofen	Other Drugs	Total
30	13,888	13,918
4551	10,481,840	10,486,391
4581	10,495,728	10,500,309

TABLE 16.8

NSAIDs with Calculated PRR, ROR, and Chi-Square Values

S. No.	Drugs	Chi-Squared with Yates' Correction (Chi-Squared Value > 4)	Proportional Reporting Ratio (PRR) and 95% Confidence Interval (PRR ≥ 2)	Reporting Odds Ratio (ROR) and 95% Confidence Interval (ROR ≥ 2)
1.	Naproxen	1.303	1.118 (0.931; 1.343)	1.118 (0.93; 1.343)
2.	Flurbiprofen	825.19	28.634 (20.375; 40.24)	29.722 (20.869; 42.33)
3.	Ketoprofen	90.552	4.949 (3.463; 7.073)	4.975 (3.473; 7.127)

16.2.2.2.1 Subgroup Analysis

Subgroup analysis was done on the basis of reporter region, age and gender. There were a total number of 32 DRESS associated with flurbiprofen. More cases were from Europe (93.75%), followed by Asia (3.125%). According to age group, more cases were found in females as compared to males. In the age group, more cases were found in adults (18–64 years) as compared to other age groups (Tables 16.7 and 16.8). A total of 30 reports of DRESS-associated ketoprofen were found. Over of cases were from Europe (73.333%) followed by South America (26.666%).

TABLE 16.9

Subgroup Analysis and Their DMA Values of Ketoprofen

Drug	Characteristics	DRESS Syndrome Number of Cases	DRESS Syndrome Total Cases	Chi-Squared with Yates' Correction (Chi-Squared Value > 4)	Proportional Reporting Ratio (PRR) and 95% Confidence Interval (PRR ≥ 2)	Reporting Odds Ratio (ROR) and 95% Confidence Interval (ROR ≥ 2)
Flurbiprofen	Male	1	6106	0.033	2.585 (0.365; 18.279)	2.591 (0.363; 18.473)
	Female	29	7018	1152.302	43.274 (30.335; 61.732)	45.565 (31.334; 66.259)
	0–11 years	1	768	1.559	9.811 (1.436; 67.036)	10.178 (1.375; 75.334)
	12–17 years	0	781	0.102	0.0 (;)	0.0 (;)
	18–64 years	28	7865	787.409	31.194 (21.813; 44.61)	33.498 (22.796; 49.225)
	65 years and above	1	3288	0.138	3.359 (0.475; 23.725)	3.372 (0.472; 24.07)
Ketoprofen	Male	13	5722	30.715	4.391 (2.553; 7.553)	4.415 (2.557; 7.622)
	Female	15	6553	48.105	5.303 (3.2; 8.788)	5.33 (3.206; 8.861)
	0–11 years	0	719	0.67	0.0 (;)	0.0 (;)
	12–17 years	0	729	0.195	0.0 (;)	0.0 (;)
	18–64 years	28	7417	82.55	4.885 (3.378; 7.066)	4.929 (3.394; 7.158)
	65 years and above	0	3003	0.902	0.0 (;)	0.0 (;)

TABLE 16.10

Subgroup Analysis on the Basis of Country and Their DMA Values

Drug	Country	No. of Cases
Flurbiprofen	Asia	1
	Africa	0
	North America	0
	Europe	30
	South America	0
	Oceania	0
	US minor outlying islands	0
Ketoprofen	Asia	0
	Africa	0
	North America	1
	Europe	22
	South America	8
	Oceania	0
	US minor outlying islands	0

TABLE 16.11

Sensitivity Analysis and Their DMA Values

Drug Class	Drugs	Chi-Squared with Yates' Correction (Chi-Squared Value > 4)	Proportional Reporting Ratio (PRR) and 95% Confidence Interval (PRR ≥ 2)	Reporting Odds Ratio (ROR) and 95% Confidence Interval (ROR ≥ 2)
Propionic acid derivative	Flurbiprofen (without ibuprofen, acetaminophen)	3.261	4.657 (3.09; 7.019)	3.571 (1.149; 11.105)
	Ketoprofen (without acetaminophen, esomeprazole, furosemide, spironolactone and ibuprofen)	61.95	4.634 (3.083; 6.968)	3.559 (1.151; 11.008)

Cases were also analyzed as per age group. The majority of ketoprofen-related DRESS cases were found in 18–64 age groups (Table 16.7). After excluding the unspecified/unknown data, females are more affected than males (15; 50% vs. 13; 43.333%) (Tables 16.9 and 16.10).

16.2.2.2.2   Sensitivity Analysis

Sensitivity analysis was performed for flurbiprofen and ketoprofen. We identified DRESS syndrome that is not associated with flurbiprofen and association with ketoprofen shown in Table 16.11.

16.3   CONCLUSION

We have identified a novel signal of MG with HMG-CoA inhibitors as well as DRESS as a novel signal associated with propionic acid derivatives.

REFERENCES

1. Sharma, A., Roy, S., Sharma, R., Kumar, A., 2023. Association of Antiviral Drugs and Their Possible Mechanisms with DRESS Syndrome Using Data Mining Algorithms. J. Med. Virol. 95. https://doi.org/10.1002/jmv.28671

2. Sharma, A., Kumar, A., 2022. Identification of Novel Signal of Clobazam-Associated Drug Reaction with Eosinophilia and Systemic Symptoms Syndrome: A Disproportionality Analysis. Acta Neurol. Scand. 146, 623–627. https://doi.org/10.1111/ane.13690

3. Berk, M., Otmar, R., Dean, O., Berk, L., Michalak, E., 2015. The Use of Mixed Methods in Drug Discovery, in: Clinical Trial Design Challenges in Mood Disorders. Elsevier, pp. 59–74. https://doi.org/10.1016/B978-0-12-405170-6.00006-3

4. Javed, F., Kumar, A., 2023. Identification of Signal of Clindamycin Associated Renal Failure Acute: A Disproportionality Analysis. Curr. Drug Saf. 18. https://doi.org/10.2174/1574886318666230228142856

5. Jain, D., Sharma, G., Kumar, A., 2023. Adverse Effects of Proton Pump Inhibitors (PPIs) on the Renal System Using Data Mining Algorithms (DMAs). Expert Opin. Drug Saf. 22, 741–752. https://doi.org/10.1080/14740338.2023.2189698

6. Blau, J.E., Tella, S.H., Taylor, S.I., Rother, K.I., 2017. Ketoacidosis Associated with SGLT2 Inhibitor Treatment: Analysis of FAERS Data. Diabetes. Metab. Res. Rev. 33, e2924. https://doi.org/10.1002/dmrr.2924

7. Drachman, D.B., 1983. Myasthenia Gravis: Immunobiology of a Receptor Disorder. Trends Neurosci. 6, 446–451. https://doi.org/10.1016/0166-2236(83)90216-3

8. Wang, S., Breskovska, I., Gandhy, S., Punga, A.R., Guptill, J.T., Kaminski, H.J., 2018. Advances in Autoimmune Myasthenia Gravis Management. Expert Rev. Neurother. 18, 573–588. https://doi.org/10.1080/14737175.2018.1491310

9. Markowska, A., Antoszczak, M., Markowska, J., Huczyński, A., 2020. Statins: HMG-CoA Reductase Inhibitors as Potential Anticancer Agents against Malignant Neoplasms in Women. Pharmaceuticals 13, 422. https://doi.org/10.3390/ph13120422

10. Kaur, B., Singh, P., 2022. Inflammation: Biochemistry, Cellular Targets, Anti-Inflammatory Agents and Challenges with Special Emphasis on Cyclooxygenase-2. Bioorg. Chem. 121, 105663. https://doi.org/10.1016/j.bioorg.2022.105663

11. Cacoub, P., Musette, P., Descamps, V., Meyer, O., Speirs, C., Finzi, L., Roujeau, J.C., 2011. The DRESS Syndrome: A Literature Review. Am. J. Med. 124, 588–597. https://doi.org/10.1016/j.amjmed.2011.01.017

12. Husain, Z., Reddy, B.Y., Schwartz, R.A., 2013. DRESS Syndrome. J. Am. Acad. Dermatol. 68, 693. e1–693.e14. https://doi.org/10.1016/j.jaad.2013.01.033

13. Shiohara, T., Kano, Y., 2016. Drug Reaction with Eosinophilia and Systemic Symptoms (DRESS): Incidence, Pathogenesis and Management. Expert Opin. Drug Saf. 1–9. https://doi.org/10.1080/14740338.2017.1270940

17 Molecular Docking and Molecular Dynamics in Signal Analysis

A Case Study

Sweta Roy, Ruchika Sharma, and Anoop Kumar

17.1 INTRODUCTION

Molecular docking is a widely employed technique in structure-based drug design, crucial for predicting how one molecule will align with another when they combine to create a stable complex. This method is highly favoured because it can anticipate the binding configuration of a small molecule ligand to the correct binding site on its target. Utilizing these methods, it becomes possible to perform an early-stage screening of molecules for their biological activity during drug development [1]. Molecular docking typically involves the interaction between a small molecule and a larger target macromolecule. It's commonly known as ligand-protein docking [2]. Molecular docking finds extensive utility and numerous applications in the realm of drug discovery. These applications encompass structure-activity investigations, lead development, virtual screening to identify promising leads, offering binding hypotheses to support predictions in mutagenesis research, aiding in the fitting of substrates and inhibitors to electron density in X-ray crystallography, conducting chemical mechanism analyses, and contributing to the design of combinatorial libraries [3].

Molecular dynamics (MDs) simulations predict how every atom in a protein or other molecular system will move over time, based on a general model of the physics governing interatomic interactions [4]. These simulations may capture a wide range of key biomolecular events, such as protein folding, conformational change, and ligand binding displaying all atom locations at femtosecond temporal precision. It can also be used to check the stability and accuracy of the designed structure [5].

17.2 CASE STUDY 1

This case study examines association of propionic acid derivative with drug reaction with eosinophilia and systemic symptoms: molecular docking and MD approaches.

17.2.1 Materials and Methods

17.2.1.1 Selection and Preparation of Target Proteins

On the basis of a literature research, we chose a variety of targets that were involved in DRESS syndrome human leukocyte antigen – HLA*B5801 (PDB ID: 5VWH), HLA*B1502 (PDB ID: 6UZM), and HLA*B5701 (3VH8). These target three-dimensional (3D) structures have been downloaded from the Protein Data Bank (PDB) at RCSB (http://www.rcsb.org/pdb/home/home.do). Downloaded structures in PDB format were imported into the Maestro programme (Schrodinger, LLC, Cambridge, USA). Using the protein preparation wizard in Maestro software (Schrodinger, LLC, Cambridge, USA), these proteins were further processed and improved. The receptor grid

DOI: 10.1201/9781032629940-17

generating job of Maestro software (Schrodinger, LLC, Cambridge, USA) was used to generate the grid for these binding pockets after using SiteMap to identify the protein's binding pocket.

17.2.2 PREPARATION OF LIGANDS

Ketoprofen's 3D structure was downloaded in .sdf format from PubChem. This format was imported into the Maestro software (Schrodinger, LLC, Cambridge, USA). Ketoprofen was prepared by utilizing LigPrep module of the software. It provides the ligand with a precise and energy-minimized structure. The force field was maintained in OPLS_2005, and throughout the ligand preparation process, the ionization state was maintained as neutral. After ligand preparation conformer of ketoprofen was selected for molecular docking.

17.2.3 MOLECULAR DOCKING

Molecular docking is the term used to describe computer-based modelling of a potential ligand interacting with a receptor or protein. Ketoprofen binding affinity to various DRESS syndrome targets was tested using molecular docking. Molecular docking was done by using glide module of the Maestro software (Schrodinger, LLC, Cambridge, USA). We used extra precision (XP) method and sampling type as it is flexible for doing molecular docking of ketoprofen against the target proteins of human leukocyte antigen – HLA*B5801 (PDB ID: 5VWH), HLA*B1502 (PDB ID: 6UZM), and HLA*B5701 (3VH8).

17.2.4 MOLECULAR DYNAMICS

17.2.4.1 Molecular Dynamics (MD) Simulations

MDs simulation is employed to investigate the characteristics of the protein, monitor internal molecular transformations within the protein, and assess the enduring stability of the protein-ligand complex over time. In this research study the MDs were done by the Desmond module (Schrodinger, LLC, Cambridge, USA) for 100 ns. The MD simulation's output file contains a simulation interaction diagram (SID), encompassing metrics such as root mean square deviation (RMSD), root mean square fluctuation (RMSF), and protein-ligand contact analysis. These metrics are employed to assess the stability of the protein-ligand complex.

17.2.5 RESULTS

17.2.5.1 Molecular Docking

The interaction of propionic acid derivative (ketoprofen) was checked towards HLA*B5801, HLA*B1502, and HLA*B1502 as described below.

17.2.6 EXTRAPOLATION OF MECHANISMS OF DRESS SYNDROME

Propionic acid derivative (ketoprofen) has shown good interaction through the active site of HLA*B5801 (PDB ID: 5VWH), HLA*B1502 (PDB ID: 6UZM), and HLA*B5701 (3VH8), as shown by the dock score in Table 17.1. The binding energy and interaction between protein targets and drug molecules are shown in Figures 17.1–17.3.

17.2.7 MOLECULAR DYNAMICS STUDIES

The MD results of ligand ketoprofen in complex with HLA-B*58:01 (5VWH) were carried out for 100 ns to determine the stability of proteins as well as ligand-protein complex.

TABLE 17.1

Molecular Docking Study of Ketoprofen against Human Leucocyte Antigen Targets

Target Name	Compound Name	H-Bonding	Hydrophobic Contact	Dock Score
5VWH	Ketoprofen		TYR74 ILE80 ALA81 TYR84 ILE95 ALA117 TYR118 TYR123	−7.657
6UZM	Ketoprofen	TYR84 THR143	TYR74 LEU81 TYR84 ILE95 TYR123 LEU156	−5.471
3VH8	Ketoprofen	ASN77	TYR9 TYR74 VAL97 TYR99 TRP133 TRP47 VAL157 LEU156 TYR159	−4.348

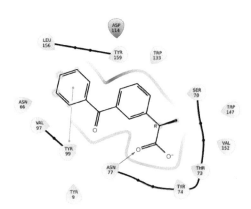

FIGURE 17.1   Interactions of ketoprofen towards HLA*B5801 (PDB ID: 5VWH).

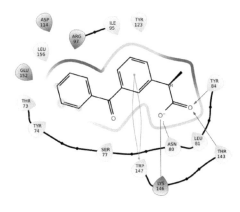

FIGURE 17.2   Interactions of ketoprofen towards HLA*B1502 (PDB ID: 6UZM).

17.2.8   ROOT MEAN SQUARE DEVIATION ANALYSIS OF THE PROTEIN

The RMSD serves as a crucial parameter for evaluating the stability, dynamic characteristics, and behaviour of the protein-ligand complex. Figure 17.4 illustrates the graph depicting the progression of protein RMSD, represented on the left Y-axis. When the protein's RMSD falls within the 1–3 Å range, it indicates the protein's stability and minimal variations in its backbone structure.

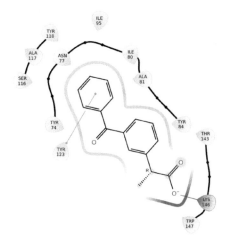

FIGURE 17.3   Interactions of ketoprofen towards HLA*B5701 (3VH8).

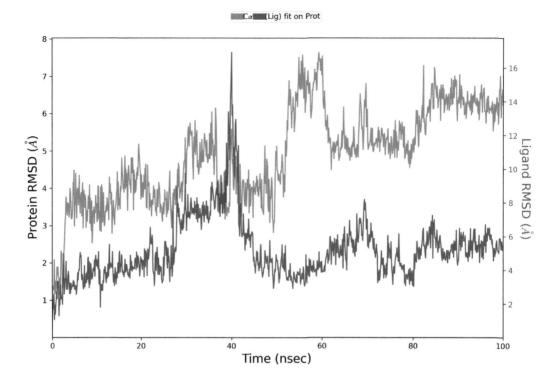

FIGURE 17.4 Structural dynamics deviation (RMSD) of protein HLA*B5801 (PDB ID: 5VWH) upon binding with ketoprofen.

The simulated complexes of the ketoprofen with HLA-B*58:01 have shown significant stability throughout the simulation having RMSD values around 8 Å, respectively. There is no stability found between the ligand and HLA-B*58:01 throughout the simulation.

17.2.9 ROOT MEAN SQUARE FLUCTUATION ANALYSIS OF THE PROTEIN

The protein's structural flexibility can be examined through RMSF, a valuable tool for detecting localized alterations within the protein sequence. Figure 17.5 highlights the regions of the protein that exhibit the greatest degree of movement during simulations, as indicated by the peaks. Region of 10–20 number amino acids has shown highest fluctuation followed by the region of 180–200 amino acids demonstrated in Figure 17.5.

17.2.10 PROTEIN-LIGAND CONTACTS

After the simulation run, a separate analysis was conducted for each system, examining the per-residue interactions by calculating the mean interaction occupancies of the binding site residues called protein-ligand contact. There is significant hydrogen bonding between ketoprofen and HLA-B*58:01 observed with LYS_143 and LYS_147, as demonstrated in Figure 17.6.

17.3 CASE STUDY 2: INTERACTION OF HMG-CoA INHIBITORS WITH MYASTHENIA GRAVIS

Myasthenia gravis (MG) is an autoimmune neuromuscular disorder in which the antibodies attack on the post-synaptic membrane components which leads to the improper neuromuscular

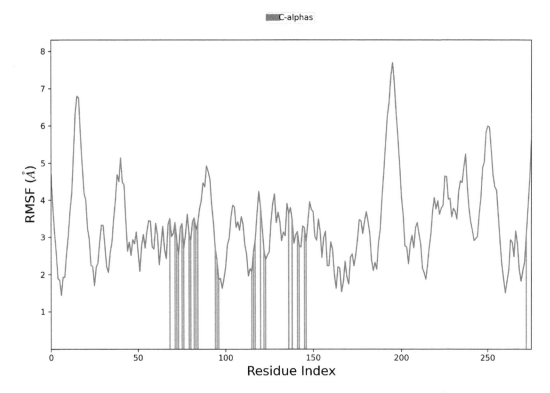

FIGURE 17.5 Backbone residual fluctuations (RMSF) plot of protein HLA*B5801 (PDB ID: 5VWH) and its complex with ketoprofen.

transmission and causes weakness and fatigue of muscles, especially skeletal muscles [6]. Most of the cases of MG are associated with the formation of antibodies against acetylcholine receptor (AChR), muscle-specific kinase (MuSK), or the protein associated to MuSK such as agrin and low-density lipoprotein receptor-related protein 4 (LRP4). MuSK is an integral membrane protein found in post-synaptic region, which plays a key role in the formation and development of the NMJ in combination with agrin and LRP4 [7].

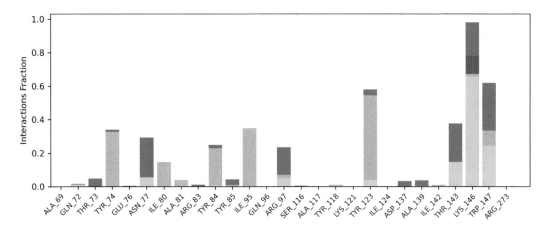

FIGURE 17.6 Histogram presentation of per residue analysis of ketoprofen in complex with HLA*B5801 (PDB ID: 5VWH) protein.

HMG-CoA inhibitors: Statins belong to the anti-hyperlipidemic class of drugs which are widely prescribed or recommended to lower the blood cholesterol level [8]. Currently, different statins are prescribed such as atorvastatin, cerivastatin, fluvastatin, lovastatin, pitavastatin, pravastatin, rosuvastatin, and simvastatin. Statins bind competitively with HMG-CoA reductase, which is an important rate limiting enzyme in mevalonic acid pathway. Blocking of the active sites prevents the conversion of HMG-CoA to mevalonic acid, and it reduces the synthesis of cholesterol in the liver [9].

17.3.1 Materials and Methods

17.3.1.1 Selection and Preparation of Target Proteins

The reported 3D crystal structure of Agrin coded with 3V64 (organism: *Rattus norvegicus*, method: X-ray diffraction, resolution value: 2.85) was obtained from PDB, available at http://www.rcsb.org/pdb/home/hoe.do. The PDB format of retrieved structures was downloaded and imported into Maestro (Schrodinger, LLC, Cambridge, USA). These proteins were further processed and refined by using protein preparation wizard in Maestro (Schrodinger, LLC, Cambridge, USA). The chemical structures of HMG-CoA were obtained from the PubChem chemistry database, and LigPrep was used for preparation and optimization of ligand using OPLS2005 force field.

17.3.2 Molecular Docking

The target protein agrin was used as the docking site for HMG-CoA using Glide (Schrodinger, LLC, Cambridge, USA). The active binding site of the protein was first located using the Schrodinger software's SiteMap module. A receptor grid generation module was then used to create a grid for these active sites. After ligand preparation, various conformers were obtained, and HMG-CoA was prepared using LigPrep. Additionally, using the XP protocol, the lowest energy conformer was docked against the target proteins. These conformers were then analysed based on the binding energy and docking score. The ability to detect binding affinity increases with decreasing glide score and binding energy. Different types of interactions between ligands and proteins can be seen through π-π, hydrophobic interactions, van der Waals, and conventional hydrogen bonding.

17.3.3 Molecular Dynamics

MD simulation is used to analyse the nature of the protein, molecular changes that occur inside the protein, and the stability of the protein-ligand complex with respect to time [10]. In this current study, MDs has been performed by using the Desmond module (Schrodinger, LLC, Cambridge, USA) using the OPLS_2005 force field. Firstly, the system setup was done for the protein-ligand complex by using the system builder module of the software in which the complex was released into the predefined solvent model TIP3P (transferable intermolecular potential 3 points), and the system was exposed into orthorhombic box. Ions such as sodium and chloride were used to neutralize the system. This build system was ready for MD simulations, which was performed by using the constant-temperature, constant-pressure ensemble (NPT), and the temperature of the system was kept at 300 kelvins with constant pressure of 1.01325 bar throughout the simulation. The MD simulation was done for 100 ns, and the system was kept relaxed before the MD simulation to get the perfect MD simulation results. The output file of the MD simulation consists of a SID which includes RMSD, RMSF, and protein-ligand contact which were used to check the stability of the protein-ligand complex.

17.3.4 Results

17.3.4.1 Molecular Docking

The interaction of HMG-CoA checked towards agrin is described below and shown in Table 17.2.

TABLE 17.2

Molecular Docking Study of Statins against Agrin Protein

Target Name	Compound Name	H-Bonding	Hydrophobic Contact	Dock Score
Agrin	Pitavastatin	LYS:1789, GLN: 1792, SER:1793	VAL:1776, ALA:1775, VAL:1925, ALA:1790, LEU:1791	−6.736
	Rosuvastatin	LYS:1789, GLN:1792, VAL:1925, ARG:1928	TYR:1772, LEU:1773, ALA:1775, VAL:1776, ALA:1790, LEU:1791, LEU:1894, VAL:1925, LEU:1930	−5.541
	Fluvastatin	GLN:1792	VAL:1776, ALA:1775, ALAP:1790, LEU:1791, PHE:1796, ILE:1812, VAL:1925	−5.119
	Atorvastatin	LYS:1789, GLN:1792, VAL:1925, ARG:1928	TYR:1772, LEU:1773, ALA:1775, VAL:1776, ALA:1790, LEU:1791, LEU:1894, VAL:1925, LEU:1930	

17.3.5 INTERACTION OF PITAVASTATIN TOWARDS AGRIN PROTEIN

Pitavastatin shows notable hydrogen bonding of LYS:1789, GLN:1792, and SER:1793, while hydrophobic contacts such as VAL:1776, ALA:1775, VAL:1925, ALA:1790, and LEU:1791 followed by the dock score −6.736 kcal/mol with agrin are demonstrated in Figure 17.7.

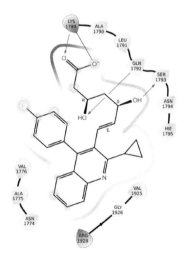

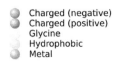

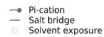

FIGURE 17.7 Interactions of pitavastatin towards agrin protein.

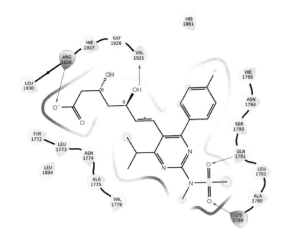

FIGURE 17.8   Interactions of rosuvastatin towards agrin protein.

17.3.6   INTERACTION OF ROSUVASTATIN TOWARDS AGRIN PROTEIN

Rosuvastatin has shown good interaction towards agrin through hydrogen bonding with LYS:1789, GLN:1792, VAL:1925, and ARG:1928 (dock score −5.541 kcal/mol). Different hydrophobic interactions such as TYR:1772, LEU:1773, ALA:1775, VAL:1776, ALA:1790, LEU:1791, LEU:1894, VAL:1925, and LEU:1930 were also observed with agrin, as shown in Figure 17.8.

17.3.7   INTERACTION OF FLUVASTATIN TOWARDS AGRIN PROTEIN

Fluvastatin showed binding affinity of −5.119 kcal/mol and interacted with agrin protein by H-bonding of GLN:1792 followed by hydrophobic interactions (VAL:1776, ALA:1775, ALAP:1790, LEU:1791, PHE:1796, ILE:1812, and VAL:1925), as shown in Figure 17.9.

17.3.8   INTERACTION OF ATORVASTATIN TOWARDS AGRIN PROTEIN

Atorvastatin has interacted with agrin through H-bonding ASN:1794, HIE:1795 followed by hydrophobic interactions such as ILE:1777, VAL:1776, ALA:1775, LEU:1773, TYR:1772, LEU:1791, ILE:1812, ILE:1925, and LEU:1930 with dock score −3.058/mol as indicated in Figure 17.10.

17.3.9   MOLECULAR DYNAMICS STUDIES

The MD results of ligands (pitavastatin, rosuvastatin, and fluvastatin) in complex with agrin were carried out for 100 ns to determine the stability of proteins as well as ligand-protein complex.

17.3.10   ROOT MEAN SQUARE DEVIATION ANALYSIS OF AGRIN

The RMSD serves as a crucial parameter for evaluating the stability, dynamic characteristics, and behaviour of the protein-ligand complex. Figures 17.11–17.13 illustrate the graph depicting the

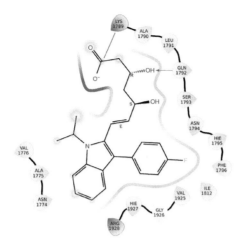

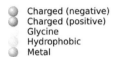

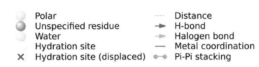

Charged (negative)   Polar   ⋯ Distance   ●— Pi-cation
Charged (positive)   Unspecified residue   ➤ H-bond   — Salt bridge
Glycine   Water   ➤ Halogen bond   ◌ Solvent exposure
Hydrophobic   Hydration site   — Metal coordination
Metal   ✕ Hydration site (displaced)   ●—● Pi-Pi stacking

FIGURE 17.9 Interactions of fluvastatin towards agrin protein.

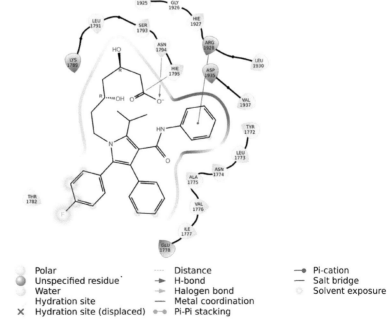

Charged (negative)   Polar   ⋯ Distance   ●— Pi-cation
Charged (positive)   Unspecified residue ˙   ➤ H-bond   — Salt bridge
Glycine   Water   ➤ Halogen bond   ◌ Solvent exposure
Hydrophobic   Hydration site   — Metal coordination
Metal   ✕ Hydration site (displaced)   ●—● Pi-Pi stacking

FIGURE 17.10 Interactions of atorvastatin towards agrin protein.

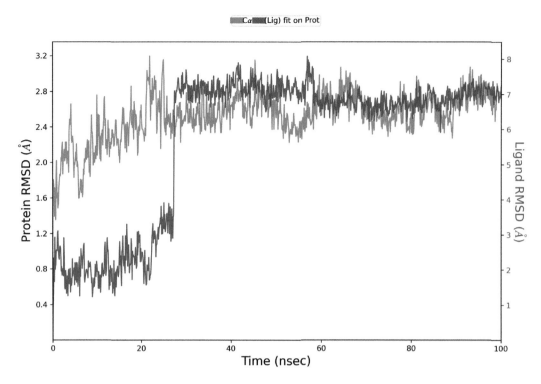

FIGURE 17.11   Structural dynamics deviation (RMSD) of protein agrin upon binding with pitavastatin.

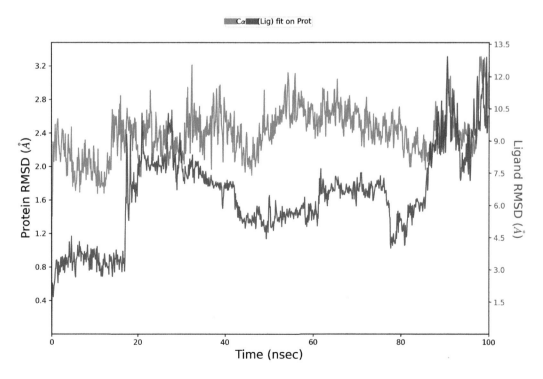

FIGURE 17.12   Backbone residual fluctuations (RMSF) plot of protein agrin and its complex with pitavastatin.

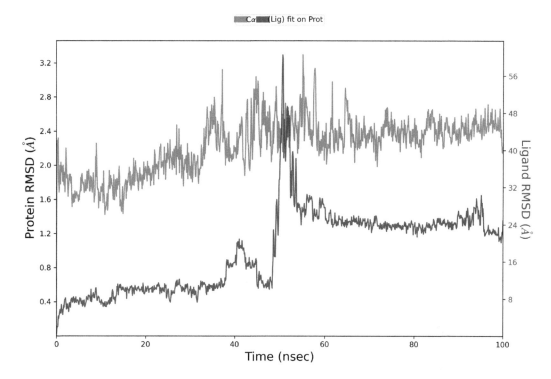

FIGURE 17.13 Histogram presentation of per-residue analysis of pitavastatin in complex with agrin protein.

progression of protein RMSD, represented on the left Y-axis. When the protein's RMSD falls within the 1–3 Å range, it indicates the protein's stability and minimal variations in its backbone structure. Among all the three ligand-protein complexes, fluvastatin with agrin has shown greater stability as compared to complex of pitavastatin and rosuvastatin with agrin. Fluvastatin has shown an RMSD value of 3.2 Å with agrin, whereas pitavastatin and rosuvastatin have shown RMSD values of 3 and 3.1 Å, as demonstrated in Figures 17.11–17.13, respectively. RMSD values of fluvastatin were found in the acceptable range, and on the basis of ligand-protein overlapping, fluvastatin was making most stable complex with agrin because fluvastatin stable overlapping with protein for more than 50 ns indicates that the protein is in stable form after binding with ligand throughout the simulation.

17.3.11 ROOT MEAN SQUARE FLUCTUATION ANALYSIS OF THE PROTEIN

The protein's structural flexibility can be examined through RMSF, a valuable tool for detecting localized alterations within the protein sequence. Figures 17.14–17.16 highlight the regions of the protein that exhibit the greatest degree of movement during simulations, as indicated by the peaks. Fluvastatin and agrin complex regions of 20–30 number amino acids have shown greatest fluctuation with RMSF value 5.5 Å, whereas pitavastatin and rosuvastatin complexes have shown highest fluctuation within the range of 10–30 and 20–35 number amino acids, respectively. It indicates that the higher the fluctuation in amino acids, the more the involvement of amino acids in the protein-ligand interaction.

17.3.12 PROTEIN-LIGAND CONTACTS

After the simulation run, a separate analysis was conducted for each system, examining the per-residue interactions by calculating the mean interaction occupancies of the binding site

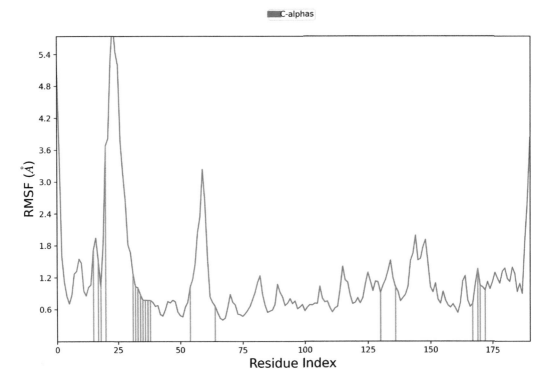

FIGURE 17.14 Structural dynamics deviation (RMSD) of protein agrin upon binding with rosuvastatin.

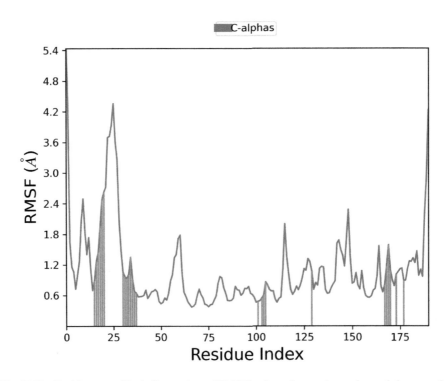

FIGURE 17.15 Backbone residual fluctuations (RMSF) plot of protein agrin and its complex with rosuvastatin.

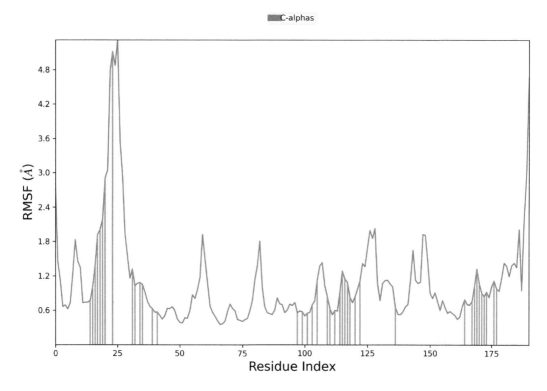

FIGURE 17.16 Histogram presentation of per residue analysis of rosuvastatin in complex with agrin protein.

residues called protein–ligand contact. Significant hydrogen bonding between ligand fluvastatin and agrin was observed with LYS_1789, SER_1793, ASN_1794, and HIS_1795, whereas pitavastatin and rosuvastatin have shown hydrogen bonding with ARG_1928, ARG_1859 and ARG_1928, SER_1793, and LYS_1789, respectively. All these interactions are demonstrated in Figures 17.17–17.19.

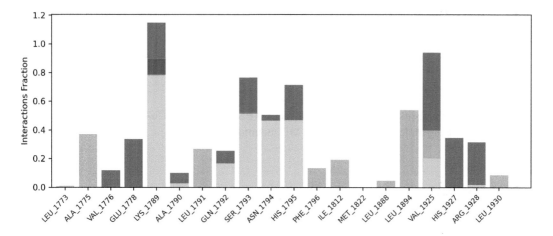

FIGURE 17.17 Structural dynamics deviation (RMSD) of protein agrin upon binding with fluvastatin.

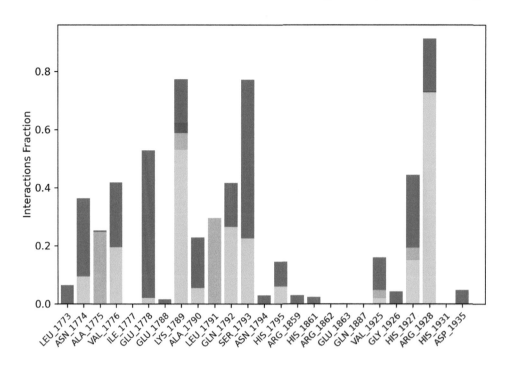

FIGURE 17.18 Backbone residual fluctuation (RMSF) plot of protein agrin and its complex with fluvastatin.

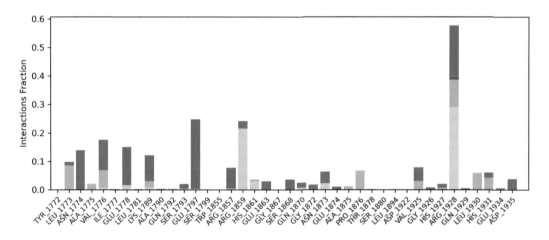

FIGURE 17.19 Histogram presentation of per residue analysis of fluvastatin in complex with agrin protein.

17.4 CONCLUSION

The possible mechanisms of ketoprofen-induced DRESS and HMG-CoA inhibitor-induced MG were predicted using molecular docking and MD techniques.

REFERENCES

1. Gupta, M., Sharma, R., Kumar, A., 2018. Docking Techniques in Pharmacology: How Much Promising? Comput. Biol. Chem. 76, 210–217. https://doi.org/10.1016/j.compbiolchem.2018.06.005
2. Kant, K., Rawat, R., Bhati, V., Bhosale, S., Sharma, D., Banerjee, S., Kumar, A., 2020. Computational Identification of Natural Product Leads that Inhibit Mast Cell Chymase: An Exclusive Plausible Treatment for Japanese Encephalitis. J. Biomol. Struct. Dyn. 0, 000. https://doi.org/10.1080/07391102.2020.1726820

3. Navyashree, V., Kant, K., Kumar, A., 2020. Natural Chemical Entities from Arisaema Genus Might Be a Promising Break-Through Against Japanese Encephalitis Virus Infection: a Molecular Docking and Dynamics Approach. J. Biomol. Struct. Dyn. 0, 000. https://doi.org/10.1080/07391102.2020.1731603

4. Hu, X., Zeng, Z., Zhang, J., Wu, D., Li, H., Geng, F., 2023. Molecular Dynamics Simulation of the Interaction of Food Proteins With Small Molecules. Food Chem. 405, 134824. https://doi.org/10.1016/j.foodchem.2022.134824

5. Stevens, J.A., Grünewald, F., van Tilburg, P.A.M., König, M., Gilbert, B.R., Brier, T.A., Thornburg, Z.R., Luthey-Schulten, Z., Marrink, S.J., 2023. Molecular Dynamics Simulation of an Entire Cell. Front. Chem. 11. https://doi.org/10.3389/fchem.2023.1106495

6. Wang, S., Breskovska, I., Gandhy, S., Punga, A.R., Guptill, J.T., Kaminski, H.J., 2018. Advances in Autoimmune Myasthenia Gravis Management. Expert Rev. Neurother. 18, 573–588. https://doi.org/10.1080/14737175.2018.1491310

7. Verschuuren, J.J.G.M., Huijbers, M.G., Plomp, J.J., Niks, E.H., Molenaar, P.C., Martinez-Martinez, P., Gomez, A.M., De Baets, M.H., Losen, M., 2013. Pathophysiology of Myasthenia gravis With Antibodies to the Acetylcholine Receptor, Muscle-Specific Kinase and Low-Density Lipoprotein Receptor-Related Protein 4. Autoimmun. Rev. 12, 918–923.

8. Ginsberg, H.N., 2006. REVIEW: Efficacy and Mechanisms of Action of Statins in the Treatment of Diabetic Dyslipidemia. J. Clin. Endocrinol. Metab. 91, 383–392. https://doi.org/10.1210/jc.2005-2084

9. Bouitbir, J., Sanvee, G.M., Panajatovic, M. V., Singh, F., Krähenbühl, S., 2020. Mechanisms of Statin-Associated Skeletal Muscle-Associated Symptoms. Pharmacol. Res. 154, 104201. https://doi.org/10.1016/j.phrs.2019.03.010

10. Bhosale, S., Kumar, A., 2021. Screening of Phytoconstituents of *Androgr11 paniculata* Against Various Targets of Japanese Encephalitis Virus: An in-Silico and in-Vitro Target-Based Approach. Curr. Res. Pharmacol. Drug Discov. 2, 100043. https://doi.org/10.1016/j.crphar.2021.100043

18 Available Databases and Tools for Signal Detection of Vaccines

Sachin Kumar and Vikas Maharshi

18.1 INTRODUCTION

Vaccines are among the safest medical treatments available today and aid in disease prevention by boosting immunity. One of the most economical ways to lower morbidity and death among children under the age of five is through regular immunization, which is a vital component of the public health system in every nation. The risk of disease for the entire population rises in the absence of comprehensive vaccination coverage. Pharmacovigilance is the study of drug safety, and it aids in preventing adverse drug effects in the general population. It began with the tragedy of the thalidomide drug, which resulted in ~10,000 birth abnormalities in children. In order to strengthen post-marketing surveillance for safer drugs, regulatory systems were quickly developed [1].

A careful oversight of vaccine's safety is crucial because the majority of vaccines are given to a group that is healthy but susceptible, i.e. children. Concerns about vaccine safety could fuel rumors, erode public faith in immunization, and ultimately have a significant impact on immunization coverage and disease incidence [2]. The safety of vaccinations must be extremely high in order for them to be accepted, as they often have delayed individual benefits but immediate adverse drug reactions (ADRs). Vaccine pharmacovigilance refers to the studies and methods used in the detection, assessment, understanding, avoidance, and reporting of adverse vaccination reactions or any other problems related to vaccination or immunization.

The complex biological products known as vaccines can contain a variety of antigens, live organisms, adjuvants, and preservatives. ADRs can result from the administration of live wild viruses, as in the case of lymphocytic meningitis following anti-mumps vaccination, or they can be non-specific and related to a substance other than the antigen [3]. Therefore, compared to other medications, each component has distinct safety consequences that require separate data to be gathered. A large population is exposed to the vaccine in a short period of time during a vaccination campaign, which increases the possibility of administration error and necessitates a different approach.

There are various factors that make vaccination evaluation separate from those of other medications. Although delayed occurrences are challenging to correlate, local or immediate adverse effects that happen as a result of administrative error or an attenuated virus can be assigned with some confidence. They take a while to respond and are affected by immunological factors in addition to pharmacological ones [4]. The causality of an event occurring after vaccination cannot be determined using generally used criteria, such as remission of the illness after treatment withdrawal and the outcome of a rechallenge [5].

The presence of databases and technologies for signal identification in relation to vaccines shows how committed the medical profession is to evaluating and enhancing vaccine safety. These tools make use of intensive data gathering and analytical methods to identify potential safety concerns that might not have been visible during clinical trials, which typically feature a small number of participants and a brief follow-up time. In this chapter, we will look at some of the important databases that are essential to vaccination safety monitoring.

DOI: 10.1201/9781032629940-18

The accessibility of databases and signal detection techniques is essential in the constantly changing environment of vaccine production and distribution in order to quickly discover and assess any potential safety issues related to vaccinations. They make it possible for stakeholders and healthcare authorities to adopt preventative steps, such as regulatory actions and communication plans, to address safety concerns and uphold public confidence in vaccination programs.

18.2   VACCINE SAFETY: WHY DOES IT NEED SEPARATE ATTENTION?

The worries concerning the safety of vaccines have grown along with the number of vaccines included in immunization schedules [6]. Numerous studies have demonstrated that there is no connection between vaccinations and a number of major adverse events following immunizations (AEFIs) that have been hypothesized to be connected to vaccinations [7, 8]. AEFI is defined by the World Health Organization (WHO) as 'any untoward medical occurrence following immunization which does not necessarily have a causal relationship to vaccine'. The safety of the vaccines now in use is backed by a significant body of research. The most widely debated safety concern of the last 20 years was perhaps the supposed link between autism and the measles, mumps, and rubella (MMR) vaccine, which was first reported in 1998 in *Lancet* by Andrew Wakefield [9]. *Lancet* issued a complete retraction of the 1998 study in 2010, citing the research's purposeful falsification [10].

The program, however, suffered a serious setback when the United Kingdom saw a drop in the MMR vaccine uptake from 91% in 1998 to 80% by 2004. Nearly 14 years after the local transmission of the disease in the UK was stopped, several measles outbreaks were reported, and the disease was once again reported to be endemic in the country in 2008 [11]. More recently, during the threat of an H1N1 pandemic in 2009–2010, there were notable vaccine refusals because people felt there was little to no risk of getting sick, they questioned the motives of the government and the pharmaceutical industry, and they remembered hearing that an earlier swine flu vaccine had reportedly been linked to Guillain-Barre syndrome. Not surprisingly, communities and even medical professionals had appallingly low adoption of the influenza A (H1N1) vaccine when it was available to control the pandemic [12].

In 2003, the oral polio vaccine (OPV) was banned in five states in northern Nigeria on the grounds that it included anti-fertility chemicals as part of a Western government initiative to reduce the population of Muslims. As a result, more than 15 previously polio-free African countries experienced a resurgence of the disease, and efforts to completely eradicate polio in India's neighboring country(ies) are still ongoing [13]. Due to rumors and erroneous information regarding the OPV, which claimed it would render people in vulnerable groups and religious minorities sterile, the effort to eradicate polio was in peril even in India [14].

The introduction of the Measles-Rubella (MR) vaccine for children between 9 months and 15 years of age, in campaign mode in the first week of February 2017 in five Indian states has recently been hampered by rumors on social media about alleged side effects like sterility/impotence, autism, and subacute sclerosing panencephalitis (SSPE) following MR vaccination [15]. Even among the educated where a sizable portion of kids had already received the required number of measles doses, the general consensus was to avoid administering extra doses of the MR vaccine during the campaign for safety reasons in various districts in southern India [16].

An insufficient immunization renders the receiver susceptible to infection, raises their risk of being ill or dying, and hinders their capacity to grow. Despite the country's long-standing government program for universal immunization [17], only 64.1% of Indian children had gotten every recommended vaccination as of 2015 [18].

Lack of access, complicated vaccination schedules needing several shots, ignorance, and real or imagined fear of vaccine adverse effects (AEFIs) are all possible causes of inadequate immunization [18]. The WHO SAGE working group defines vaccination reluctance as 'delay in acceptance or refusal despite the availability of vaccine services'. It is complex and context-specific, varying depending on the time, location, and immunizations. Influencing elements include complacency, practicality, and confidence [19].

18.3 VARIOUS DATABASES FOR VACCINE-RELATED ADVERSE EVENTS ACROSS THE GLOBE

18.3.1 GLOBAL SCENARIO

Many nations maintain an efficient national AEFI surveillance system to monitor AEFI.

18.3.1.1 Adverse Event Reporting of Vaccines in the United States

A national initiative to monitor the safety of vaccines approved for use in the United States is the Vaccine Adverse Event Reporting System (VAERS). VAERS gathers and examines reports of unfavorable outcomes following immunization that is supervised by the Food and Drug Administration (FDA) and the Centers for Disease Control and Prevention (CDC). The FDA and CDC jointly financed this program when it first started in 1990. Anyone (vaccine manufacturers, healthcare professionals, vaccine recipients, and parents) may report any clinically significant adverse events (AEs) that develop after vaccination through this spontaneous reporting system. Manufacturers are required to report all the suspected adverse events, whereas healthcare professionals are required to report certain types of them. VAERS is a passive reporting system that relies on reporting of adverse events by individuals to FDA and CDC. VAERS may indicate toward a likely safety concern with vaccine(s) by detecting unusual patterns of reporting which may further be investigated by regulatory authorities. Primary objectives of VAERS are to detect unusual, rare, and/or new adverse events, detect unusual rise in reporting of known adverse events with vaccines, identify potential risk factors associated with adverse events, identify possible reporting clusters like a particular region or a particular vaccine or its batch/lot, etc., and thus in turn evaluate the safety of vaccine(s) to serve as a safety monitoring system. Though VAERS can suggest a safety concern with a particular vaccine, it cannot evaluate and tell that the adverse event is certainly caused by a particular vaccine or not [20, 21].

18.3.1.2 Adverse Event Reporting of Vaccines in Canada

The Canadian Adverse Events Following Immunization Surveillance System (CAEFISS) is the name of the post-marketing safety monitoring system for commercially marketed vaccinations in Canada [22]. It was established through collaboration between the federal, provincial, and public health authorities of Canada and is managed by the agency's division of vaccine safety. Data on AEFI are gathered through both passive and active surveillance from a variety of collaborators, including provincial and territory health departments, medical practitioners, and pharmaceutical companies. CAEFISS includes both passive and active surveillance. CAEFISS detects reporting of previously unknown adverse events as well as increased reporting of a known adverse event following immunization and thus may suggest to regulatory authority on further evaluation of a particular issue to monitor the safety of vaccines in Canada. Consumers may report the adverse event after receiving vaccines to the healthcare professionals (vaccine provider) form where the report is sent to local public health units. The latter transfer the information to provincial/federal/territorial immunization authorities who incorporates the information in the database. Thereafter, advisory committee on causality assessment evaluates the causal relation between reported adverse event and the suspected culprit vaccine using a standard procedure developed by WHO [23].

18.3.1.3 Adverse Event Reporting of Vaccines in the United Kingdom

It is the responsibility of the UK's Medicines and Healthcare Products Regulatory Agency (MHRA) to ensure the security of all commercially available drugs, including vaccines. Through the Yellow Card Scheme, vaccine-related suspected AEs can be reported to the MHRA [24].

18.3.1.4 Adverse Event Reporting of Vaccines in the European Union

The European Medicines Agency (EMA) collects reports and keeps track of AEFI in the European Union (EU). The pharmacovigilance system is the main system for AEFI reporting and surveillance within the EU. About 1.7 million safety reports pertaining to COVID-19 vaccines were evaluated by the EMA in the EudraVigilance (EV) database in 2021, and more than 900 possible signals were found [25].

18.3.2 INDIAN PERSPECTIVE

One of the biggest producers and exporters of vaccines worldwide is India. Additionally, India delivers the Universal Immunization Program (UIP), one of the largest immunization programs in the world, to approximately 27 million babies and 30 million expecting mothers each year in an effort to prevent a number of diseases that can be prevented by vaccines [26, 27]. The Indian government started the AEFI surveillance program in 1986, not long after UIP was introduced in 1985 [18]. This program was started with the primary goals of detecting AEFI, managing it quickly, and taking the necessary steps to stop such incidents from happening again [28].

The Government of India released another updated AEFI guideline in 2015 that was produced in accordance with the WHO/CIOMS guideline in order to forward the agenda of effective implementation and surveillance on vaccination safety [29, 30].

18.3.2.1 Adverse Event Reporting

The Government of India owns and manages Co-WIN (COVID Vaccine Intelligence Network), a web-based infrastructure that enables COVID-19 immunization of citizens. The Co-WIN system is a tool for collecting all the vaccination-related information for the UIP [31]. Co-WIN is integrated with the SafeVac system. By selecting the AEFI button in Co-WIN, which will take the user to the appropriate SafeVac form, vaccine providers can report any adverse events they notice at the vaccination facility, on SafeVac.

18.4 SIGNALS IN RELATION TO VACCINE SAFETY

The COVID-19 pandemic has made the function of vaccines in combating infectious diseases around the world more crucial than ever. Vaccine development has always placed a high priority on safety; however, like other medical treatments, there is a chance that side effects could occur. A strong post-market safety surveillance system is necessary, together with properly executed clinical studies, to guarantee safety throughout the vaccine lifecycle. The effectiveness and reliability of signal detecting systems as well as responsible communication are crucial to the safety of vaccines.

As the world battles to put a stop to the coronavirus disease 2019 (COVID-19) pandemic, it is crucially necessary to grasp the risks linked to COVID-19 vaccines. One of numerous systems used to document AEs that occur after immunization, including the COVID-19 vaccinations, is the VAERS, which is jointly run by the US-FDA and the CDC. Similar to other safety surveillance systems, VAERS offers the chance to swiftly identify potential risks associated with vaccinations through a technique known as signal detection.

A safety signal is described by the WHO as information that has been made public regarding a possible causal relationship between an AE and a product, even while the specifics of that relationship are unknown or only partially understood [32]. Signal detection, at its most fundamental level, is the proactive search for safety signs. As a multifaceted and interdisciplinary process, signal detection can be applied at many levels of evidence and data as well as using a range of approaches.

A 'signal' is a crucial early warning mechanism used to identify and look into potential safety issues related to vaccinations in the field of vaccine safety. These signals are observable patterns, trends, or clusters of reports of adverse events that stand out from the background noise of anticipated negative reactions after vaccination. They act as warning signs that demand more investigation because they might point to a potential connection between a particular vaccine or group of vaccines and unfavorable health outcomes. The emergence of new or unexpected adverse events, temporal clustering, multiple reports of the same event from different sources, a higher-than-expected rate compared to baseline, or an increased incidence of a specific adverse event are just a few examples of how signals can appear [33].

For instance, if reports of severe allergic responses to a flu vaccination suddenly increase noticeably soon after administration, it would be a warning sign that further research is necessary. These signals are regarded seriously, and regulatory agencies, public health authorities, and vaccine producers perform thorough investigations, including epidemiological research and in-depth analysis, to ascertain whether a causal association between the vaccine and the adverse event exists. Given that many reported events may be coincidental or unrelated to vaccination, it is important to note that the majority of signals do not ultimately result in the identification of real safety concerns. Nevertheless, they are essential for preserving the public's confidence in vaccination programs and ensuring the continued safety of vaccines [34].

The most significant incident was in 2010, when there was a rise in fever and febrile seizures after seasonal influenza vaccinations [35, 36]. This was later determined to be caused by one trivalent influenza vaccine brand's (Fluvax® bioCSL) enhanced reactogenicity. Western Australian emergency department doctors issued the initial warning. Ten days later, the vaccine was pulled off the market, but not before further cases had emerged, including one in which an 11-month-old baby suffered a protracted febrile seizure that left the child profoundly disabled [37]. The multi million-dollar compensation judgment took into account the reporting delays identified by an independent review [37, 38], as well as the state and federal response times.

Retrospective investigations of AEFI with seasonal influenza vaccines demonstrated a potential increase in allergy-related AEFI in 2015, which was a separate signal event [39]. Fortunately, the severity in this case was not great enough to require changing the program. The Australian passive surveillance system's persistent inability to quickly recognize AEFI signal events was disclosed [40]. The implementation of active surveillance systems has been a positive development, but they can be costly, time-consuming, and frequently focus on known potential issues (new immunizations or high-risk groups) [41]. It is essential that passive systems are used to their utmost extent for quick signal identification because they continue to form the foundation of AEFI surveillance.

A 'signal' is detected when AEFI happens more frequently than is generally expected. Signal detection in vaccine monitoring requires a comprehensive approach since AEFI might range from an uncommon occurrence of a severe AEFI to a higher incidence or increased severity of a recognized, often occurring AEFI [42, 43]. The responsibility of ensuring the safety of vaccines is intricate and shared. It can be done in a variety of ways, one of which is by documenting specific instances of adverse responses that are believed to be related to vaccinations. The task is challenging since it can be challenging to assign causality to a specific case report. In order to streamline the process, an expert advisory council created the causality assessment form, a standardized evaluation tool [44]. Establishing the likelihood of a causal relationship between a side effect and the vaccine of interest is the goal of signal evaluation in the context of vaccine safety. Depending on the findings, the regulatory authorities may decide to take no further action, request additional testing or monitoring, update the product's information and/or risk management strategy, or implement immediate safety restrictions like suspending or revoking a marketing authorization [45]. Finding an AEFI with a greater incidence is frequently challenging and has been compared to 'finding a needle in a haystack' [46]. Through analysis of each complaint individually or through active surveillance, rare but extremely significant instances can be found.

18.5 METHODS OF SIGNAL DETECTION

Signal detection requires information on adverse event following vaccine administration. According to International Council for Harmonization of Technical Requirements of Pharmaceuticals for Human Use (ICH), sources of information could be unsolicited (spontaneous reports, medical literature, online and offline media), solicited (clinical trials, registries, health programs, and surveys), contractual agreements (inter-industry exchange of safety information), and regulatory authority sources (individual case safety reports originated from regulatory authorities). A number of tools and methods (called data mining) have been developed to sift spontaneous reporting data. Two types of approaches have been described for signal detection, namely traditional and statistical approaches [47].

18.5.1 TRADITIONAL APPROACHES

Traditional methods of signal detection are sub-classified into qualitative and simple quantitative approaches. Traditional qualitative approaches involve analysis of spontaneously reported adverse events by (a) review of individual case reports and/or case series which could be available in pharmacovigilance databases or in medical literature. Simple quantitative methods of signal detection involve (b) aggregate analysis through individual case count, measuring simple reporting rates or exposure adjusted reporting rates [47].

18.5.1.1 Qualitative Approach

18.5.1.1.1 Review of Individual Cases or Case Series

Cases (reported) are reviewed by trained specialists, and during the process, signals are detected. The most common approach for signal detection used here is the 'index case' or 'striking case' approach, in which the first well-documented reported case of an adverse event following administration of a vaccine may be considered a signal. However, clinically, a series of similar reported cases (called clustering) are required to strongly suspect an association between vaccine administration and development of an adverse event [47].

In the review process, special attention is paid to the 'designated medical events' (which are rare, serious, and have high vaccine attributable risk), 'events of special interest' (which may vary from region to region and organization to organization), 'targeted medical events' (whose classification is based on a vaccine), and 'hyperacute events'. Usually, one to three similarly reported cases are considered sufficient to generate a signal in these situations.

Analysis of periodic safety review information, e.g. periodic safety update reports (PSUR, in India and Europe), is an important tool in signal detection [47].

18.5.1.2 Quantitative Approaches

18.5.1.2.1 Aggregate or Simple Analysis

In this method, a number of adverse events reported are considered. Signals are detected when the value of the following for a reported event is higher than expected:

a. *Counts* (numbers of a particular adverse event report)
b. *Proportion*: This includes the following:
   i. Numbers of a particular adverse event/total number of adverse events with that vaccine)
   ii. Numbers of a particular adverse event/estimated exposure to the drug

One limitation of this method that can be pointed out is what exactly is meant with 'expected' in case of a spontaneous reporting system.

Though traditional methods have higher subjectivity and require expertise and experience, they provide high clinical information value for detection of signals and create the foundation for the signal detection process [47].

18.5.2 STATISTICAL APPROACHES FOR SIGNAL DETECTION

Most of these methods rely on relative reporting (of adverse events) frequencies (disproportionality analysis). Statistical approaches also called data mining algorithms are not to replace traditional methods, rather to support them to analyze large volume of datasets in a more efficient and systematic way. Statistical methods make the following assumptions related to adverse event reporting:

a. Adverse event is reported more frequently with the vaccine of interest than other vaccines/drugs.
b. For same adverse event, magnitude of reporting (underreporting) is same across the vaccines and/or drugs.
c. Reporting rate of adverse events or overall reporting pattern could be considered a reference against which reporting of a particular adverse event-vaccine pair of interest can be compared [47].

18.5.2.1 Disproportionality Analysis

Objective of disproportionality analysis is to identify statistically prominent reporting associations between a particular vaccine and adverse event pair within spontaneous reporting database(s). In statistical methods a statistic is calculated called 'statistic of disproportionate reporting' (SDR). Finding SDR (statistically prominent reporting relationship) does not necessarily establish causality between adverse event-vaccine pair and does not necessarily mean that the signal exists and vice versa because of the presence of inherent biases and artifacts in pharmacovigilance data. Therefore, decision about existence of signal should not solely be made on the basis of statistical prominence rather needs full clinical evaluation as well.

Though many threshold values are used, currently there is no 'gold standard' approach to determine a threshold for defining SDR. Too high threshold may lead to high false negatives (not being able to detect signals when they really exist), and too low threshold value may result in high false positivity (detecting existence of signals when they actually do not exist).

Methods of signal detection using disproportionality analysis are classified under two headings, *viz.* Frequentist's approaches (here the probabilities are viewed as long-term frequency with assumption of repeatable experiment) and Bayesian approaches (here probability is viewed as degree of belief which includes prior beliefs and knowledge updated with availability of new information) [47, 48].

18.5.2.1.1 Frequentist's Approach

Data mining methods convert information from a large database into 2×2 contingency tables for each vaccine-adverse event pair [47, 49]. A 2×2 contingency table formed based on reports of a particular vaccine-adverse event pair in a database is mentioned in Table 18.1.

Various measures of strength of association can be calculated from this table, *viz.* 'relative reporting', 'reporting odds ratio (ROR)', 'proportional reporting ratio (PRR)', and 'information component (IC)'. For these measures confidence interval is calculated based on Frequentist's notion of repeated sampling. ROR and PRR are two most commonly used Frequentist methods [47, 48, 50]. Measures of association and formulae (based on 2×2 table above) for their calculation are provided in Table 18.2.

TABLE 18.1

A 2×2 Contingency Table Formed Based on Reports of a Particular Vaccine-Adverse Event Pair in a Database

	Reports of Event of Interest	Reports of All Other Events	Total
Reports for vaccine of interest	A	B	A + B
Reports for all other vaccines	C	D	C + D
Total	A + C	B + D	A + B + C + D

TABLE 18.2

Measures of Association and Formulae to Calculate (with Reference to 2×2 Contingency Table)

Measures of Association	Formulae to Calculate
ROR	AD/BC
PRR	A × (C + D)/C × (A + B)
RR	A (A + B + C + D)/(A + C) (A + B)
IC	$\log_2 IC = A (A + B + C + D)/(A + C)(A + D)$

18.5.2.1.2 *Bayesian Methods*

In a large dataset, for each vaccine-event combination millions of 2×2 tables may be formed, and for each table calculation of measures of association may be tedious job. Therefore, there was a need to develop newer methods of signal detection. Bayesian methods of signal detection in pharmaco-vigilance were pioneered by WHO in the 1990s. Two major Bayesian methods used for signal detection are 'Bayesian Confidence Propagation Neural Network (BCPNN)' and 'Multi-item Gamma Poisson Shrinker (MGPS)'.

In a large and sparse database, cell 'A' of 2×2 table for a rare vaccine-event combination may have very less or sometimes nil value. Variances of such vaccine-event combination (when prior scientific knowledge and/or biological plausibility is poor) will be relatively higher (highly variable 'Observed/Expected' ratio or 'O/E' for cell 'A' in 2×2 table). Frequentist's approach addresses this issue by calculating statistically significant threshold or confidence interval. Multiplicity correction is also employed for such instances in Frequentist's approach [47, 48].

Bayesian approaches in such cases (less frequently reported vaccine-event combination and highly variable O/E) calculate an association metric called reported ratio (RR) but averaged over whole database or set to 'one'. This serves as a null RR or O/E value for all vaccine-event combinations and is then combined (by weighting scheme) with calculated association metric for 2×2 tables of individual vaccine-event combination. Thus, the calculated association metric for individual vaccine-event combination table is a composite value that ranges between null (overall average) and the association metric calculated purely based on individual 2×2 table of that vaccine-event combination. For rare/less frequently (or none) reported vaccine-event combination, the composite value of association metric will be toward null (larger values due to chance fluctuation are reduced toward null or one and is called 'Bayesian Shrinkage' of crude O/E). Similarly, O/E values that are less than null can be pulled up toward the null. This approach minimizes overall error [47, 48].

Shrinkage matrix is known by different names, *viz.* 'information component' in BCPNN and 'Empirical Bayes Geometric Mean' in MGPS. These shrinkage matrices have their credibility interval.

Both of the Bayesian methods follow the above-described principle similarly but may be implemented in different ways.

Bayesian methods make an assumption that, because of sampling variability in a noisy database, the true O/E value for most vaccine-event associations lies near to one.

In summary statistical/data mining approaches are developed as complementary to the traditional methods. In large databases of spontaneously reported adverse events, the bulk of reports is so large to be beyond the capability analysis through traditional approaches. Using the traditional methods for analyzing such a large volume of complex datasets may delay the detection of signal which sometimes could be high public health impact. These methods of signal detection identify a drug-event combination that is disproportionally reported as compared to the distribution of other drug-event combinations in the background dataset. However, disproportionately high reporting is not sufficient to generate a signal rather it indicates toward a need for further evaluation [47].

TABLE 18.3

Method of Signal Detection According to Methods of Data Collection [47]

Method of Data Collection	Method of Signal Detection
Passive surveillance	
• Spontaneous reporting (in VAERS in the US, Yellow Card system in the UK, and EudraVigilance in EU)	• Review of designated/targeted medical events • Review of other medical events reports for 'striking' feature (e.g. positive rechallenge)
• Targeted collection (exposure based or outcome based)	• Periodic aggregate analysis of spontaneous reports • Data mining approach for pattern of disproportionate reporting
Active surveillance	
• Data collection (from patient or prescriber) through surveys	• Maximized sequential probability ratio testing (MxSPRT) • Analysis of higher relative risk of event in treated versus
• From electronic patient records	untreated (control) individuals

18.6   CHOICE OF METHOD FOR SIGNAL DETECTION

Choice of method for signal detection depends on the data analyzed that in turn depends on the method of data collection. Data can be collected through two approaches in pharmacovigilance, *viz.* passive and active surveillance [47]. The method of signal detection according to method of data collection is compiled in Table 18.3.

18.7   ONCE A SIGNAL IS DETECTED

Once the signal is detected through individual or aggregate analysis, it is evaluated through some sequential steps like signal triage, clarification, early evaluation, and, if required, formal evaluation using additional information from sources like mechanistic studies about the adverse event, non-clinical studies, observational studies, registry-based studies, clinical trials, population databases like electronic patient record, background event (adverse event of interest) rate in the similar population, and the existing knowledge of vaccines/drugs from the same class. After confirmation, signal may be disseminated to the stakeholders, including regulatory authorities, healthcare professionals, and general public (Figure 18.1).

18.8   DIFFERENCES BETWEEN SIGNAL DETECTION OF VACCINES AND THAT FOR DRUGS

There is significant overlap between signal detection methods for drugs and vaccines. Nevertheless, few noteworthy differences do exist. These differences are covered under the following headings.

18.8.1   UNIVERSAL IMMUNIZATION AND IMPACT OF PUBLIC COMMUNICATION ABOUT SIGNALS

In contrast to medicines, many vaccines are administered on a mass scale called universal immunization because a wider coverage is required to generate a herd immunity and overcome the disease burden. Communication of a signal generated by weak scientific evidence may adversely affect the vaccine coverage and the associated benefits at population level and not at individual level [47].

18.8.2   IMPLICATIONS OF AGE IN SIGNAL DETECTION

Many of the times the age of administration of a vaccine and onset of a particular disease coincide with each other. In this situation, the vaccine may falsely be interpreted as culprit for occurrence of

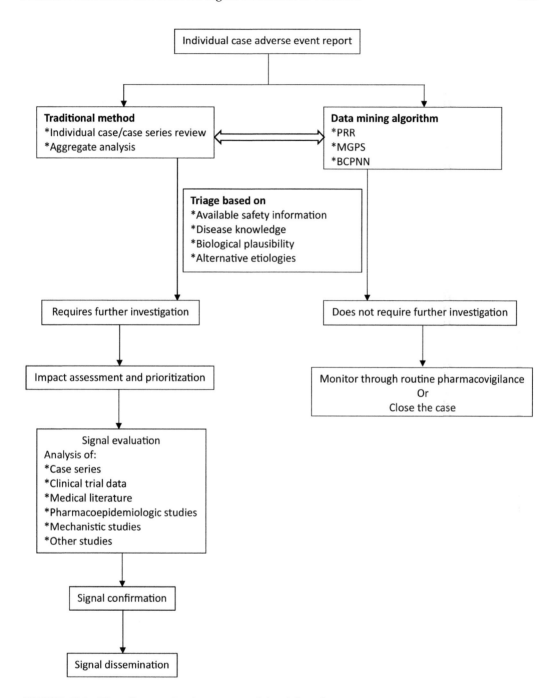

FIGURE 18.1   Flow diagram showing process of signal detection.

that disease because of pseudo-temporal association even when disease occurred as a part of its natural course without any relation to vaccine administration. However, on the other hand, the condition could have occurred as an adverse event following vaccine administration, but association between the two may be neglected considering the occurrence of condition as a part of its natural course.

Therefore, temporal association between vaccine administration and occurrence of a particular condition does not necessarily establish a causal association but simultaneously does not rule out the causal association between the two and requires further investigation [47].

18.8.3   SETTINGS OF VACCINE ADMINISTRATION

Most of the time, a clinician is not present at the site of vaccine administration which may affect the quantity and quality of reporting of adverse events. For example, at the mass vaccination site, clusters of vasovagal-like symptoms may occur which may mistakenly be reported as severe adverse events without medical confirmation. On the other hand, sometimes, important vaccine-related adverse event is identified during mass vaccination, e.g. Guillain-Barre syndrome identified as an adverse event following swine flu vaccination [47].

18.8.4   LIVE ATTENUATED VACCINES

Live attenuated vaccines contain viruses and bacteria which are supposed to cause mild infection to enhance immunity. Sometimes, instead of mild infection, the microorganisms in such vaccines may cause related pathologic disease which requires confirmation whether disease has occurred because of vaccine strain of wild type strain of the microorganism, before assessment of causality [47].

18.8.5   ANTIGENIC AND NON-ANTIGENIC COMPONENTS IN A VACCINE

A vaccine may have two types of components, *viz.* antigenic (which are expected to induce a desired immune response) and non-antigenic (like adjuvant to induce immune boosting effect of antigenic component, stabilizing agent, and sterilizing agent). Both of these components may be involved individually or in combination to produce an adverse event. For example, antigenic component may induce an unintended immune response (e.g. autoimmunity) and allergy due to non-antigenic components [47].

18.8.6   COMBINATION OF VACCINES AND SIMULTANEOUS ADMINISTRATION OF MULTIPLE VACCINES

Many vaccines are available in fixed-dose combinations (e.g. DPT: diphtheria, pertussis, and tetanus toxoid), whereas many vaccines are administered at the same time at different body sites. If an adverse event appears in such circumstances, it becomes difficult to identify a single culprit vaccine, and sometimes a non-culprit vaccine may falsely be suspected as a causative factor, e.g. development of polio following administration of DPT and OPV and linking the causality with DPT vaccine [47].

18.8.7   DATA ANALYTICAL CHALLENGES

Some spontaneous adverse event reporting databases contain reports purely because of vaccines (e.g. VAERS in the U.S.), and others may have reports for both vaccines and drugs (e.g. EV in EU). In vaccine-only databases reports of one particular vaccine may be significantly higher than others which may skew the database. In mixed databases, usually reports because of drugs outnumber those because of vaccines. Inclusion of such disparate and dissimilar databases for signal detection may put challenges in data analysis and require consideration of potential biases and confounders as well [47].

For signal detection, unvaccinated persons are frequently used as control or comparator. But because they did not receive the vaccine, they may have some characteristics that could have precluded them to receive the vaccine and make them different from those who received vaccine. However, these characteristics may also be related to development of adverse events that may act as confounding factors.

Confounding by indication is more for drugs (which are usually given to sick persons) than for vaccines (which are usually given to healthy persons). Vaccines are mostly given in children (for prophylaxis of diseases), whereas drugs are mostly given to older people (for treatment of diseases) [47].

These differences may affect choice of an appropriate comparison group and analytical approach.

18.8.8 POSSIBLE ANALYSIS BY CLASS, BRAND, OR BATCH/LOT

Analysis may be done for whole class of same vaccines (e.g. all inactivated influenza vaccines) or by brand or by batch/lot. For example, the association between administration of inactivated influenza vaccine and development of Guillain-Barre syndrome may be analyzed altogether for all inactivated influenza vaccines, or it may be done independently for different brands of influenza vaccines or for different batches of the same brand. Decision regarding the type of analysis is crucial [47].

18.8.9 FREQUENCY OF ADMINISTRATION

Vaccines are largely administered infrequently (e.g. BCG vaccine for prophylaxis of severe TB disease is administered only at birth, DPT vaccine for three or four times only and that also at an interval of many days to months), whereas drugs are administered more frequently (e.g. antihypertensive drugs are administered lifelong on daily basis).

Vaccines, once administered, imparts a long-term immunity; therefore, dechallenge is usually not applicable. Moreover, because they are used infrequently (like prophylactic BCG vaccine, which is administered only once as aforementioned), opportunity to have rechallenge is also much less for vaccines than that for drugs [47].

18.8.10 AUTOMATED SIGNAL DETECTION

Some databases have reports for both the vaccines and drugs. In such circumstances, choice of an appropriate comparator is vital because choosing a comparator including drugs may identify false signals (e.g. sudden infant death syndrome) or may identify already known, mild, and expected reactions (e.g. local injection site reactions).

It may be appropriate to do some analyses for vaccine(s) alone and others including the drugs as well.

18.9 CONCLUSION

The exploration of databases and the utilization of sophisticated tools for signal detection of vaccine adverse events represent crucial advancements in vaccine safety monitoring and public health protection.

Databases, encompassing a wide array of healthcare information sources, have emerged as invaluable assets for tracking and analyzing vaccine adverse events. These repositories offer a rich source of real-world data critical for signal detection.

The development and implementation of advanced signal detection tools, such as data mining algorithms and statistical methods, have greatly improved our ability to identify potential safety concerns associated with vaccines. These tools empower us to sift through vast datasets efficiently and effectively.

The integration of diverse data streams, including electronic health records (EHRs), vaccine registries, and spontaneous reporting systems, is fundamental to achieving comprehensive vaccine safety surveillance. This integration provides a holistic view of vaccine safety and enhances our capacity to detect adverse events.

The practical application of databases and signal detection tools has led to tangible benefits for public health. It has allowed for the early detection of rare adverse events, leading to rapid responses, modifications to vaccination strategies, and, ultimately, the safeguarding of population health.

As with any complex field, challenges exist, including data quality issues and the need to address privacy concerns. Future endeavors should focus on addressing these challenges, expanding data sources, refining signal detection methods, and strengthening collaboration among stakeholders.

In light of these observations, it is evident that databases and signal detection tools represent indispensable components of vaccine safety surveillance. Their continued evolution and integration into public health systems are essential for maintaining and enhancing the safety and efficacy of vaccination programs. As we move forward, it is paramount to foster collaboration among researchers, healthcare providers, regulatory agencies, and the pharmaceutical industry to ensure the ongoing effectiveness of these tools in protecting global public health. By doing so, we can continue to build trust in vaccines and work toward a healthier and safer world for all.

REFERENCES

1. Ronald DM, Elizabeth BA. The basis of pharmacovigilance introduction. In: Ronald DM, Elizabeth BA, editors. Pharmacovigilance. 2nd ed. Great Britain: John Wiley Sons Inc; 2007. pp. 1–3.
2. European Medicines Agency. Concept Paper for a Guideline on the Conduct of Pharmacovigilance for Vaccines. European Medicines Agency. 2005. Available from: http://www.sefap.it/farmacovigilanza_news_200512/EMEA_conceptpaper.pdf [Last accessed on 04 June 2024].
3. Autret-Leca E, Bensouda-Grimaldi L, Jonville-Béra AP, Beau-Salinas F. Pharmacovigilance of vaccines. Arch Pediatr. 2006;13:175–180.
4. Clajus C, Spiegel J, Bröcker V, Chatzikyrkou C, Kielstein JT. Minimal change nephrotic syndrome in an 82 year old patient following a tetanus-diphteria-poliomyelitis-vaccination. BMC Nephrol. 2009;10:21.
5. World Health Organization. Causality assessment of an adverse event following immunization. Available from: https://iris.who.int/bitstream/handle/10665/340802/9789241516990-eng.pdf?sequence=1 [Last accessed on 04 June 2024].
6. Baxter P. Pertussis vaccine encephalopathy: 'oh! Let us never, never doubt. Dev Med Child Neurol. 2010;52:883–884.
7. Chen RT, Shimabukuro TT, Martin DB, Zuber PLF, Weibel DM, Sturkenboom M. Enhancing vaccine safety capacity globally: A lifecycle perspective. Vaccine. 2015;33:D46–D54.
8. Institute of Medicine. Adverse Effects of Vaccines: Evidence and Causality. Washington, DC: The National Academies Press. Available from: https://doi.org/10.17226/13164 [Accessed on 20 September 2023].
9. Wakefield AJ, Murch SH, Anthony A, et al. Ileal-lymphoid-nodular hyperplasia, non-specific colitis, and pervasive developmental disorder in children. Lancet. 1998;351:637–641.
10. Anonymous. Retraction—ileal-lymphoid-nodular hyperplasia, nonspecific colitis, and pervasive developmental disorder in children. Lancet. 2010;375:445. https://doi.org/10.1016/S0140-6736(10)60175-4
11. Fitzpatrick MMMR. Risk, choice, chance. Br Med Bull. 2004;69:143–153.
12. Giannattasio A, Mariano M, Romano R, et al. Sustained low influenza vaccination in healthcare workers after H1N1 pandemic: A cross sectional study in an Italian healthcare setting for at-risk patients. BMC Infect Dis. 2015;15:329–336.
13. Jegede AS. What led to the Nigerian boycott of the polio vaccination campaign? PLoS Med. 2007;4:0417–0422. https://doi.org/10.1371/journal.pmed.0040073
14. Chaturvedi S, Dasgupta R, Adhish V, et al. Deconstructing social resistance to pulse polio campaign in two north Indian districts. Indian Pediatr. 2009;46:963–974.
15. Media Report. WhatsApp Rumours about Vaccinations Hamper India's Drive to Halt Measles and Rubella, The Government Plans to Extend the Campaign by a Week to Achieve Targets: February 24, 2017. Available from: https://scroll.in/pulse/830129/rumours-aboutmeasles-rubella-vaccine-hit-coverage [Accessed on 20 September 2023].
16. States Seek to Allay Fears over Measles-Rubella Vaccine. January 29, 2017. Available from: http://www.thehindu.com/news/national/Statesseek-to-allay-fears-over-measles-rubella-vaccine/article17110296.ece [Accessed on 20 September 2023].
17. Integrated Child Health and Immunization Survey (INCHIS). Report – Rounds 1 & 2. 2016. Immunization Technical Support Unit, Ministry of Health and Family Welfare, Government of India, New Delhi. Available from: http://www.itsu.org.in/integrated-child-healthimmunization-survey-inchis-report-rounds-1-2 [Accessed on 20 September 2023].
18. Hardt K, Schmidt-Ott R, Glismann S, Adegbola RA, Meurice FP. Sustaining vaccine confidence in the 21st century. Vaccines (Basel). 2013;1:204–224.
19. World Health Organization. Report of the Sage Working Group on Vaccine Hesitancy. Geneva. 2014. Available from: http://www.who.int/immunization/sage/meetings/2014/october/1_Report_WORKING_GROUP_vaccine_hesitancy_final.pdf [Accessed on 20 September 2023].
20. Understanding the Vaccine Adverse Event Reporting System (VAERS). Available from: https://www.fda.gov/files/vaccines,%20blood%20&%20biologics/published/Understanding-the-Vaccine-Adverse-Event-Reporting-System-(VAERS).pdf [Accessed on 20 September 2023].

21. Varricchio F, Iskander J, Destefano F, Ball R, Pless R, Braun MM, et al. Understanding vaccine safety information from the vaccine adverse event reporting system. Pediatr Infect Dis J. 2004;23:287–294.
22. Canadian Adverse Events Following Immunization Surveillance System (CAEFISS). Available from: https://www.canada.ca/en/public-health/services/immunization/canadian-adverse-events-following-immunization-surveillance-system-caefiss.html [Last accessed on 04 June 2024].
23. MacDonald NE, Law BJ. Canada's eight-component vaccine safety system: A primer for healthcare workers. Paediatr Child Health. 2017;22:e13–e16.
24. Medicines and Medical Devices Regulation. What You Need to Know. Available from: https://www.adam-aspire.co.uk/wp-content/uploads/2011/02/mhra-medicines-and-medical-devices-regulation.pdf [Last accessed on 04 June 2024].
25. European Medicines Agency. Annual Report on EudraVigilance for the European Parliament, the Council and the Commission. Reporting Period: 1 January to 31 December 2021. 2021. Available online: https://www.ema.europa.eu/en/documents/report/2021-annual-report-eudravigilance-european-parliament-council-commission_en.pdf [Accessed on 20 September 2023].
26. National Quality Assurance Standards for AEFI Surveillance Program. 2016. Available from: https://mohfw.gov.in/sites/default/files/National%20Quality%20Assurance%20Standard%20AEFI%20on%2022-11-16%20B.pdf [Last accessed on 12 October 2018].
27. Kalaiselvan V, Thota P, Singh GN. Pharmacovigilance programme of India: Recent developments and future perspectives. Indian J Pharmacol. 2016;48:624–628.
28. Chitkara AJ, Thacker N, Vashishtha VM, Bansal CP, Gupta SG. Adverse event following immunization (AEFI) surveillance in India, position paper of Indian academy of pediatrics, 2013. Indian Pediatr. 2013;50:739–741.
29. Joshi J, Das MK, Polpakara D, Aneja S, Agarwal M, Arora NK. Vaccine safety and surveillance for adverse events following immunization (AEFI) in India. *Indian J Pediatr*. 2018;85:139–148.
30. Revised AEFI Guidelines: Executive Summary. Available from: https://mohfw.gov.in/sites/default/files/Revised%20AEFI%20Guidelines%20Execute%20Summary.pdf [Accessed on 20 September 2023].
31. Arjun MC, Singh AK, Parida SP. CoWIN: The future of universal immunization program in India. Indian J Community Med. 2023;48(4):514–517.
32. Edwards IR, Aronson JK. Adverse drug reactions: Definitions, diagnosis, and management. Lancet. 2000;356(9237):1255–1259.
33. Van Holle L, Zeinoun Z, Bauchau V, Verstraeten T. Using time-to-onset for detecting safety signals in spontaneous reports of adverse events following immunization: A proof of concept study. Pharmacoepidemiol Drug Saf. 2012;21(6):603–610.
34. Chung EH. Vaccine allergies. Clin Exp Vaccine Res. 2014;3(1):50–57.
35. Gold MS, Effler P, Kelly H, Richmond PC, Buttery JP. Febrile convulsions after 2010 seasonal trivalent influenza vaccine: Implications for vaccine safety surveillance in Australia. Med J Aust. 2010;193(9):492–493.
36. Australian Government Department of Health and Ageing Therapeutic Goods Administration. Overview of Vaccine Regulation and Safety Monitoring and Investigation into Adverse Events Following 2010 Seasonal Influenza Vaccination in Young Children Canberra: 8 October 2010. Available from: www.tga.gov.au/sites/default/files/alerts-medicine-seasonal-flu-101008.pdf [Accessed on 20 September 2023].
37. Moulton Emily. Parents of Saba Button who was victim of flu debacle received payout from WA Government. Perth Now. 2014. Available from: www.perthnow.com.au/news/wa/parents-of-saba-button-who-was-victim-of-flu-vaccine-debacle-receive-payout-from-wa-government-ng-bcb-f1a145ec457cc170ef5692f3b691d [Accessed on 20 September 2023].
38. Stokes B. Ministerial review into the public health response into the adverse events to the seasonal influenza vaccine. In: Government of Western Australia Department of Health. Perth; 2010. Available from: www.health.wa.gov.au/publications/documents/Stokes_Report.pdf. [Accessed on 20 September 2023].
39. Clothier HJ, Crawford N, Russell MA, Buttery JP. Allergic adverse events following 2015 seasonal influenza vaccine, Victoria, Australia. Euro Surveill. 2017;22(20):30535.
40. Clothier HJ, Crawford NW, Russell M, Kelly H, Buttery JP. Evaluation of 'SAEFVIC', a pharmacovigilance surveillance scheme for the spontaneous reporting of adverse events following immunization in Victoria, Australia. Drug Saf. 2017;40(6):483–495.
41. Crawford NW, Clothier H, Hodgson K, Selvaraj G, Easton ML, Buttery JP. Active surveillance for adverse events following immunization. Expert Rev Vaccines. 2014;13(2):265–276.
42. Waldman EA, Luhm KR, Monteiro SA, Freitas FR. Surveillance of adverse effects following vaccination and safety of immunization programs. Rev Saude Publica. 2011;45(1):173–184.

43. Harpaz R, DuMouchel W, LePendu P, Bauer-Mehren A, Ryan P, Shah NH. Performance of pharmacovigilance signal-detection algorithms for the FDA adverse event reporting system. Clin Pharmacol Ther. 2013;93(6):539–546.

44. Collet JP, MacDonald N, Cashman N, Pless R. Monitoring signals for vaccine safety: The assessment of individual adverse event reports by an expert advisory committee. Advisory Committee on Causality Assessment. Bull World Health Organ. 2000;78(2):178–185.

45. European Medicines Agency Guideline on Good Pharmacovigilance Practices (GVP) Module IX-Signal Management (Rev 1). Available online: https://www.ema.europa.eu/en/documents/scientific-guideline/guideline-good-pharmacovigilance-practices-gvp-module-ix-signal-management-rev-1_en.pdf (Accessed on 22 December 2022).

46. Evans SJ, Waller PC, Davis S. Use of proportional reporting ratios (PRRs) for signal generation from spontaneous adverse drug reaction reports. Pharmacoepidemiol Drug Saf. 2001;10(6):483–486.

47. CIOMS Working Group VIII. Practical Aspects of Signal Detection in Pharmacovigilance. 2010. Available from: https://cioms.ch/wp-content/uploads/2018/03/WG8-Signal-Detection.pdf (Accessed on 29 August 2023).

48. European Medicines Agency. Guideline on the Use of Statistical Signal Detection Methods in the EudraVigilance Data Analysis System. 2006. Available from: https://www.ema.europa.eu/en/documents/regulatory-procedural-guideline/draft-guideline-use-statistical-signal-detection-methods-eudravigilance-data-analysis-system_en.pdf (Last accessed on 04 June 2024).

49. Zeinoun Z, Seifert H, Verstraeten T. Quantitative signal detection for vaccines: Effects of stratification, background and masking on GlaxoSmithKline's spontaneous reports database. Hum Vaccin. 2009;5(9):599–607.

50. Cai Y, Du J, Huang J, Ellenberg SS, Hennessy S, Tao C, Chen Y. A signal detection method for temporal variation of adverse effect with vaccine adverse event reporting system data. BMC Med Inform Decis Mak. 2017;17(Suppl 2):76.

19 Identification and Validation of Signals Associated with Vaccines

Juny Sebastian

19.1 INTRODUCTION

Vaccines have revolutionized the field of public health by effectively preventing and controlling the spread of infectious diseases. Immunization is widely regarded as one of the most cost-effective preventive health interventions available. Childhood vaccination averts approximately 4 million global fatalities annually. Immunization has potential to avert more than 50 million deaths between 2021 and 2030. The complete eradication of smallpox, the remarkable progress toward polio eradication, and the significant declines in illness and death from diseases like measles, whooping cough, and flu are all powerful stories of vaccines' success in protecting public health.[1]

The societal impact of vaccines is undeniable, resulting in healthier population and substantial economic benefits. There exist over 25 vaccines that are both safe and efficacious, serving as preventive measures against diseases, supporting overall well-being across all stages of life, and contributing to the prevention and containment of outbreaks. However, the success of the vaccination program hinges not only on their potential benefits but also on their safety and effectiveness.[2]

19.1.1 IMPORTANCE OF VACCINE SAFETY AND EFFECTIVENESS

Vaccine safety and effectiveness are inseparable components of the public health endeavor. Vaccines must not only provide protection against targeted diseases but must also undergo rigorous scrutiny to ensure that they do not pose undue risks. Regulatory agencies mandate for the testing and evaluation of each of the vaccine before its introduction into public use. Pre-licensure studies, which encompass thousands of participants Adverse Events Following Immunization (AEFI), are noted, and post-licensure vaccine recipients typically number in the millions. Consequently, extremely rare AEFI cases might only come to light after licensing, necessitating robust monitoring mechanisms. Effective monitoring systems facilitate the identification of safety signals. When such signals are detected, it prompts the initiation of epidemiological studies aimed at assessing potential risks and quantifying their extent.[3,4]

19.2 SIGNALS (SAFETY SIGNALS)

19.2.1 DEFINITION AND DESCRIPTION

The Council for International Organizations of Medical Sciences (CIOMS) has defined a signal as "information (from one or multiple sources) which suggests a new potentially causal association, or a new aspect of a known association, between an intervention and an event or set of related events, either adverse or beneficial, that is judged to be of sufficient likelihood to justify verificatory action." A signal is therefore a hypothesis, together with data and arguments.[5]

As per the WHO-UMC definition, a safety signal pertains to information concerning a potential side effect, whether new or known that could be linked to a product. Typically, such signals arise from more than a single report of a suspected side effect. It's important to clarify that a signal doesn't directly establish a causal link between a side effect and a product. Instead, it serves as a hypothesis that warrants further investigation, supported by data and arguments. The information conveyed by a signal may be novel or supplementary and encompasses both adverse and favorable effects of an intervention. It can further illuminate a previously recognized correlation between a product and an adverse drug effect. Since a signal's information isn't definitive, it can significantly evolve with accumulating data over time. Once a signal surfaces, it triggers a causality assessment, which delves into the connection between a product and the occurrence of a side effect.[6]

19.2.2 STRUCTURE AND PROCESS OF SIGNAL ASSESSMENT

19.2.2.1 Sources of Data and Information

Signals about the safety and effectiveness of medications can be gleaned from a variety of data sources, including pharmacovigilance and pharmacoepidemiological data. These sources, such as spontaneous reporting systems (SRSs), active surveillance systems, and research studies, provide comprehensive scientific information on medication use and outcomes. Periodic analysis of databases containing suspected adverse reactions is key to identifying these signals. These databases can differ in terms of size and scope, ranging from marketing authorization holder databases and national databases to EudraVigilance and the database maintained by the WHO Programme for International Drug Monitoring (VigiBase).[5–7]

The clinical safety data sources outlined in ICH E2D guidelines during post-approval phase[8] are compiled in Table 19.1.

19.2.2.2 Signal Detection

Effective signal detection requires a flexible methodology that considers the nature of the data (e.g., spontaneous reports, clinical trials), relevant vaccine characteristics (e.g., age group, novel vs. established), and the specific product being evaluated. It is imperative to incorporate data from all relevant sources and exercise clinical judgment throughout this process. Identifying potential safety issues in vaccines can involve a variety of methods, including reviewing individual reports of side effects (Individual Case Safety Reports [ICSRs]), conducting statistical analyses of large datasets, or a combination of both. The most appropriate approach depends on the amount and availability of data. When it is not relevant or feasible to assess each individual case (e.g., signals detected from published studies, healthcare record data), assessment of aggregated data should be considered.[5,7,8]

TABLE 19.1

Clinical Safety Data Sources Outlined in ICH E2D Guidelines during Post-Approval Phase[8]

Sources of Individual Case Safety Reports	Description of the Source
I. Unsolicited sources	Spontaneous reports; Internet; Literature; Other sources (press or other social media).
II. Solicited sources	Organized data collection systems (include registries, clinical trials, post-approval safety surveillance, consumer support and disease management programs, surveys among patients or healthcare professionals, information gathering on efficacy of the product or patient outcome; some of these may involve record-linkage, i.e. finding entries that refer to the same entity in two or more files).
III. Contractual agreements	Exchange of safety information between different companies.
IV. Regulatory authority sources	Individual case safety reports, such as Suspected Unexpected Serious Adverse Reactions (SUSARs), that originate from regulatory authorities.

19.2.2.3 Evaluation during Signal Validation and Further Assessment

During the process of signal validation rooted in the assessment of Individual Case Safety Report (ICSR) data, the ensuing factors warrant careful consideration:

- Previous awareness, which includes factors such as:
  - Existing knowledge of the adverse reaction in product information such as summary of product characteristics (SmPC) and package leaflet.
  - Does the identified signal correspond to an adverse reaction already documented in the SmPC for other medicinal products containing the same active substance? It's crucial to recognize that certain signals might be unique to a specific medicinal product or its formulation.
  - Has the potential association between the identified signal and the medicinal product been previously assessed during the initial marketing authorization application, within the risk management plan (RMP), periodic safety update reports (PSURs), or other regulatory procedures? This assessment relies on information held by and readily accessible to the relevant organizations.
- The robustness of the evidence demands an evaluation of various aspects, encompassing but not limited to:
  - The aggregate number of cases (excluding duplicates), specifically emphasizing supportive instances. These might involve cases exhibiting a coherent temporal relationship, affirmative de- or re-challenge outcomes, absence of alternative causative factors, professional assessments of possible relevance, and corroborative results from pertinent investigations.
  - Number of exposed patients.
  - Associated terms and complications such as clinical impact or different stages of the same reaction.
  - Consistency of evidence across cases in terms of onset, pattern, and association strength.
  - Data quality and documentation.
  - Alignment with globally accepted case definitions (e.g., Brighton Collaboration for vaccines).
  - A dose-response correlation
  - Plausibility grounded in biological and pharmacological mechanisms.
  - The disproportional tendency in reporting, when relevant.
- Regarding clinical relevance and contextual considerations:
  - The gravity and severity of the reaction.
  - The reaction's outcome and its potential for reversal.
  - Insights that augment existing awareness of an adverse reaction, encompassing its intensity, duration, result, frequency, or management.
  - Side effects arising from interactions with other medications.
  - Occurrences in specific groups with increased risk (such as elderly, children, pregnant women, persons with pre-existing conditions).
  - Distinctive usage patterns (side effects associated with misuse, overdose, off-label use, counterfeit products, or medication errors).
- Supplementary resources could enhance evidence in favor of or against causal associations or potentially shed light on novel facets of established associations. The decision to consider these sources during further signal assessment hinges on their pertinence to the signal and availability to respective organizations. Such sources might encompass the following:
  - Clinical trial data
  - Corresponding cases documented in scientific literature, including data pertaining to substances within the same pharmacological class:

- Details on the spread and impact of both the side effect and the associated disease
- Findings from in vitro and in vivo studies
- Prioritize extensive databases, especially if the signal emerged from national or product-specific ones
- Analyzing healthcare databases to reveal demographics, health conditions, and medication patterns of patients exposed to a specific factor
- Information from regulatory authorities across the globe[5,7]

The potential determinations regarding the process of signal detection are shown in Figure 19.1.

19.2.2.4 Signal Prioritization

Every organization involved in the signal management process should continuously evaluate whether identified signals suggest important risks impacting patient or public health or requiring a reassessment of the medicinal product's risk-benefit balance.

Evaluating the impact of a potential safety signal requires careful consideration of the following factors:

- Seriousness, severity, outcome, and reversibility of the adverse reaction, including its preventability.
- Patient exposure to the drug (number of users) and the estimated frequency of the suspected adverse reaction.
- Patient exposure among vulnerable populations (e.g., pregnant women, children, elderly, individuals with pre-existing medical conditions) and/or populations with atypical usage patterns, where applicable.
- Consider the consequences of treatment discontinuation on the progression of the disease under treatment and the availability of alternative therapeutic options.
- Anticipated level of regulatory action, encompassing potential modifications to product information (adverse reactions, warnings, contraindications), implementation of additional risk minimization strategies, or even suspension or revocation of marketing authorization.
- The potential applicability of the identified signal to other medicinal products within the same class should be evaluated to assess the broader public health implications.

Signals with potential for significant media attention and public concern (e.g., adverse events following mass immunization) may warrant expedited signal management, with the timeframe dictated by prioritization. At any stage, if available information suggests potential risks requiring timely prevention or minimization, appropriate measures should be considered, even before formal signal assessment is complete. Clinical judgment and flexibility should be applied throughout the process.[5,7]

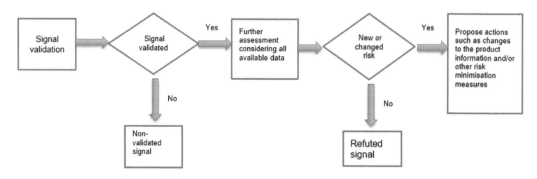

FIGURE 19.1 Potential determinations regarding the process of signal detection.

19.3   APPROACHES TO SIGNAL DETECTION

19.3.1   TRADITIONAL APPROACHES

Post-licensure monitoring of medicinal product safety has historically relied on SRSs. Following the devastating thalidomide birth defects in the 1960s, various countries implemented passive public health surveillance systems to monitor medication safety. Despite the introduction of alternative methods aiming for more proactive hazard identification following initial authorization, SRSs retain a crucial role.[5]

Traditional pharmacovigilance methods or analyzing spontaneous reports involve the following:

* Delving into individual adverse event reports and series collected in pharmacovigilance databases and published sources to explore potential drug safety signals.
* Conducting aggregated analyses of case reports using: absolute case counts (raw numbers of reported events), simple reporting rates (proportion of patients with reported events), or exposure-adjusted reporting rates (adjusting rates for factors like medication usage).

Conventional pharmacovigilance methodologies hold particular significance when evaluating designated medical events (DMEs) or infrequent occurrences. In such scenarios, individual case assessments carry greater importance, with heightened sensitivity outweighing specificity.[5]

Upon identification of a safety signal through individual case reports or aggregated analyses of spontaneous adverse event reports, a well-defined sequence of investigative steps is initiated. These entail signal triage, elucidation, early assessment, and, if necessitated, formal evaluation via autonomous datasets like hypothesis-testing research studies. This investigative journey mandates an integrated, holistic approach grounded in biological plausibility and the encompassing body of scientific evidence available.[5]

While not essential for every investigation, the following data sources can offer valuable additional information to support signal evaluation:

* *Population-based databases*: such as insurance claim or electronic patient record databases, providing insights into broader usage patterns and event prevalence.
* *Non-interventional studies*: including pharmacoepidemiological studies and patient registry studies, offering real-world evidence on medication safety in diverse populations.
* *Knowledge regarding drugs in the same pharmacologic class*: leveraging existing information on similar drugs to assess potential shared risks and mechanisms.
* *Background rates of the event under investigation in patients with relevant underlying disease conditions*: establishing the baseline incidence of the event in the target population to differentiate from drug-related effects.
* *Non-clinical and pharmacology studies*: exploring preclinical data on drug mechanisms and potential toxicity to inform further investigation.
* *Mechanistic studies of the adverse effect*: elucidating the biological pathways and processes underlying the suspected side effect to establish causality.
* *Clinical trials*: data from controlled studies providing information on safety and efficacy in specific patient populations.
* *Industry data on product complaints*: utilizing reports of adverse events submitted by patients and healthcare professionals to identify potential real-world safety signals.[5]

19.3.2   STATISTICAL DATA MINING

In the late 1990s, statistical data mining techniques emerged as a complement to conventional signal detection methods for the regular assessment of spontaneous adverse reaction data. Originally devised for systematic signal detection in expansive databases of adverse event information within health authorities and drug monitoring centers' SRSs, these statistical methods found application in the WHO International Drug Monitoring program, the United States FDA Adverse Event Reporting System (AERS), the United Kingdom's Medicines Control Agency's ADROIT (now Sentinel) database, and the European Medicines Agency's (EMA) EudraVigilance.[5]

Large SRSs, such as these, pose challenges to conventional pharmacovigilance approaches due to their vast numbers of adverse event reports. Relying solely on traditional methodologies can elevate the risk of overlooking early signals related to some drug-induced adverse drug reactions (ADRs), potentially bearing significant public health consequences.[5]

A distinguishing feature of extensive SRSs administered by health authorities and monitoring centers is the substantial heterogeneity and diversity present across individual drugs and drug categories. The wider diversity within these databases contributes to richer background data, reducing vulnerability to phenomena like masking and cloaking of drug-event associations compared to narrower or less varied repositories maintained by pharmaceutical companies. Nevertheless, acknowledging these limitations, pharmaceutical companies are progressively incorporating statistical data mining techniques within their proprietary spontaneous reporting databases. This might run concurrently with data mining of databases held by health authorities or drug monitoring centers.[5]

In suitable contexts, data mining can improve the efficiency of a pharmacovigilance program by identifying signals that might remain undetected or manifest significantly later when reliant solely on traditional methods (though this might not hold true for all signals). Data mining algorithms sift through large datasets of spontaneous adverse event reports, identifying drug-event pairs that significantly differ from the expected patterns based on the overall background data. However, it's essential to note that disproportionate frequency of reporting alone doesn't establish causality. Instead, it accentuates the need for additional investigation, often involving clinical assessment. The selection of the background dataset for disproportionality analysis has a profound impact on the outcomes, influencing the identification and characterization of potential drug-event associations. The dataset's size and its diversity concerning the represented products (drugs) and adverse events play pivotal roles in this analysis.[5]

19.3.2.1 Statistical Methods of Signal Detection

19.3.2.1.1 Basic Descriptive Analysis: Identifying Unusual Patterns

Basic descriptive analysis involves examining frequencies and proportions of adverse events in vaccinated populations. This method highlights deviations from expected baseline rates and prompts further investigation when specific adverse events occur more frequently than anticipated.[5,9]

19.3.2.1.2 Disproportionality Assessment: Highlighting Associations

Disproportionality analysis focuses on detecting disproportional reporting of adverse events compared to other events in pharmacovigilance databases. Methods like the reporting odds ratio (ROR) and the proportional reporting ratio (PRR) help identify potential signals by assessing the strength of association between specific adverse events and vaccines.[5]

19.3.2.1.3 Bayesian Data Mining: Unveiling Hidden Patterns

Bayesian data mining leverages Bayesian statistical principles to uncover hidden patterns in large datasets. It assesses the probability of observed adverse events given vaccination status, highlighting events with higher-than-expected occurrences and contributing to the identification of signals that might not be apparent through other methods.[5,10]

19.3.2.1.4 Sequential Methods: Real-Time Surveillance

Sequential methods enable real-time surveillance of vaccine safety data. Techniques like the sequential probability ratio test (SPRT) adjust analysis parameters based on accumulating data, allowing for the timely detection of signals. These methods ensure that emerging signals are promptly investigated and addressed.[11]

19.3.2.2 Conceptual Framework of the Data Management Process

Figure 19.2 outlines a comprehensive framework for signal detection program, detailing the sequential steps involved in identifying, prioritizing, and evaluating potential drug safety concerns. This framework emphasizes the integration of signal evaluation with risk management activities, recognizing that the choice of data sources and analytical approaches should be tailored to the specific needs of each investigated signal (for example, not all cases require a pharmacoepidemiological study).[5]

19.3.2.3 Comprehending Data Mining through an Integral Approach

When employing disproportionality analysis, it's vital to acknowledge the inherent limitations of spontaneous AERSs. Quantitative signal detection techniques cannot fully eliminate biases like confounding by indication or other inherent biases within the data from spontaneous adverse event reports. These methods are also unable to rectify notable deficiencies or inconsistencies in individual case-level data, or challenges within the overall data acquisition mechanism.[5,7]

The fundamental aspect of integrating statistical data mining methodologies with traditional signal detection methods within pharmacovigilance lies in the comprehensive scientific evaluation of disproportionality analysis outcomes. The interpretation of data mining results for drug safety should be informed by a comprehensive analysis of relevant safety data from diverse sources. This approach factors in knowledge of a medicinal product's established safety profile, its pharmacology, the treated patient populations, biological plausibility, and potential alternative causes for suspected ADRs.[5,7]

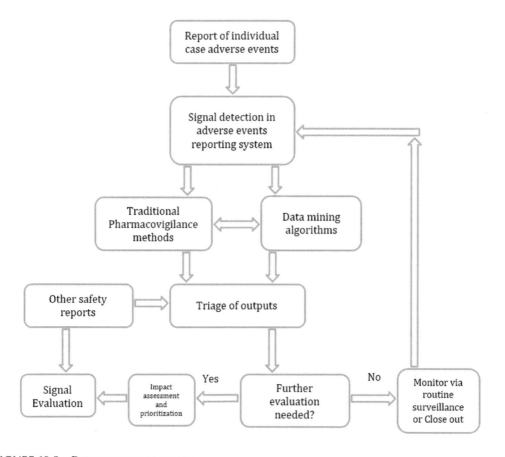

FIGURE 19.2 Data management process.

Conducting a thorough disproportionality analysis requires a series of thoughtful decisions. This includes choosing the best data sources, statistical methods, and filtering criteria, then using your pharmacovigilance knowledge and clinical expertise to interpret the results accurately. Documenting these decisions and the steps taken throughout the assessment of findings derived from disproportionality analysis is vital to keep track of signal detection activities and to comprehend the progression of emerging signals.[5,7]

The interpretation and further evaluation of drug-event associations identified through quantitative signal detection necessitate a multidisciplinary team comprising qualified professionals. This team should ideally include drug safety scientists, epidemiologists, statisticians, data analysts, and medical professionals. Their collective expertise in pharmacovigilance, statistics, data analysis, and clinical medicine is instrumental in comprehending the nature, strength, and potential implications of the identified associations.[5,7]

19.4   CHALLENGES IN SIGNAL ASSESSMENT

While the process of signal identification is crucial for maintaining vaccine safety, it is not without its challenges. Navigating these challenges is essential to ensure accurate and meaningful interpretation of data. Let's explore some of the key challenges that arise in the context of identifying signals associated with vaccines.[12–16]

19.4.1   RARITY OF ADVERSE EVENTS: THE SIGNAL-TO-NOISE RATIO

Many adverse events following vaccination are rare occurrences, making it challenging to distinguish genuine signals from random noise. Statistical methods must be robust enough to detect these infrequent events while accounting for the larger background of non-vaccine-related events.[12–16]

19.4.2   REPORTING BIASES AND UNDER-NOTIFICATION: DATA LIMITATIONS

Pharmacovigilance databases rely on voluntary reporting, leading to under-notification of adverse events. Healthcare professionals and patients may not always report events, leading to an incomplete picture. This reporting bias can impact the accuracy of signal detection and hinder the identification of associations.[12–16]

19.4.3   TEMPORAL ASSOCIATION VS. CAUSALITY: UNRAVELING COMPLEXITY

Temporal associations between vaccination and adverse events do not necessarily imply causality. Disentangling true causal relationships from coincidental temporal patterns is a challenge. Proper validation methodologies, such as well-designed observational studies, are required to establish causality definitively.[12–16]

19.4.4   CONFOUNDING FACTORS AND BIAS: ENSURING VALIDITY

Observational studies and real-world evidence are susceptible to confounding factors and bias that can distort signal detection. Factors like patient demographics, underlying health conditions, and concomitant medications can introduce confounders that need to be carefully addressed in analysis.[12–16]

19.4.5   INTERPRETING SIGNAL STRENGTH: RISK ASSESSMENT

Identifying a signal is just the first step. Assessing the clinical significance and potential risks associated with the identified signal requires a nuanced approach. The magnitude of the risk, the population affected, and the overall benefit-risk balance must be considered when interpreting signal strength.[12–16]

19.5 CASE STUDIES FOR SIGNAL IDENTIFICATION AND VALIDATION

Real-world case studies offer valuable insights into the practical application of signal identification and validation methodologies in the realm of vaccine safety. These examples illustrate the challenges faced, methodologies employed, and outcomes observed in the process of identifying and validating signals associated with vaccines.

19.5.1 Case Study 1: Narcolepsy and Pandemic Influenza Vaccine

The association between an increased risk of narcolepsy and the AS03-adjuvanted pandemic influenza A (H1N1) vaccine was identified in several European countries. Signal detection algorithms analyzed pharmacovigilance databases, revealing an unexpected clustering of narcolepsy cases following vaccination. Subsequent epidemiological studies, including case-control analyses and cohort studies, confirmed the association, leading to regulatory actions and changes in vaccine recommendations.[17]

19.5.2 Case Study 2: Rotavirus Vaccine and Intussusception

A potential link between rotavirus vaccination and intussusception, a rare bowel condition, prompted thorough signal validation. Disproportionality analyses detected a higher reporting rate of intussusception after vaccination. Observational studies, including large cohort analyses, were conducted to assess causality and risk magnitude. While the signal was validated, the overall benefit-risk profile of the vaccine favored continued vaccination due to the substantial reduction in severe rotavirus disease.[18]

19.5.3 Case Study 3: Measles-Mumps-Rubella (MMR) Vaccine and Autism

The widely debunked association between the measles-mumps-rubella (MMR) vaccine and autism serves as a cautionary tale for the complexities of signal validation in pharmacovigilance. Initial studies suggesting an association were subsequently debunked by large-scale epidemiological studies. Rigorous research, including cohort studies across diverse populations, consistently demonstrated no causal association between the MMR vaccine and autism. This case underscores the importance of thorough validation and its impact on public health decisions.[19]

19.6 ETHICAL CONSIDERATIONS IN SIGNAL IDENTIFICATION AND VALIDATION

The process of signal identification and validation in vaccine safety is not only a scientific endeavor but also one fraught with ethical considerations. Ensuring the ethical conduct of these activities is crucial to maintaining public trust, safeguarding individual rights, and making informed public health decisions. Let's explore some key ethical considerations that arise in the context of identifying and validating signals associated with vaccines.[20–22]

19.6.1 Transparency and Informed Consent

Transparency in the communication of vaccine safety findings is paramount. Individuals have the right to be informed about potential risks and benefits associated with vaccines. This information empowers them to make informed decisions about their health and the health of their families.[20–22]

19.6.2   Balancing Public Health and Individual Rights

Balancing the need for public health protection with individual rights is a delicate ethical dilemma. Decisions based on signal identification and validation may impact public health policies and vaccination recommendations. Striking the right balance between minimizing potential risks and respecting individual autonomy is crucial.[20–22]

19.6.3   Timely Reporting and Action

Timely reporting of potential signals is an ethical imperative. Delayed reporting could lead to continued exposure to risks. Ethical responsibilities extend to promptly investigating and taking appropriate actions based on validated signals to prevent harm and ensure public safety.[20–22]

19.6.4   Avoiding Alarmism and Panic

Ethical considerations extend to communication strategies. While transparency is vital, responsible communication is necessary to avoid creating unnecessary alarmism or panic. Misleading interpretations or exaggerated claims can erode public trust in vaccines and undermine vaccination efforts.[20–22]

19.6.5   Protection of Vulnerable Populations

Special ethical attention is needed when dealing with vulnerable populations, such as elderly, children, pregnant women, and those with compromised immune status. Signal identification and validation must prioritize the safety and well-being of these groups and ensure they are not unduly exposed to potential risks.[20–22]

19.7   Summary

Identifying and validating potential vaccine safety issues are critical processes that protect public health by ensuring safe and effective immunization programs. Vaccines, crucial in disease prevention and outbreak mitigation, undergo rigorous evaluation prior to and post-approval. The gravity of vaccine safety is emphasized by the annual prevention of millions of childhood deaths through vaccination.

This chapter delves into the complex realm of signal identification and validation linked to vaccines. The landscape of signal detection is dynamic, encompassing various data sources like SRSs and advanced statistical data mining methods. The historical foundation of SRSs, arising from the thalidomide tragedy, has evolved with contemporary statistical techniques, reaffirming their role in pharmacovigilance. Crucial elements of signal validation are explored, encompassing evidence strength and clinical context. These parameters guide the evaluation of adverse events and facilitate informed decision-making. Integrating statistical data mining with conventional methods enhances signal detection, unveiling insights often overlooked. However, the scientific interpretation of disproportionality analysis remains paramount, factoring in pharmacology, patient profiles, and existing safety data.

This chapter serves as a comprehensive guide, shedding light on the complexities of signal identification and validation in the vaccine safety landscape. It navigates diverse methodologies, underlining the collaborative effort among professionals like epidemiologists, statisticians, and healthcare practitioners. Ultimately, the chapter underscores the constant need for vigilance and adaptable methodologies to ensure vaccine safety and public health.

REFERENCES

1. Centers for Disease Control and Prevention (CDC). 2023. Fast facts on Global Immunization. https://www.cdc.gov/globalhealth/immunization/data/fast-facts.html (accessed August on 18, 2023)
2. World Health Organization. 2018.10 facts on immunization. https://www.who.int/mongolia/health-topics/vaccines/10-facts-on-immunization (accessed on August 18, 2023)
3. World Health Organization. 2021.Vaccine efficacy, effectiveness and protection https://www.who.int/news-room/feature-stories/detail/vaccine-efficacy-effectiveness-and-protection (accessed on August 20, 2023)
4. World Health Organization. 2023. Global Vaccine Safety Blueprint 2.0 (GVSB2.0) 2021–2023. https://www.who.int/publications/i/item/9789240036963 (accessed on September 12, 2023)
5. Council for International Organizations of Medical Sciences (CIOMS). Geneva. 2021. Practical Aspects of Signal Detection in Pharmacovigilance Report of CIOMS Working Group VIII. https://cioms.ch/wp-content/uploads/2018/03/WG8-Signal-Detection.pdf (accessed on August 23,2023)
6. Uppsala Monitoring Centre. 2022. What is a signal? https://who-umc.org/signal-work/what-is-a-signal/ (accessed on August 25, 2023)
7. European Medical Agency. 2017. Guideline on good pharmacovigilance practices (GVP) Module IX – Signal management (Rev 1). https://www.ema.europa.eu/en/documents/scientific-guideline/guideline-good-pharmacovigilance-practices-gvp-module-ix-signal-management-rev-1_en.pdf (accessed on 25 August 2023)
8. European Medical Agency. 2004. ICH Topic E 2 D Post Approval Safety Data Management. https://www.ema.europa.eu/en/documents/scientific-guideline/international-conference-harmonisation-technical-requirements-registration-pharmaceuticals-human-use_en-12.pdf (accessed on 26 August 2023)
9. Bate, Andrew, and S. J. W. Evans. "Quantitative signal detection using spontaneous ADR reporting." *Pharmacoepidemiology and Drug Safety* 18, no. 6 (2009): 427–436.
10. Bate, Andrew, Marie Lindquist, I. Ralph Edwards, Sten Olsson, Roland Orre, Anders Lansner, and R. Melhado De Freitas. "A Bayesian neural network method for adverse drug reaction signal generation." *European Journal of Clinical Pharmacology* 54 (1998): 315–321.
11. Maro, Judith C., Jeffrey S. Brown, Gerald J. Dal Pan, and Martin Kulldorff. "Minimizing signal detection time in postmarket sequential analysis: Balancing positive predictive value and sensitivity." *Pharmacoepidemiology and Drug Safety* 23, no. 8 (2014): 839–848.
12. Bizimungu, Christelle, Martine Sabbe, Françoise Wuillaume, Jamila Hamdani, Philippe Koch, and Jean-Michel Dogné. "Challenges in assessing COVID-19 vaccines safety signals—The case of ChAdOx1 nCoV-19 vaccine and corneal graft rejection." *Vaccines* 11, no. 5 (2023): 954.
13. Harpaz, Rave, William DuMouchel, Robbert van Manen, Alexander Nip, Steve Bright, Ana Szarfman, Joseph Tonning, and Magnus Lerch. "Signaling COVID-19 vaccine adverse events." *Drug Safety* 45, no. 7 (2022): 765–780.
14. Brattig Correia, Rion, Ian B. Wood, Johan Bollen, and Luis M. Rocha. "Mining social media data for biomedical signals and health-related behavior." *arXiv e-Prints* 3 (2020): 433–458.
15. Miotto, Riccardo, Li Li, Brian A. Kidd, and Joel T. Dudley. "Deep patient: An unsupervised representation to predict the future of patients from the electronic health records." *Scientific Reports* 6, no. 1 (2016): 1–10.
16. Pathak, Jyotishman, Kent R. Bailey, Calvin E. Beebe, Steven Bethard, David S. Carrell, Pei J. Chen, and Dmitriy Dligach et al. "Normalization and standardization of electronic health records for high-throughput phenotyping: The SHARPn consortium." *Journal of the American Medical Informatics Association* 20, no. e2 (2013): e341–e348.
17. Nohynek, Hanna, Jukka Jokinen, Markku Partinen, Outi Vaarala, Turkka Kirjavainen, Jonas Sundman, and Sari-Leena Himanen et al. "AS03 adjuvanted AH1N1 vaccine associated with an abrupt increase in the incidence of childhood narcolepsy in Finland." *PloS One* 7, no. 3 (2012): e33536.
18. Yih, W. Katherine, Tracy A. Lieu, Martin Kulldorff, David Martin, Cheryl N. McMahill-Walraven, Richard Platt, Nandini Selvam, Mano Selvan, Grace M. Lee, and Michael Nguyen. "Intussusception risk after rotavirus vaccination in US infants." *New England Journal of Medicine* 370, no. 6 (2014): 503–512.
19. Taylor, Luke E., Amy L. Swerdfeger, and Guy D. Eslick. "Vaccines are not associated with autism: An evidence-based meta-analysis of case-control and cohort studies." *Vaccine* 32, no. 29 (2014): 3623–3629.
20. National Academies of Sciences, Engineering, and Medicine. (2017). The National Vaccine Injury Compensation Program: A Reference Guide. National Academies Press. https://www.ncbi.nlm.nih.gov/books/NBK236425/

21. Moodley, Keymanthri, Kate Hardie, Michael J. Selgelid, Ronald J. Waldman, Peter Strebel, Helen Rees, and David N. Durrheim. "Ethical considerations for vaccination programmes in acute humanitarian emergencies." *Bulletin of the World Health Organization* 91 (2013): 290–297.

22. Henao-Restrepo, Ana Maria, Anton Camacho, Ira M. Longini, Conall H. Watson, W. John Edmunds, Matthias Egger, and Miles W. Carroll et al. "Efficacy and effectiveness of an rVSV-vectored vaccine in preventing Ebola virus disease: Final results from the Guinea ring vaccination, open-label, cluster-randomised trial (Ebola Ça Suffit!)." *Lancet* 389, no. 10068 (2017): 505–518.

20 Neurological and Behaviour Alterations with COVID-19 Vaccines Using the Vaccine Adverse Event Reporting System (VAERS)

A Retrospective Study

Mohd Amir, Saquib Haider and Anoop Kumar

20.1 INTRODUCTION

According to the World Health Organisation (WHO), there were approximately 767,518,723 confirmed cases of COVID-19 as of June 14, 2023, with a total of 6,947,192 deaths globally [1]. Vaccination is an efficient and well-tolerated way to reduce the prevalence of infection and eradicate the coronavirus pandemic. Fortunately, various types of vaccines have been developed and have received emergency approval from the regulatory authorities [2]. The U.S. Food and Drug Administration (FDA) has approved Pfizer-BioNTech Comirnaty, Moderna, and Novavax COVID-19 vaccines, etc. for emergency use for the prevention of SARS-CoV-2 infections [3]. Till now, a total of 11 COVID-19 vaccines have been approved across the globe, and the details are compiled in Table 20.1 [4, 5].

Despite being usually safe, COVID-19 vaccinations have been linked to several adverse events such as headache, fatigue, fever, injection site pain, abdominal pain, dizziness, nausea vomiting, and body aches [6]. One of the most common side effects of the COVID-19 vaccination is headache, which is reported in around half of the patients [7]. And in other studies, Bell's palsy was the most often reported neurological problem [8]. In one study, headache is the most common neurological side effect of the SARS-CoV-2 vaccine, with myelitis, Guillain-Barre syndrome (GBS), and venous sinus thrombosis (VST) coming in second and third [9]. Overall, the correlation of COVID-19 vaccination with neurological events is unclear so far. To the best of our knowledge, one study [3] has been conducted so far to find out the association of COVID-19 vaccines with neurological events using the Vaccine Adverse Event Reporting System (VAERS) database. Guo et al. have used the VAERS database to find out the association of COVID-19 vaccines with various adverse events including neurological; however, the data included were only up to December 31, 2021 [3]. Therefore, there is a need for updated analysis to determine the exact association of COVID-19 vaccines with neurological events. Thus, in the current investigation, we use the data available in the VAERS database up to July 15, 2022, to provide an updated association of COVID-19 vaccines with neurological events.

20.2 MATERIALS AND METHODS

20.2.1 DATABASES

The COVID-19 Vaccine Knowledge Base (Cov19VaxKB) tool (https://violinet.org/) was used to query the VAERS, which is available at https://vaers.hhs.gov/about.html [10–12].

The relevant data were extracted for neurological events related to available COVID-19 vaccines, i.e., Pfizer-BioNTech Comirnaty, Moderna, Johnson & Johnson's Janssen, and unknown vaccines up to July 15, 2022.

20.2.2 SUBGROUP ANALYSIS

Subgroup analysis was done to find out the cases of neurological events with COVID-19 vaccines in different gender and age groups. The age group was categorized as 18–64 (adult) and 65 and above (senior), whereas gender was categorized as male and female. The cases were extracted using the following query in the Cov19VaxKB tool.

- *Vaccine Name*: Select the COVID-19 vaccine
- *USA State/Territory*: Select any
- *Age*: Select the age limit
- *Sex*: Select the gender
- *Click query* for the results

20.2.3 STATISTICAL ANALYSIS

The Cov19VaxKB tool was used to compute the number of cases > 0.2% of total reports and the proportional reporting ratio, i.e., PRR ($\geq$ 2) with associated Chi-squared value (> 4) to identify a possible signal. The methodology is summarized in a flow chart in Figure 20.1. Using the following query in the Cov19VaxKB tool, statistical analysis data was extracted.

- Click on *vaccine safety* to query the data
- Now click on statistical analysis
- *Adverse Event*: Name of adverse event selected
- *COVID-19 Vaccine Name*: COVID-19 vaccine selected
- Click *Run* for the result

20.3 RESULTS

20.3.1 NEUROLOGICAL EVENTS WITH COVID-19 VACCINES

A total of 49,603 cases of neurological events were found with COVID-19 vaccines in the VAERS database. A total of 5,405 cases of ageusia were found with Moderna, Johnson & Johnson's Janssen, and Pfizer-BioNTech Comirnaty vaccines. Out of 5,405 cases, 475 cases were found with Johnson & Johnson's Janssen vaccine and 2,204 cases with Moderna, whereas the remaining 2,726 cases were found with Pfizer-BioNTech Comirnaty.

Cases of anosmia were also found in the VAERS database associated with COVID-19 vaccines. A total of 378 cases of anosmia were found with Johnson & Johnson's Janssen vaccine, whereas 2,326 cases were found with the Pfizer-BioNTech Comirnaty vaccine. The cases of anxiety (4890), dysgeusia (2,684), and Bell's palsy (1,926) were found with Pfizer-BioNTech Comirnaty vaccine. The cases of facial paralysis (293) were found with Johnson & Johnson's Janssen vaccine and (1,473) with Pfizer-BioNTech Comirnaty vaccine.

Cases of seizures (701), chills (11,717), headaches (16,531), sleep disorders (1,119), and parosmia (160) were found with Johnson & Johnson's Janssen vaccine. Furthermore, no cases of neurological events were observed with the Novavax vaccine.

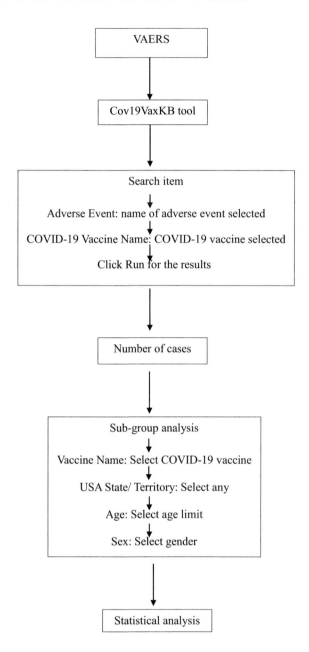

FIGURE 20.1 Flow chart of the general methodology.

20.3.2 ASSOCIATION OF NEUROLOGICAL EVENTS WITH COVID-19 VACCINES

The association of COVID-19 vaccines with neurological events was determined by calculating PRR with the associated Chi-squared value. The strong signal of anosmia was found to be associated with the Pfizer vaccine as indicated by PRR, i.e., 3.317. The disproportionality measure indicated a significant association of COVID-19 vaccines with ageusia, anosmia, anxiety, dysgeusia, Bell's palsy, facial paralysis, seizures, chills, headaches, sleep disorders, and parosmia. The details are compiled in Table 20.1.

TABLE 20.1

Statistical Analysis of Behaviour and Neurological AEs with COVID-19 Vaccines

Serial No.	Cardiovascular Event	Vaccine	PRR	Chi-Squared	Number of Cases	% of Event	Statistically Significant
1.	Ageusia	Janssen	2.007	220.79	475	0.68	Yes
		Moderna	2.011	688.32	2204	0.57	Yes
		Pfizer	2.834	1688.9	2726	0.69	Yes
2.	Anosmia	Janssen	2.027	180.66	378	0.54	Yes
		Pfizer	3.317	1815.5	2326	0.59	Yes
3.	Anxiety	Pfizer	2.057	1603.6	4890	1.23	Yes
4.	Dysgeusia	Pfizer	2.334	1168	2684	0.68	Yes
5.	Bell's palsy	Pfizer	2.954	1272.2	1926	0.48	Yes
6.	Facial paralysis	Janssen	2.053	145.14	293	0.42	Yes
		Pfizer	2.310	625.96	1473	0.37	
7.	Seizure	Janssen	2.436	536.92	701	1	Yes
8.	Chills	Janssen	2.228	7821.8	11,717	16.65	Yes
9.	Headache	Janssen	2.161	10,611	16,531	23.49	Yes
10.	Sleep disorder	Janssen	2.134	620.24	1119	1.59	Yes
11.	Parosmia	Janssen	2.234	99.212	160	0.23	Yes

20.3.3 SUB-GROUP ANALYSIS

Subgroup analysis was done to find cases of neurological events in different gender and age groups. The cases of ageusia (male = 172, female = 287, unknown gender = 16), anosmia (male = 135, female = 229, unknown gender = 14), facial paralysis (male = 146, female = 145, unknown gender = 2), seizures (male = 385, female = 294, unknown gender = 22), chills (male = 3393, female = 7,576, unknown gender = 748), headache (male = 4572, female = 11,053, unknown gender = 906), sleep disorders (male = 350, female = 758, unknown gender = 11), and parosmia (male = 50, female = 104, unknown gender = 6) were found with Johnson & Johnson's Janssen vaccine, whereas the cases of ageusia (male = 823, female = 1853, unknown gender = 50), anosmia (male = 700, female = 1,588, unknown gender = 38), anxiety (male = 1238, female = 3571, unknown gender = 81), dysgeusia (male = 471, female = 2164, unknown gender = 1), Bell's palsy (male = 842, female = 1,044, unknown gender = 40), and facial paralysis (male = 521, female = 922, unknown gender = 30) were found with Pfizer-BioNTech Comirnaty vaccine. A total of 656 cases of ageusia were found in males, 1,484 cases were found in females, and 64 cases were found in unknown gender with the Moderna vaccine.

Most of the neurological events were found in the 18–64-year and 65-year and above age groups. The cases of ageusia (18–64 = 357, 65 and above = 64, unknown age = 54), anosmia (18–64 = 298, 65 and above = 49, unknown age = 31), facial paralysis (18–64 = 226, 65 and above = 52, unknown age = 15), seizures (18–64 = 603, 65 and above = 37, unknown age = 61), chills (18–64 = 9162, 65 and above = 670, unknown age = 1885), headaches (18–64 = 12,831, 65 and above = 1,087, unknown age = 2,613), sleep disorders (18–64 = 994, 65 and above = 100, unknown age = 25), and parosmia (18–64 = 125, 65 and above = 14, unknown age = 21) were found with Johnson & Johnson's Janssen vaccine, whereas the cases of ageusia (18–64 = 1,900, 65 and above = 527, unknown age = 299), anosmia (18–64 = 1,690, 65 and above = 373, unknown age = 263), anxiety (18–64 = 4,284, 65 and above = 605, unknown age = 1), Bell's palsy (18–64 = 1,350, 65 and above = 360, unknown age = 216), and facial paralysis (18–64 = 1,012, 65 and above = 326, unknown age = 135) were found with Pfizer-BioNTech Comirnaty vaccine, and the cases of ageusia (18–64 = 1,399, 65 and above = 629, unknown age = 2028) were associated with the Moderna vaccine. The details of the subgroup analysis are compiled in Table 20.2.

TABLE 20.2

Subgroup Analysis

Serial No.	Neurological Events	Vaccine	Characteristics	Number of Cases (AE)	Total Cases
1.	Ageusia	Janssen	Male	172	475
			Female	287	
			Unknown gender	16	
			18–64 years	357	475
			65 years and above	64	
			Unknown age	54	
		Moderna	Male	656	2,204
			Female	1,484	
			Unknown gender	64	
			18–64 years	1,399	2,204
			65 years and above	629	
			Unknown age	2,028	
		Pfizer	Male	823	2,726
			Female	1,853	
			Unknown gender	50	
			18–64 years	1,900	2,726
			65 years and above	527	
			Unknown age	299	
2.	Anosmia	Janssen	Male	135	378
			Female	229	
			Unknown gender	14	
			18–64 years	298	378
			65 years and above	49	
			Unknown age	31	
		Pfizer	Male	700	2,326
			Female	1,588	
			Unknown gender	38	
			18–64 years	1,690	2,326
			65 years and above	373	
			Unknown age	263	
3.	Anxiety	Pfizer	Male	1,238	4,890
			Female	3,571	
			Unknown gender	81	
			18–64 years	4,284	4,890
			65 years and above	605	
			Unknown age	1	
4.	Dysgeusia	Pfizer	Male	471	2,684
			Female	2,164	
			Unknown gender	49	
			18–64 years	2,126	2,684
			65 years and above	331	
			Unknown age	227	
5.	Bell's palsy	Pfizer	Male	842	1,926
			Female	1,044	
			Unknown gender	40	
			18–64 years	1,350	1,926
			65 years and above	360	
			Unknown age	216	

(Continued)

TABLE 20.2 *(Continued)*

Subgroup Analysis

Serial No.	Neurological Events	Vaccine	Characteristics	Number of Cases (AE)	Total Cases
6.	Facial paralysis	Janssen	Male	146	293
			Female	145	
			Unknown gender	2	
			18–64 years	226	293
			65 years and above	52	
			Unknown age	15	
		Pfizer	Male	521	1473
			Female	922	
			Unknown gender	30	
			18–64 years	1,012	1473
			65 years and above	326	
			Unknown age	135	
7.	Seizures	Janssen	Male	385	701
			Female	294	
			Unknown gender	22	
			18–64 years	603	701
			65 years and above	37	
			Unknown age	61	
8.	Chills	Janssen	Male	3,393	11,717
			Female	7,576	
			Unknown gender	748	
			18–64 years	9,162	11,717
			65 years and above	670	
			Unknown age	1,885	
9.	Headaches	Janssen	Male	4,572	16,531
			Female	11,053	
			Unknown gender	906	
			18–64 years	12,831	16,531
			65 years and above	1,087	
			Unknown age	2,613	
10.	Sleep disorders	Janssen	Male	350	1,119
			Female	758	
			Unknown gender	11	
			18–64 years	994	1,119
			65 years and above	100	
			Unknown age	25	
11.	Parosmia	Janssen	Male	50	160
			Female	104	
			Unknown gender	6	
			18–64 years	125	160
			65 years and above	14	
			Unknown age	21	

20.4 DISCUSSION

A total of 49,603 cases of neurological events were found with COVID-19 vaccines in the VAERS database found with Moderna, Johnson & Johnson's Janssen, and Pfizer-BioNTech Comirnaty vaccines. A signal is new information related to the safety of medicines. The most used methods for signal detection are frequency-based and Bayesian methods [13]. In the present study, we used the frequency-based method in the identification of neurological events as novel signals with COVID-19 vaccines. Researchers across the globe have also tried to identify the various signals associated with COVID-19 vaccines. Noseda et al. have conducted a disproportionality analysis to find out the association of acute inflammatory neuropathies with COVID-19 vaccines using VigiBase; however, more data is required to confirm the signals [14]. Very few studies have been conducted so far to identify signals related to the neurological system with COVID-19 vaccines.

Guo et al. have used the VAERS database to find an association of COVID-19 vaccines with various adverse events including neurological; however, data was included up to December 31, 2021 [3]. Therefore, the number of cases reported earlier is different from the cases reported by the current study. For example, Guo et al. [3] stated that in the combination of the Pfizer, Moderna, and Janssen vaccines, headaches (17.5%) were the most often reported side effects. The safety of co-administration of mRNA COVID-19 and seasonal inactivated influenza vaccines was checked by Moro et al. using the VAERS database; however, no unique or unexpected patterns of AEs were observed [15]. The safety profile of COVID-19 mRNA vaccines using VAERS was also checked by Laurini et al. [16]. Moro et al. have also used a VAERS database to examine the adverse events related to COVID-19 vaccines in pregnant women. No significant patterns of maternal or infant-foetal outcomes were detected concerning COVID-19 vaccinations [17]. Emerging case reports and case series have reported the association of COVID-19 vaccines with neurological events [18–20]. Recently, Amir et al. [21] have also identified a signal of various cardiovascular events with the COVID-19 vaccines using Vaccine Adverse Event Reporting System (VAERS) database. The results of the current investigation have also indicated the association of COVID-19 vaccines with neurological events.

20.5 CONCLUSION

In conclusion, the study identified neurological events linked to COVID-19 vaccinations, including ageusia, anosmia, anxiety, dysgeusia, Bell's palsy, facial paralysis, seizure, chills, headache, sleep disorder and parosmia, and other complications. A sub-group analysis was carried out based on age and gender, showing cases from various age groups and genders. This implies a potential signal for unfavourable neurological health problems. The results indicated a statistically significant association between neurological events and COVID-19 vaccines. However, further causality assessment is required to confirm the association.

20.6 LIMITATIONS

The current study used the data of VAERS, which is based on passive surveillance systems. Therefore, underreporting is one of the limitations. Furthermore, it contains only data from Pfizer-BioNTech Comirnaty, Johnson & Johnson's Janssen, and the Moderna vaccine. Therefore, other COVID-19 vaccines were not assessed.

ACKNOWLEDGEMENTS

The authors would like to acknowledge the Cov19VaxKB tool for providing clean data of VAERS.

REFERENCES

1. World Health Organisation (WHO) [Internet]. WHO Coronavirus (COVID-19) Dashboard [cited 2023 June 14]. Available from: https://covid19.who.int/
2. Singh C, Naik BN, Pandey S, Biswas B, Pati BK, Verma M, Singh PK. Effectiveness of COVID-19 vaccine in preventing infection and disease severity: a case-control study from an Eastern State of India. Epidemiol Infect. 2021;149:e224. https://doi.org/10.1017/S0950268821002247.
3. Guo W, Deguise J, Tian Y, Huang PC, Goru R, Yang Q, Peng S, Zhang L, Zhao L, Xie J, He Y. Profiling COVID-19 vaccine adverse events by statistical and ontological analysis of VAERS case reports. Front Pharmacol. 2022;13. https://doi.org/10.3389/fphar.2022.870599.

4. Koirala A, Joo YJ, Khatami A, Chiu C, Britton PN. Vaccines for COVID-19: the current state of play. Paediatr Respir Rev. 2020;35:43–49. https://doi.org/10.1016/j.prrv.2020.06.010.

5. Han X, Xu P, Ye Q. Analysis of COVID-19 vaccines: types, thoughts, and application. J Clin Lab Anal. 2021;35(9):e23937. https://doi.org/10.1002/jcla.23937.

6. Corrêa DG, Cañete LA, Dos Santos GA, de Oliveira RV, Brandão CO, da Cruz Jr LC. Neurological symptoms and neuroimaging alterations related with COVID-19 vaccine: cause or coincidence? Clin Imaging. 2021 Dec 1;80:348–352. https://doi.org/10.1016/j.clinimag.2021.08.021.

7. Garg RK, Paliwal VK. Spectrum of neurological complications following COVID-19 vaccination. Neurol Sci. 2022 Jan;43(1):3–40. https://doi.org/10.1007/s10072-021-05662-9.

8. Castillo RA, Castrillo JM. Manifestations associated with COVID-19 vaccine. Neurología (English Edition). 2022 Oct 23. https://doi.org/10.1016/j.nrleng.2022.09.007.

9. Finsterer J. Neurological side effects of SARS-CoV-2 vaccinations. Acta Neurol Scand. 2022 Jan;145(1):5–9. https://doi.org/10.1111/ane.13550.

10. Huang PC, Goru R, Huffman A, Lin AY, Cooke MF, He Y. Cov19VaxKB: a web-based integrative COVID-19 vaccine knowledge base. Vaccine: X. 2022;10:100139. https://doi.org/10.1016/j.jvacx.2021.100139.

11. Cai Y, Du J, Huang J, Ellenberg SS, Hennessy S, Tao C, Chen Y. A signal detection method for temporal variation of adverse effect with vaccine adverse event reporting system data. BMC Med Inform Decis Mak. 2017;17(2):93–100.

12. Shimabukuro TT, Nguyen M, Martin D, DeStefano F. Safety monitoring in the vaccine adverse event reporting system (VAERS). Vaccine. 2015;33(36):4398–405. https://doi.org/10.1016/j.vaccine.2015.07.035.

13. Park G, Jung H, Heo SJ, Jung I. Comparison of data mining methods for the signal detection of adverse drug events with a hierarchical structure in postmarketing surveillance. Life. 2020;10(8):138. https://doi.org/10.3390/life10080138.

14. Noseda R, Ripellino P, Ghidossi S, Bertoli R, Ceschi A. Reporting of acute inflammatory neuropathies with COVID-19 vaccines: subgroup disproportionality analyses in VigiBase. Vaccines. 2021 Sep 14;9(9):1022. https://doi.org/10.3390/vaccines9091022.

15. Moro PL, Zhang B, Ennulat C, Harris M, McVey R, Woody G, Marquez P, McNeil MM, Su JR. Safety of co-administration of mRNA CoVID-19 and seasonal inactivated influenza vaccines in the vaccine adverse event reporting system (VAERS) during July 1, 2021–June 30, 2022. Vaccine. 2023;41(11):1859–1863. https://doi.org/10.1016/j.vaccine.2022.12.069.

16. Laurini GS, Montanaro N, Broccoli M, Bonaldo G, Motola D. Real-life safety profile of mRNA vaccines for COVID-19: an analysis of VAERS database. Vaccine. 2023;41(18):2879–2886. https://doi.org/10.1016/j.vaccine.2023.03.054.

17. Moro PL, Olson CK, Clark E, Marquez P, Strid P, Ellington S, Zhang B, Mba-Jonas A, Alimchandani M, Cragan J, Moore C. Post-authorization surveillance of adverse events following COVID-19 vaccines in pregnant persons in the vaccine adverse event reporting system (VAERS), December 2020–October 2021. Vaccine. 2022;40(24):3389–3394. https://doi.org/10.1016/j.vaccine.2022.04.031.

18. Yang Y, Huang L. Neurological disorders following COVID-19 vaccination. Vaccines. 2023 Jun 19;11(6):1114. https://doi.org/10.3390/vaccines11061114.

19. Tondo G, Virgilio E, Naldi A, Bianchi A, Comi C. Safety of COVID-19 vaccines: spotlight on neurological complications. Life. 2022 Aug 29;12(9):1338. https://doi.org/10.3390/life12091338.

20. Rosenblum HG, Hadler SC, Moulia D, Shimabukuro TT, Su JR, Tepper NK, Ess KC, Woo EJ, Mba-Jonas A, Alimchandani M, Nair N. Use of COVID-19 vaccines after reports of adverse events among adult recipients of Janssen (Johnson & Johnson) and mRNA COVID-19 vaccines (Pfizer-BioNTech and Moderna): update from the Advisory Committee on Immunization Practices—United States, July 2021. Morb Mortal Wkly Rep. 2021 Aug 8;70(32):1094. 10.15585/mmwr.mm7032e4.

21. Amir M, Latha S, Sharma R, Kumar A. Association of cardiovascular events with COVID-19 vaccines using vaccine adverse event reporting system (VAERS): A retrospective study. Curr Drug Saf. 2023 Nov 28. doi: 10.2174/0115748863276904231108095255.

21 Pharmacovigilance-Based Drug Repurposing via Inverse Signals

Simran Ohra and Anoop Kumar

21.1 INTRODUCTION

21.1.1 SIGNALS

A signal is any reported information on a new side effect (or adverse event [AE]) that may be associated with the use of a drug or it can be on a previously known side effect with new information such as a change in a dose range in which the effect occurs and change in severity, indicating a more susceptible population. However, a signal is only a hypothesis and it does not indicate a direct causal relationship between a drug and a side effect, and further causality assessment is required to validate a signal. Generally, a signal is considered as the harmful effect of a drug; however, a signal can be both a harmful and a beneficial effect of a drug. This can be viewed as a concept for drug repurposing to explore a drug's side effects, which eventually turned out to be beneficial for the patients. A signal can be obtained from various sources such as clinical reports, scientific literature, and post-marketing data [1–3]. A signal can be identified using disproportionality analysis, in which a drug and AE association is considered a signal when it satisfies Evans' criteria, i.e., ROR > 2, the lower limit of 95% CI (confidence interval) > 1.0, and no. of cases > 3 (Figure 21.1) [4].

21.1.1.1 Inverse Signals

Based on the approaches followed by researchers, an inverse signal can be defined as follows:

In terms of disproportionality analysis, an inverse signal follows the criteria ROR < 1 and upper limit of 95% CI < 1 as validated by a few researchers. For example, a researcher needs to identify potential drugs for the treatment of viral infections using the AE search approach. First, the researcher would search in the pharmacovigilance database or tools (e.g., FAERS, OpenVigil 2.1) for the AE terms related to the viral infection such as viral upper respiratory tract infections and influenza viral infection, using MedDRA terminologies. Then apply the inverse signal criteria of disproportionality analysis, i.e., ROR < 1, and upper limit of 95% CI < 1 to search for the drugs associated with these AE terms. Drugs that fulfill this criterion could be useful for generating the hypothesis for drug repurposing against viral infections [5, 6].

Another approach is by finding the inverse association with the disease, i.e., using the AE terms that are inversely related to the disease (or opposite of the disease pathogenesis) for which one wants to propose a hypothesis for the treatment. For example, Raynaud's phenomenon (RP) is a condition in which vascular blood flow is decreased to the extremities resulting in abrupt color changes of fingers and toes. Drugs were identified for repurposing against RP by searching for the condition opposite to the vascular pathophysiology of RP, i.e., erythromelalgia, increased blood flow, and vasodilation. These AE terms (erythromelalgia etc.) were searched in the pharmacovigilance databases to identify the drugs associated with these terms and could be further used for hypothesis generation for drug repurposing for the treatment of RP [7]. The flow of inverse signal detection is presented in Figure 21.2.

DOI: 10.1201/9781032629940-21

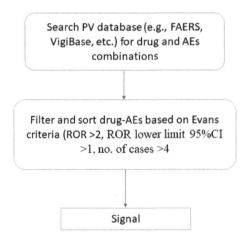

FIGURE 21.1   Flow to identify a signal from a pharmacovigilance database.

21.2   DRUG REPURPOSING

Traditional drug discovery is a long and expensive process that can take up to 10–15 years [8], and the total average cost ranges from $2 to $3 billion to develop and approve a new drug [9]. Drug repurposing is a strategy that is used to find an already approved drug for a novel indication for which it has not been explored yet. The drug repurposing approach offers various advantages over traditional methods by reducing the time frame, cost of development, and risk of failure because the safety of the repurposed drug is sufficiently established; hence, most of the preclinical and safety testing can be bypassed, and the drug can enter into the late phase of the drug development process [10]. Some examples of repurposed drugs are given in Table 21.1.

Various databases are available that provide information on drugs, diseases, genes, proteins, and drug-target interactions that can be used for drug repurposing. Examples include DrugComb, Ensembl, UniProt, Search Tool for the Retrieval of Interacting Genes (STRING), Dependency Map (DepMap), Protein Data Bank (PDB), repoDB, and VigiAccess [20]. Apart from these, clinical trial databases (e.g., clinicalTrials.gov) [21], social media [22], FDA labels [23], etc. can also be used to obtain the information for drug repurposing.

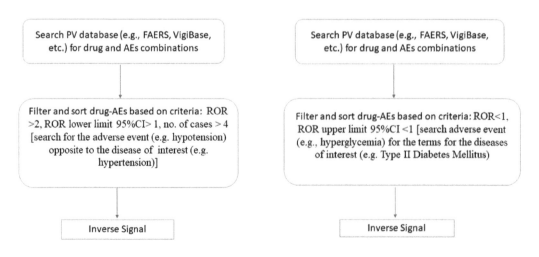

FIGURE 21.2   Methods of drug repurposing.

TABLE 21.1

List of Some Successful Repurposed Drugs with Their Future Repurposing Potential

S no.	Drug	Original Indication	Repurposed Indication	Further Exploration	Year of Repurposing	Reference
1.	Raloxifene	Osteoporosis	Breast cancer prevention		2007	[11]
2.	Thalidomide	Hypnotic sedative, anti-emetic for morning sickness	Plasma cell myeloma (in combination therapy)	Nano-sized thalidomide analog as anticancer	2006	[12, 13]
3.	Minoxidil	Anti-hypertensive	Androgenetic alopecia and alopecia areata	–	1988	[14, 15]
4.	Sildenafil	Anti-hypertensive	Erectile dysfunction and pulmonary arterial hypertension	Premature ejaculation, female sexual dysfunction, cardiovascular disease, myocardial infarction and ischemia / reperfusion injury, neurological research, pulmonary arterial hypertension	1988	[4, 5, 16]
5.	Aspirin	Analgesic	Antiplatelet aggregation drug	Colorectal cancer	1980	[17]
6.	Amantadine	Influenza	Parkinson's disease	COVID-19	1973	[18, 19]

21.2.1 TECHNIQUES FOR DRUG REPURPOSING

The drug repurposing approach is widely used by research institutes, pharmaceutical companies, repurposing technology companies, and academicians. Their strategic approach for drug repurposing may differ from each other such as academicians usually focus on *in silico* or high-throughput screening (HTS) methods, whereas pharmaceutical companies mainly focus on the late-stage development process of the molecule. These strategic approaches can be broadly classified into two main categories (Figure 21.3). First is the experimental approach in which molecules with known safety and mechanism of action are screened and serve as a source of hits for repurposing [24], for example, target-based screening, drug-centric approach [25], and phenotypic approach [26].

Target-based screening: Experimental target-based screening involves systematically testing existing drugs against specific biological targets or pathways associated with a particular disease. This approach aims to identify potential drug candidates based on their ability to interact with the intended target [25].

Drug-centric approach: The experimental drug-centric approach in drug repurposing focuses on exploring existing drugs, often with known safety profiles, for new therapeutic indications. Instead of targeting specific biological pathways or targets, this approach assesses the broader pharmacological effects of drugs and identifies potential candidates based on their overall impact on cellular and physiological processes [25].

Phenotypic approach: The experimental phenotypic approach in drug repurposing focuses on evaluating the observable effects of existing drugs on disease-relevant phenotypes. Unlike target-centric approaches, this method aims to identify compounds that exhibit desirable changes in cellular or physiological behaviors, offering a holistic perspective on potential therapeutic applications [26].

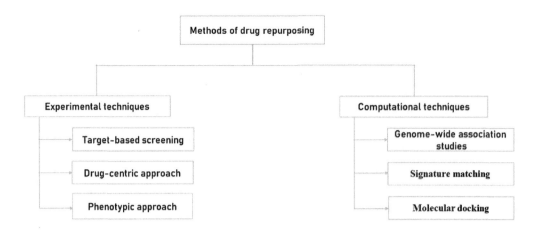

FIGURE 21.3   Methods of drug repurposing.

Second is the computational approach, which uses complex, standardized data from various sources such as clinical trial reports, gene expression profiles, and protein networks to further use this information for drug repurposing [27]. Examples include genome-wide association studies (GWAS) [28], signature matching [29], and molecular docking [30].

GWAS: GWAS are a computational genetics approach used to identify genetic variants associated with specific traits or diseases across the entire genome. In the context of drug repurposing, GWAS can provide insights into the genetic basis of diseases, helping to identify potential drug targets and repurposing candidates [28].

Signature matching in drug repurposing: Signature matching is a computational approach in drug repurposing that involves comparing the molecular or phenotypic signatures of diseases or patients with those of existing drugs. The goal is to identify drugs that exhibit patterns or characteristics similar to those associated with a particular condition, suggesting potential repurposing opportunities [29].

Molecular docking in drug repurposing: Molecular docking is a computational technique used in drug repurposing to predict and analyze the interactions between small molecules (drugs) and their target proteins at the atomic level. This approach plays a crucial role in understanding the binding mechanisms and potential efficacy of existing drugs against new targets [30].

21.3   PHARMACOVIGILANCE-BASED APPROACHES

Pharmacovigilance includes the activities to detect, assess, and understand the adverse effects or any other problem related to the drug [31] to further analyze the underlying cause of the occurrence of these effects or events. Pharmacovigilance is conventionally used for the sole purpose of detecting and preventing the adverse drug reactions (ADRs) that are associated with the use of a drug. However, recent advances in the technology and the potential of the data obtained from the real-world population in a pharmacovigilance study have taken this discipline to a newer level. Now, with the use of computational tools and technologies pharmacovigilance can be used for repurposing drugs [32]. There are various companies such as Biovista (https://www.biovista.com/), Numedicus (https://www.numedicus.co.uk/methods-and-output/), CureHunter (https://www.cure-hunter.com/public/showTopPage.do), Cures Within Reach (https://www.cureswithinreach.org/), and Aragen (https://www.aragen.com/) that are purposefully working on drug repurposing synergistically with pharmacovigilance [32, 33].

The adverse effects information provides pharmacological information about the drug which can be used to develop drug profiles and compare the similarity between molecules. This is one of the approaches to exploring an already approved drug for a newer indication, which is based on the idea that the drug that has a common adverse reaction profile should work by a similar mechanism of action. This approach requires post-marketing data to obtain the adverse effect profile of the drugs. This similarity in drug adverse effects profile could be used to compare, analyze, and predict drug's actions and applied for repurposing of drugs [34].

21.3.1 CASE STUDIES

Recently, few studies have used a pharmacovigilance-based approach in the identification of potential repurposing drugs against various diseases. The details of these studies are mentioned below.

Battini et al. have identified angiotensin-converting enzyme inhibitors (ACEi), adrenergic beta-antagonists (beta-blockers), dipeptidyl-peptidase-IV inhibitors (DPP-4i) in neurological patients using pharmacovigilance-based approach [35].

Liu et al. identified various drug classes (antidepressants, antiepileptics, and drugs for urinary frequency and incontinence) that could be used for the treatment of hyperhidrosis using the FAERS database and inverse signal approach [36].

Battini et al. [35] focused on exploring the potential antidepressant effects of antidiabetic drugs using extensive population data derived from the FAERS and VigiBase. The methods involved the identification of cases (depressed patients experiencing therapy failure) and non-cases (depressed patients encountering any other AE) from primary cohorts of patients treated with antidepressants in both databases. Various disproportionality scores, including Reporting Odds Ratio (ROR), Proportional Reporting Ratio (PRR), Empirical Bayes Geometric Mean (EBGM), and Empirical Bayes Regression-Adjusted Mean (ERAM), were then computed to compare cases to non-cases regarding concurrent exposure to specific antidiabetic agents, such as biguanides, sulfonylureas, thiazolidinediones, DPP4-inhibitors, GLP-1 analogs, and SGLT2 inhibitors—agents supported by preliminary evidence from the literature. The results, particularly for GLP-1 analogs, consistently demonstrated statistically significant decreases in all disproportionality scores in both analyses [37].

Ko et al. [38] have used pharmacovigilance-based drug repurosing techniques in the identification of marketed drugs against psoriasis. The drug and AE were extracted from FAERS and ROR, information component (IC), and empirical Bayes geometric mean (EBGM) were used to calculate inverse signals [38].

Paci et al. have developed an algorithm SAveRUNNER to reposition drugs for cardiovascular diseases with fewer ADRs [39].

Nakagawa et al. [40] have identified a repurposing drug against rheumatoid arthritis (RA) using a pharmacovigilance-based approach. The drug and AE data were extracted from FAERS and JMDC. The ROR and IC were calculated. The authors have reported a potential inverse association between the antipsychotic haloperidol and RA. Furthermore, the possible antirheumatic mechanism of haloperidol was identified using bioinformatics tools [40].

Böhm et al. [6] searched for inverse signals for drugs against viral respiratory infection using the OpenVigil 2.1 tool-MedDRA17. They have used pharmacovigilance data from the FAERS and query the OpenVigil 2.1 tool to extract the information for the association of drugs and AEs that occur less frequently than expected to provide hypotheses for drug repurposing. The study showed gestagens and tyrosine kinase inhibitors as new putative therapeutic concepts for viral respiratory infections [6].

The other databases like VigiBase were also used in the identification of repurposing candidates. Chrétien et al. [41] have identified four potential drugs that can be repurposed against Alzheimer's disease (AD) using disproportionality analysis. Furthermore, the potential anti-AD activity was also confirmed experimentally [41].

Yokoyama et al. have identified an alternative use of digoxin as an anticancer agent using FAERS and JMDC integrated with bioinformatic tools [42].

Zaza et al. [7] suggested the WHO pharmacovigilance database as a powerful tool for repurposing drugs and found that fumaric acid could be an effective drug for the treatment of secondary RP using the AE signature matching approach. They have used the AE signature (unique characteristics) of drugs to generate the hypothesis for drug repurposing [7].

de Anda-Jáuregui et al. [43] have explored the relationship between anti-inflammatory drugs and functional pathways of ADRs through network models. The information related to ADRs was extracted from the US Food and Drug Administration (FDA) Adverse Event Reporting System (FAERS). Finally, they have identified naproxen, meloxicam, etodolac, tenoxicam, flufenamic acid, fenoprofen, and nabumetone as candidates for drug repurposing with lower ADR risk [43].

Hosomi et al. [5] have performed an analysis of real-world data from the FAERS and Japan Medical Data Center (JMDC) database. The cross-validation was done using bioinformatic tools to analyze gene expression data. The potential drug reposition signals were identified using disproportionality analysis (DPA) and sequence symmetry analysis (SSA). The analysis of the FAERS and JMDC database identified haloperidol, diazepam, and hydroxyzine as potential repurposing candidates for the treatment of inflammatory bowel disease (IBD) [5].

21.4   CURRENT CHALLENGES AND FUTURE PERSPECTIVE

Navigating through the pharmacovigilance and drug repurposing confronts an array of challenges in the pharmaceutical research. Integrating data from various sources, including electronic health records, clinical trials, and spontaneous reporting systems, poses a challenge due to differences in formats and standards. Ensuring the accuracy and completeness of the data is crucial for meaningful signal detection and analysis. Differentiating meaningful signals of potential drug repurposing opportunities from background noise or confounding factors requires advanced analytical methods and expertise. There is a need for a skilled workforce with expertise in both pharmacovigilance and drug repurposing, which may be limited. Adequate funding for research in pharmacovigilance-based drug repurposing is needed, especially for large-scale studies and data analytics.

In the future, the field of pharmacovigilance and drug repurposing is set to undergo transformative changes, leveraging advanced technologies and a wealth of data for more efficient and effective healthcare solutions. Artificial intelligence and machine learning will play a pivotal role in identifying potential drug repurposing candidates. These technologies will sift through vast amounts of patient data, genetic information, and clinical trial results to discover novel uses for existing drugs. This will significantly expedite the drug development process and reduce costs. Real-time monitoring systems will become more sophisticated, allowing for continuous surveillance of drug safety. Integrated with wearable devices, these systems will provide immediate feedback on how patients respond to repurposed drugs, ensuring a proactive approach to pharmacovigilance. Collaboration between pharmaceutical companies, research institutions, and regulatory bodies will be streamlined through interconnected databases and shared platforms. This collaborative ecosystem will facilitate quicker identification of repurposing opportunities and enable faster regulatory approvals. Ethical considerations and privacy safeguards will be paramount, ensuring that the increased use of patient data is done responsibly and transparently. Regulatory frameworks will evolve to accommodate the dynamic nature of drug repurposing and the integration of advanced technologies.

Overall, the future perspective of pharmacovigilance-based drug repurposing is marked by agility, collaboration, and a patient-centric approach, driving innovation in healthcare and improving treatment outcomes.

REFERENCES

1. European Medicine Agency Science Medicine and Health, Signal Management. https://www.ema.europa.eu/en/human-regulatory/post-authorisation/pharmacovigilance/signal-management (accessed September 01, 2023).
2. Practical Aspects of Signal Detection in Pharmacovigilance, Report of CIOMS Working Group VIII. https://cioms.ch/wp-content/uploads/2018/03/WG8-Signal-Detection.pdf (accessed September 01, 2023).

3. Uppsala Monitoring Centre. What is signal? https://who-umc.org/signal-work/what-is-a-signal/ (accessed September 01, 2023).

4. Kumar, Vijay, Anand Prakash Singh, Nicholas Wheeler, Cristi L. Galindo, and Jong-Joo Kim. "Safety profile of D-penicillamine: A comprehensive pharmacovigilance analysis by FDA adverse event reporting system." *Expert Opinion on Drug Safety* 20, no. 11 (2021): 1443–1450.

5. Hosomi, Kouichi, Mai Fujimoto, Kazutaka Ushio, Lili Mao, Juran Kato, and Mitsutaka Takada. "An integrative approach using real-world data to identify alternative therapeutic uses of existing drugs." *PLoS One* 13, no. 10 (2018): e0204648.

6. Böhm, Ruwen, Claudia Bulin, Vicki Waetzig, Ingolf Cascorbi, Hans-Joachim Klein, and Thomas Herdegen. "Pharmacovigilance-based drug repurposing: The search for inverse signals via OpenVigil identifies putative drugs against viral respiratory infections." *British Journal of Clinical Pharmacology* 87, no. 11 (2021): 4421–4431.

7. Zaza, Putkaradze, Roustit Matthieu, Cracowski Jean-Luc, and Khouri Charles. "Drug repurposing in Raynaud's phenomenon through adverse event signature matching in the World Health Organization pharmacovigilance database." *British Journal of Clinical Pharmacology* 86, no. 11 (2020): 2217–2222.

8. Parvathaneni, Vineela, Nishant S. Kulkarni, Aaron Muth, and Vivek Gupta. "Drug repurposing: A promising tool to accelerate the drug discovery process." *Drug Discovery Today* 24, no. 10 (2019): 2076–2085.

9. Park, Kyungsoo. "A review of computational drug repurposing." *Translational and Clinical Pharmacology* 27, no. 2 (2019): 59–63.

10. Pushpakom, Sudeep, Francesco Iorio, Patrick A. Eyers, K. Jane Escott, Shirley Hopper, Andrew Wells, and Andrew Doig et al. "Drug repurposing: Progress, challenges and recommendations." *Nature Reviews Drug Discovery* 18, no. 1 (2019): 41–58.

11. Policy brief, Repurposing of medicines in oncology – the underrated champion of sustainable innovation. https://www.gov.si/assets/ministrstva/MZ/DOKUMENTI/pomembni-dokumenti/Policy-brief_Repurposing-of-medicines-in-oncology-the-underrated-champion-of-sustainable-innovation.pdf (accessed July 31, 2023).

12. Rehman, Waqas, Lisa M. Arfons, and Hillard M. Lazarus. "The rise, fall and subsequent triumph of thalidomide: Lessons learned in drug development." *Therapeutic Advances in Hematology* 2, no. 5 (2011): 291–308.

13. Ali, Imran, Waseem A Wani, Kishwar Saleem, and Ashanul Haque. "Thalidomide: A banned drug resurged into future anticancer drug." *Current Drug Therapy* 7, no. 1 (2012): 13–23.

14. Massey, Thomas H., and N. P. Robertson. "Repurposing drugs to treat neurological diseases." *Journal of Neurology* 265 (2018): 446–448.

15. Bhar, Shanta. "Role of Drug Repurposing in Sustainable Drug Discovery." (2023).

16. Jackson, Graham, H. Gillies, and I. Osterloh. "Past, present, and future: A 7-year update of Viagra®(sildenafil citrate)." *International Journal of Clinical Practice* 59, no. 6 (2005): 680–691.

17. Jourdan, Jean-Pierre, Ronan Bureau, Christophe Rochais, and Patrick Dallemagne. "Drug repositioning: A brief overview." *Journal of Pharmacy and Pharmacology* 72, no. 9 (2020): 1145–1151.

18. Abreu, GE Aranda, ME Hernández Aguilar, D. Herrera Covarrubias, and F. Rojas Durán. "Amantadine as a drug to mitigate the effects of COVID-19." *Med Hypotheses* 140, no. 109755 (2020): 10–1016.

19. Stott, Simon RW, Richard K. Wyse, and Patrik Brundin. "Drug repurposing for Parkinson's disease: The international linked clinical trials experience." *Frontiers in Neuroscience* 15 (2021): 653377.

20. Tanoli, Ziaurrehman, Umair Seemab, Andreas Scherer, Krister Wennerberg, Jing Tang, and Markus Vähä-Koskela. "Exploration of databases and methods supporting drug repurposing: A comprehensive survey." *Briefings in Bioinformatics* 22, no. 2 (2021): 1656–1678.

21. Su, Eric Wen. "Drug repositioning by mining adverse event data in ClinicalTrials.gov." *Computational Methods for Drug Repurposing* 1903 (2019): 61–72.

22. Nugent, Timothy, Vassilis Plachouras, and Jochen L. Leidner. "Computational drug repositioning based on side-effects mined from social media." *PeerJ Computer Science* 2 (2016): e46.

23. Fang, Hong, Stephen Harris, Zhichao Liu, Shraddha Thakkar, Junshuang Yang, Taylor Ingle, Joshua Xu, Lawrence Lesko, Lilliam Rosario, and Weida Tong. "FDALabel for drug repurposing studies and beyond." *Nature Biotechnology* 38, no. 12 (2020): 1378–1379.

24. Cha, Y., T. Erez, I. J. Reynolds, D. Kumar, J. Ross, G. Koytiger, and R. Kusko et al. "Drug repurposing from the perspective of pharmaceutical companies." *British Journal of Pharmacology* 175, no. 2 (2018): 168–180.

25. Ng, Yan Ling, Cyrill Kafi Salim, and Justin Jang Hann Chu. "Drug repurposing for COVID-19: Approaches, challenges and promising candidates." *Pharmacology & Therapeutics* 228 (2021): 107930.

26. Reaume, Andrew G. "Drug repurposing through nonhypothesis driven phenotypic screening." *Drug Discovery Today: Therapeutic Strategies* 8 (2011): 85–88.
27. Hodos, Rachel A., Brian A. Kidd, Khader Shameer, Ben P. Readhead, and Joel T. Dudley. "In silico methods for drug repurposing and pharmacology." *Wiley Interdisciplinary Reviews: Systems Biology and Medicine* 8, no. 3 (2016): 186–210.
28. Reay, William R., and Murray J. Cairns. "Advancing the use of genome-wide association studies for drug repurposing." *Nature Reviews Genetics* 22, no. 10 (2021): 658–671.
29. Shukla, Rammohan, Nicholas D. Henkel, Khaled Alganem, Abdul-rizaq Hamoud, James Reigle, Rawan S. Alnafisah, and Hunter M. Eby et al. "Signature-based approaches for informed drug repurposing: Targeting CNS disorders." *Neuropsychopharmacology* 46, no. 1 (2021): 116–130.
30. Kumar, Shivani, and Suresh Kumar. "Molecular docking: a structure-based approach for drug repurposing." In *In Silico Drug Design*, pp. 161–189. Academic Press, 2019.
31. New Drugs and Clinical Trial Rules, 2019. https://cdsco.gov.in/opencms/export/sites/CDSCO_WEB/Pdf-documents/NewDrugs_CTRules_2019.pdf (accessed on October 05, 2023).
32. Flower, D. R. "Pharmacovigilance, drug repositioning, and virtual screening." *J Pharmacovigilance* 1, no. 1 (2013): 100–103.
33. Naylor, David M., D. M. Kauppi, and J. M. Schonfeld. "Therapeutic drug repurposing, repositioning and rescue." *Drug Discovery* 57 (2015).
34. Vilar, Santiago, and George Hripcsak. "The role of drug profiles as similarity metrics: Applications to repurposing, adverse effects detection and drug–drug interactions." *Briefings in Bioinformatics* 18, no. 4 (2017): 670–681.
35. Battini, Vera, Sara Rocca, Greta Guarnieri, Anna Bombelli, Michele Gringeri, Giulia Mosini, and Marco Pozzi et al. "On the potential of drug repurposing in dysphagia treatment: New insights from a real-world pharmacovigilance study and a systematic review." *Frontiers in Pharmacology* 14 (2023): 1057301.
36. Liu, Yi, Rongrong Fan, Nurmuhammat Kehriman, Xiaohong Zhang, Bin Zhao, Lin Huang. "Pharmacovigilance-based drug repurposing: Searching for putative drugs with hypohidrosis or anhidrosis adverse events for use against hyperhidrosis." *European Journal of Medical Research* 28 (2023): 95.
37. Battini, Vera, Robbert P. Van Manen, Michele Gringeri, Giulia Mosini, Greta Guarnieri, Anna Bombelli, and Marco Pozzi et al. "The potential antidepressant effect of antidiabetic agents: New insights from a pharmacovigilance study based on data from the reporting system databases FAERS and VigiBase." *Frontiers in Pharmacology* 14 (2023): 1128387.
38. Ko, Minoh, Jung Mi Oh, and In-Wha Kim. "Drug repositioning prediction for psoriasis using the adverse event reporting database." *Frontiers in Medicine* 10 (2023): 1159453.
39. Paci, Paola, Giulia Fiscon, Federica Conte, Rui-Sheng Wang, Diane E. Handy, Lorenzo Farina, and Joseph Loscalzo. "Comprehensive network medicine-based drug repositioning via integration of therapeutic efficacy and side effects." *npj Systems Biology and Applications* 8, no. 1 (2022): 12.
40. Nakagawa, Chihiro, Satoshi Yokoyama, Kouichi Hosomi, and Mitsutaka Takada. "Repurposing haloperidol for the treatment of rheumatoid arthritis: An integrative approach using data mining techniques." *Therapeutic Advances in Musculoskeletal Disease* 13 (2021): 1759720X211047057.
41. Chrétien, Basile, Jean-Pierre Jourdan, Audrey Davis, Sophie Fedrizzi, Ronan Bureau, Marion Sassier, and Christophe Rochais et al. "Disproportionality analysis in VigiBase as a drug repositioning method for the discovery of potentially useful drugs in Alzheimer's disease." *British Journal of Clinical Pharmacology* 87, no. 7 (2021): 2830–2837.
42. Yokoyama, Satoshi, Yasuhiro Sugimoto, Chihiro Nakagawa, Kouichi Hosomi, and Mitsutaka Takada. "Integrative analysis of clinical and bioinformatics databases to identify anticancer properties of digoxin." *Scientific Reports* 9, no. 1 (2019): 16597.
43. de Anda-Jáuregui Guillermo, Kai Guo, Brett A. McGregor, and Junguk Hur. "Exploration of the anti-inflammatory drug space through network pharmacology: Applications for drug repurposing." *Frontiers in Physiology* 9 (2018): 151.

22 Prospective Role of Preclinical Experimentation in Validating Safety Signals

Chandragouda Raosaheb Patil, Chandrakant Gawli, and Shvetank Bhatt

22.1 INTRODUCTION

The elixir sulphanilamide disaster in the late 1930s, followed by the thalidomide tragedy, raised serious public concerns regarding drug safety. In response to these incidents, the Federal Food, Drug and Cosmetic Act (FD&C Act) was enacted in 1938. Subsequently, the Kefauver-Harris Drug Amendments were introduced in 1962 to further address and enhance drug safety regulations. Over the past decades, there has been a significant increase in the issuance of regulatory guidance documents aimed at improving drug safety evaluation.

Following these, several guidelines have been developed by both national and international organizations, such as the International Conference on Harmonization, the Council for International Organizations of Medical Sciences, the Food and Drug Administration (FDA), and the European Commission. The purpose of these guidelines is to establish standardized and comprehensive approaches to assessing the safety of pharmaceutical products, ensuring that they meet rigorous safety standards before being approved for use in the market. These efforts reflect the ongoing commitment to safeguarding public health and preventing the occurrence of drug-related disasters like the ones that prompted the implementation of the FD&C Act and the Kefauver-Harris Drug Amendments in the past (1–5).

During clinical development programmes, it is expected that both adverse events associated with and not associated with the drug's mechanism of action may occur. Identification of the adverse events not associated with the mechanism of action of a drug under development or in the post-marketing surveillance phase is referred to as the 'safety signal identification'. When the number of adverse events associated with a product's use exceeds the expected rate, it is known as a safety signal (6). In pharmacovigilance a safety signal is considered information that arises from one or multiple sources, including observations and experiments, and suggests a new potentially causal association or a new aspect of a known association between its administration and an event or a set of related events (7).

Signals can arise at any time during the life-course of a drug, from the preclinical phase through the post-marketing phase. Signals are generated through the intentional, but hypothesis-free, comparison of the number of events observed in a population with the number expected. The determination of whether an excess of an adverse event represents a true causal relationship between the drug and the event is a challenge faced by drug developers and clinical researchers and is particularly difficult with infrequent adverse events (8) (Figure 22.1).

In the current scenario, safety signals are only detected with the help of clinical trials. However, with the present paradigm of drug discovery, even the preclinical pharmacological evaluation has remained the mainstay of initial determination of drug efficacy, toxicity, and safety of all drugs entering human trials. For identifying the efficacy, in vitro animal-free testing, along with the in vivo and ex vivo experimental animal-based testing, is performed. In this chapter, we hypothesize that the safety signals may be validated with the help of even the preclinical phase and this validation

DOI: 10.1201/9781032629940-22

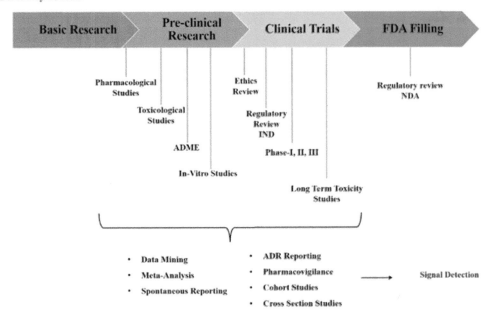

FIGURE 22.1   Generation of signals during drug discovery and the development process.

will prevent undue exposure of the patients to a drug with safety concern. Here, we describe a case study of on the significance of preclinical data in validating the safety signal, along with the narration of complexities and challenges in refining and evaluating safety signals.

22.2   IMPORTANCE OF PRECLINICAL DATA IN DRUG DISCOVERY AND DEVELOPMENT

The early drug discovery process involves extensive testing of the prospective druggable chemicals through in vitro and in vivo preclinical testing. On average, 2.7 years are needed to establish the pharmacodynamics, pharmacokinetics, toxicity, and safety for a drug candidate (9). During the basic research related to any drug discovery programme, a systematic data is generated to substantiate the efficacy, mechanism of action, and the molecular pathways targeted by the test drug. During these initial testing, the high throughput assays based on a highly specific drug-target interactions are used to select the effective candidates. Through the basic research, hits acting against the predefined pathogenic molecular pathways are identified for further development. The hit identification is based on the in silico and in vitro screening of the compounds against the selected biological targets involved in the pathogenesis of target disease. As the assays used in this type of research are focussed on finding hits interacting with the specific molecular pathways of interest, the outcomes provide precise information about their potency and kinetics of the interaction with the biological targets, probable biological activities, and target selectivity of the hits (9).

Further development of hits to leads involves structural modifications in the hits to improve their molecular interactions with the biological targets. During this phase of drug discovery, the objectives are to improve the selectivity of the developed leads as well as minimization of the non-specific interactions with the other irrelevant biological targets. The leads are then further improved in terms of their pharmacokinetic properties, reduced toxicities, and scaled up synthetic feasibilities (10).

The optimized leads undergo exhaustive pharmacological, toxicological, and safety screening through validated in vitro and in vivo experimental models under GLP specifications (11). The preclinical screening of the test compounds is regulated as per the guidelines prescribed by ICH, OECD, and country specific regulations like CDSCO guidelines for preclinical testing of toxicity and safety of the test drugs. This data is used for submission of the Investigational New Drug Application. This preclinical data plays a cardinal role in regulatory decision-making on suitability of the selected drug candidates for clinical investigations. The first in human trial is exclusively based on the preclinical screening of test drugs (1, 12).

22.3   EXTRAPOLATION OF PRECLINICAL SIGNALS TO THE CLINICAL CONTEXT

Preclinical data generated through the pharmacodynamics, pharmacokinetics, toxicity, and safety testing provides a wealth of useful data that is extrapolatable to humans. The shorter life spans of experimental animals provide a window to the effects of drugs across weaning, puberty, adulthood, pregnancy, and geriatric stages (13).

While translating the preclinical data to a clinical context, the following factors may be considered.

22.3.1   DOSE CONVERSION

The major purpose of the toxicity and safety testing studies is to determine the 'first in human' dose of the test drug (14). We have reviewed the approaches used in converting animal doses to human doses. Such dose conversion is performed considering the body surface areas, pharmacokinetics, physiological time, and clinical safety. As per the US FDA, the approach to estimate a human dose is to allometrically scale the NOAEL (no observable adverse effect level) and determine the maximum recommended dose (15). Preclinical toxicity testing involves a systematic plan to estimate the NOAEL, as the highest dose that does not induce any significant adverse effects. This dose level is then scaled to the human equivalent dose (HED) based on the weight ratios considering the average rate of humans to be 60 kg. The body surface area that represents the metabolic rate of a species is also considered in this extrapolation (16). The animal species that provides a minimum HED is considered a sensitive species, and it is selected for the estimation of the MRSD – maximum recommended starting dose (17). The HED is further divided by 10 to increase the safety of the first human dose. All these dose conversions assure that the participants of the clinical trials are exposed to most safe doses of the test drug. This indicates that the animal species used for the drug testing are being exposed to sufficiently higher doses of the test drug, including the lethal doses. Hence, a great wealth of data is generated on the adverse effects of a test drug during the animal studies and may provide valuable clues for validation of the safety signals encountered during the clinical studies. In the safety pharmacology follow-up studies as per the ICH guidelines, there are provisions to carry out additional safety studies in experimental animals which may have a significant role in validating the signals and in determining the mechanisms of adverse events.

Pugsley et al. (18) have reviewed the principles of safety pharmacology and have stated that an adverse event encountered during a preclinical safety study does not necessarily end the progression of the test drug through the development programme. However, a proper interrogation of the adverse effect signal generally ensues and triggers a mechanistic investigation of the observed signal through not only the core battery tests but also follow-up testing to generate data on pharmacodynamics, pharmacokinetics, and drug interaction studies. Thus, the preclinical studies provide very important data on dose-dependence of the adverse event signal ranging from NOAEL up to the lethal dose. This data is generated under strictly controlled variables and hence provides advantages over the patient-related trials where a number of confounding factors interfere with the validation of the signal.

22.3.1.1 Duration of Dosing

The important feature of preclinical data is the feasibility of exposure of subjects to test drugs across time spans ranging from single-dose exposure, exposure during a specific phase of development (from embryonic to old-age), and even life-long administration. Furthermore, the life span of experimental animals used in preclinical screening is considerably shorter, and thus, the effects of test drugs during various developmental phases can easily be established.

An editorial note by Quinn (19) has elaborated on the conversion of age from rat days to human years. According to Quinn, there is no simple formula to convert rat days to human years. He has emphasized how during different stages of life, the rat day to human year conversion varies and has concluded that an average 16.7 rat days corresponds to 1 human year. However, there is a need to consider the exact stage of human life during which the effects of drugs are to be elucidated and calculate the converting factor for rat days to a human year. Based on the suggestions by Quinn (19), recently, the adverse effects of SGLT-2 inhibitors were reported in rats and mice (20). In this study, an exposure to 4-week treatment of rats was considered to be equivalent to an exposure for 2.5 human years. There are several other studies reporting similar treatment durations to determine whether exposure of rats over a period of a few weeks can correlate with the clinical outcomes of an exposure of multiple human years.

22.3.1.2 Route of Administration

Apart from the extrapolation of doses and duration of treatment from animals to humans, there is an important role involved with the route of administration and establishment of pharmacokinetics. Pharmacokinetic and toxicokinetic studies are an integral part of regulatory testing.

The regulatory testing of drugs under investigation involves toxicokinetic assessment substantiating at least the kinetic parameters like area under curve (AUC), and peak plasma levels (C_{max}) along with establishing the maximum tolerated dose (MTD) (21). In certain cases, the evidence of toxicity in preclinical studies is unveiled at lower doses than their clinical counterpartd. Hence, the preclinical toxicokinetic data and estimates like NOAEL and NOEL may provide valuable information for dose escalations in clinical studies and in validating the signals encountered during post-marketing surveillance.

22.3.1.3 Use of Specific Age Groups (Juvenile/Geriatric)

Occurrence of adverse events during the trials in special populations like juveniles may pose additional challenges in being validated through additional clinical evaluations. In such a scenario, the preclinical data generated in the juveniles may support the evidence encountered during paediatric exposure. As per the ICH M 3 (R2) guideline (22, 23), it is recommended that preclinical studies related to the paediatric populations should be considered while designing clinical trials for drugs intended to be used in paediatrics. Furthermore, it is also suggested that paediatric clinical trials should be planned in paediatric patients suffering from the targeted diseases. These facts underscore the challenges in validating the adverse events in paediatric populations where the nuisance of the confounding factors becomes more prominent if the signal validation is to be planned in additional patients. However, regulatory toxicity and safety pharmacology studies including follow-up studies in neonatal and juvenile animals may prove to be indispensable in validating the signals without interference of confounding factors. Though such studies may necessitate testing of human metabolites of the test drug and evaluation in more than one species, certain studies have revealed that the age and sex differences in toxicokinetic drugs observed in juvenile and adult animals provide hints about the age differences in kinetics, repeated-dose scenarios, and human pharmacokinetics.

Apart from this, the preclinical investigations on the toxicity of drugs are reflective of the species and sex variations encountered in the clinical setting. The toxicokinetic drugs in certain cases may vary with gender due to the involvement of multiple factors like hormones, composition of body compartments, and stages of oestrous cycles. There are even possibilities that other toxicokinetic issues arising from the enterohepatic circulation, saturation kinetics, induction or inhibition of metabolizing enzymes, number of half-lives to achieve the steady-state concentration, and so on, are well extrapolated from preclinical studies to clinical investigations. These studies may play a significant role in the validation of the clinical safety signals.

22.3.1.4 Preclinical Studies Provide Numerous Other Advantages

These advantages include access to physiological, biochemical, and histopathological parameters during drug exposure to validate the safety signals. During the safety studies using telemetry systems, a continuous monitoring of blood pressure, ECG, and biochemical parameters is feasible using implanted sensors and transducers. This is a unique advantage, and hence the data collected during preclinical investigations might provide added advantages over the clinical validation of the safety signals (24). Such access to the biological parameters is impossible in clinical investigations. Hence, the preclinical data may provide unique advantages if used for the validation of the safety signals.

22.3.1.5 In Vitro Evaluations of Drug Effects

The modern paradigm of drug discovery involves extensive target-based evaluation of the pharmacodynamic profiling of the druggable candidates. Such evaluations include cell-free assays for target binding and enzyme activation/inhibition assays; regulatory toxicity and safety assays using in vitro interactions with biological targets (25), e.g. hERG assay for cardiotoxicity; cell-based screening of specific toxic responses, e.g., in vitro micronucleus assays; and comet assays (26). Such in vitro investigations may provide important insights in validating the safety signals.

Recent advances in cell culture technology like 3D cell culture provide unprecedented assay capabilities to establish drug effects and safety issues. The 3D cell culture-based detailed screening may now provide opportunities to use more human-relevant screening of drug effects using human cell-based and organoid-based assay protocols.

22.3.1.5.1 Advantages of In Vitro Studies over Preclinical In Vivo Studies

In the field of drug discovery, in vitro studies are recommended for initial screening of NCEs or herbal molecules. It saves the drug discovery time, which is usually 14 years or more. By doing in vitro studies, we can reduce the number of animals used as well as save manpower, resources, and time nowadays. The US FDA recommends some in vitro studies for regulatory approvals. In vitro studies have various advantages over in vivo:

- In vitro studies are easy to perform over in vivo studies as they take less time to perform and are easily reproducible.
- Less manpower, resources, and space are required to perform in vitro studies.
- Ethical clearance is not required to perform in vitro studies using cell lines.
- In vitro studies are generally more sensitive over in vivo studies, and a smaller number of drugs are required to get the results.
- Maintenance of laboratory conditions such as temperature, pressure, and humidity is easy to maintain for in vitro studies over chronic in vivo studies.
- Signalling inside the cell can easily be demonstrated and studied using in vitro studies.
- Long-term in vitro studies are cheaper as compared to in vivo studies.

22.3.1.5.2 Shortcomings of In Vitro Techniques over In Vivo

We cannot consider results obtained from in vitro studies 100%, as drug molecules behave totally in different ways inside the body due to the interaction of various systems with each other. Sometimes a molecule gives very good results in in vitro studies but fails inside the body. Hence to prove the effect of a drug, in vivo studies are also warranted. In in vivo studies we can study the effect of various systems on a single cell which is not possible in the case of in vitro studies. In a similar way, behavioural studies cannot be performed in vitro. We can get actual results using in vivo animal models. In addition, long-term effects of drugs like for months and years cannot be studied using in vitro techniques. In addition, cost of various chemicals, cell lines, and equipment used in various in vitro techniques is higher and increases the overall cost of the drug discovery process.

22.4   DIFFICULTIES IN CLINICAL TRANSLATION OF EXPERIMENTAL PRECLINICAL DATA

In drug discovery, drugs will be evaluated in human beings i.e. clinics after successful completion of all preclinical studies. Dose is generally extrapolated based on body weight from animals to human beings. However, a very small number of drugs going to clinical trials have shown the same efficacy as they have shown in preclinical setup. Some drugs have shown less effects, and some may show more effects. Because of these variations we cannot directly extrapolate the dose from animals to human beings, and more parameters need to be included to reduce these variations while extrapolating the dose from animals to human beings. Selection of animal models is also important in this aspect as the models with all the validities, namely, face, construct, and predictive validities have accurate results. These results can be extrapolated in human beings more precisely as compared to the models which have only one validity.

Nowadays, in some instances, animal testing is replaced with alternative in vitro methods, including in vitro tests using cell lines, tissue samples, and the use of alternative organisms such as bacteria, three-dimensional modelling and bioprinting, in silico tests, organ-on-chip technologies such as three-dimensional organoids, computer modelling, and phase-0 in human microdosing trials. The variations can be reduced using in vitro studies as laboratory conditions are easy to maintain throughout the experiments.

22.5   DOSE EXTRAPOLATION

Extrapolation of doses from one species to other species is important in the drug discovery process. The extrapolation of doses from animal species to humans is a very critical parameter. Scaling of doses is important for identifying first human doses based on the doses used in any preclinical studies. The approaches mainly used to convert the dose from preclinical to clinical are either based on surface area or body weight. Algometric methodology considers the differences in body surface area, which is linked with the weight of the animal while extrapolating the doses of various drugs among the species (14). However, weight-to-surface area is not consistent, and it should be noted that various pharmacokinetic parameters and pharmacodynamic properties of molecules should be taken into account while scaling as physiology of animals and human beings is different from each other. One should study all the parameters such as plasma protein binding, metabolism, and drug-binding characteristics, which should be monitored carefully (27).

22.6   BODY COMPARTMENT AND COMPOSITION VARIATIONS

It is known that more than 97% genes of human beings and rats or mice are similar. However, differences are observed with respect to body compartments and physiology of various organ systems. For example, some animals have ruminant digestive systems, i.e., sheep have four-chambered stomachs. These animals distribution of drugs will follow different patterns, and in a similar way other pharmacokinetic parameters will also change. Similarly, the human brain is a very complex system with more logical, expressive, and emotional quotients as compared to animals. Various physiological parameters such as breathing rate, enzymatic composition, and hormonal parameters vary slightly in human beings as compared to animals. Cell and extracellular space composition is slightly varied between animals and human beings. The presence of various ions, gases, waste products, and inflammatory markers are also not exactly similar in animals and human beings. These variations in composition should be taken into account for extrapolation of doses from animals to human beings. Compared to human RBCs, rat RBCs had decreased deformability, membrane rigidity, aggregability, and micro-vesiculation after component manufacturing processes (28). Similarly, gut microbiota of mice and non-human primates appears to be closer to human beings as compared to rats (29).

22.7 METABOLIC VARIATIONS

Animal models are commonly utilized in the preclinical discovery of new drugs to predict the metabolic behaviour of new compounds in humans. Larger animals have a high metabolic rate as compared to small animals of same species. Metabolism is the process of chemical conversion of lipophilic drugs to hydrophilic drugs. The liver is our main metabolizing organ. Biotransformation or metabolism is divided into two phases: Phase-1 or non-conjugative reactions include oxidation, reduction, and hydrolytic reactions. Phase-2 or conjugative or synthetic reactions include conjugation of phase-1 metabolites with glucuronide, sulphate, glutathione, glycine, etc. Most of the oxidation reactions and some reductive reactions and glucuronide conjugations are mainly performed by cytochrome P450 enzyme, which is present in liver microsomes. The metabolic rate of rodents is much faster as compared to human beings. For example, BMR is ~6.4 times faster in the rodent. Heart rate is on average ~4.7 times faster in the rat than in humans (260–400 versus 60–80 beats/min), and respiratory rate is ~6.3 times faster in the rat than in humans (75–115 versus 12–18/min) (30). It is important to understand that animals and humans are different with respect to isoform composition, expression, and catalytic activities of drug-metabolizing enzymes. Human CYP1A1, -1A2, -2B6, -2C8, -2C9, -2C19, and -3A4 are considered to be inducible isoforms of the CYP enzyme, whereas CYP2D6 is not. As reported in a paper by Martignoni et al. (31), CYP2E1 shows no large differences in metabolism patterns between species, and extrapolation between species appears to hold quite well. In contrast, the species-specific isoforms of CYP1A, -2C, -2D, and -3A show appreciable interspecies differences in terms of catalytic activity, and some caution should be applied when extrapolating metabolism data from animal models to humans. The substrate specificity of human CYP2A6 is considerably different from CYP2A enzymes in animal species. The CYP2D family shows genetic polymorphism resulting in variations in functional activity in drug metabolism in humans. Quinidine inhibits CYP2D in humans, dogs, and monkeys, but not in rats and mice (31).

22.8 VARIATIONS DUE TO THE ABSENCE OF CERTAIN ANATOMIC STRUCTURES IN ANIMALS

Rats do not have chemoreceptor trigger zones, gall bladder, and tonsils as compared to humans who possess all the above-mentioned structures. So if we want to evaluate any anti-emetic drug in rats, it is not possible as rats don't vomit. Similarly, studies of gall stones cannot be performed in rats. The pancreas is also very diffused in rats, and hence pancreatomy cannot be performed easily in rats. In most of the cases, diabetes is induced in rats by chemicals.

22.9 VARIATIONS IN THE SENSITIVITY BETWEEN ANIMALS AND HUMAN BEINGS

Animal models are used in drug discovery to predict human toxicity, and yet analysis suggests that animal models are poor predictors of drug safety in humans (13). Drugs that showed safety in preclinical settings have shown toxicity when used in clinics. Sensitivity for certain drugs in animals and human beings is different. These drugs behave differently in animals as compared to human beings and is based on their sensitivity parameter changes. For example, sensitivity of guinea pigs towards allergens is much more as compared to human beings, so extrapolation of results directly may not be feasible, and some additional parameters are required to be considered. Animal species which are more sensitive to certain drugs is preferable for toxicity studies as compared to less sensitive species. For example, female rats are more sensitive and preferable for toxicity studies as compared to male animals. Mice are less sensitive to thalidomide than other species like non-human primates and rabbits (32).

22.10   VARIATIONS IN THE DIURNAL CYCLE

The diurnal cycle varies between animals and human beings. Circadian rhythms are physical, mental, and behavioural changes that follow a 24-hour cycle. Generally, laboratory animals are considered nocturnal species, their activities occur more at night as compared to daytime, while human beings are more active in daytime as compared to night. There are variations observed with respect to levels of various enzymes, hormones, cytokines, oxidative stress parameters, and temperature with day and night. The study of these variations and circadian rhythms is done under chrono-pharmacology and chrono-pharmacokinetics. Different animal species have different food habits, and their activities also change within a day. With these variation levels, microbiomes also vary during the day in animals. Change in microbiota is associated with different neurological and gastrointestinal changes. These variations are also required to be considered for extrapolation of dose from animals to human beings.

22.11   BEHAVIOUR SIGNALS IN ANIMALS AND HUMAN BEINGS

Lots of variations are observed in behaviour studies using animals. It's difficult to validate behaviour parameters in animals. To reduce these variabilities a greater number of animals are included in the study. In the case of human beings, these variations are less as the physiological system is well developed and responds well to external stimulus as compared to animals. An animal behaviour model should have predictive, construct, and face validity to extrapolate perfectly for the human being. Changes in animal behaviour can be done by traumatic brain injury, olfactory bulbectomy, which does not primarily or perfectly mimic that of human beings and variations are observed. Extrapolation of behaviour signals from animals to human beings is a difficult task. The change in behavioural signals can be minimized if we perform the animal and human experiments/evaluations at the same time using similar experimental conditions such as temperature, pressure, humidity, and so on. These behaviour signals are different from species to species. Change in the strain of animal also leads to change in behavioural signals (33).

22.12   CONCLUDING REMARKS AND FUTURE PROSPECTS

The vast preclinical data that is generated in the initial phases of drug discovery contains useful information on the pharmacodynamics, pharmacokinetics, toxicities, and safety issues related to the drug candidates. Generally, the regulatory toxicity and safety studies are carried out to generate the data for submissions of the regulatory requirements. Though this data may appear discrete and inconclusive about the clinical safety aspects, specific data points from the preclinical studies may provide important hints about the clinically observed adverse events. We suggest that the preclinical data might be revisited once a safety concern is raised about the drug.

REFERENCES

1. OECD. Available at https://www.oecd.org/india/. Accessed on 20th July 2023.
2. International Council for Harmonisation of Technical Requirements for Pharmaceuticals for Human Use. Available at www.ich.org. Accessed on 25th August 2023.
3. European Medicines Agency EMA. Available at https://www.ema.europa.eu/en/homepage. Accessed on 31st August 2023.

4. FDA. Available at https://www.fda.gov/. Accessed on 30th September 2023.

5. Council for International Organizations of Medical Sciences. Available at: https://cioms.ch/ Accessed on 31st August 2023.

6. Pharmacovigilance G. Final guidance: good pharmacovigilance practices and pharmacoepidemiologic assessment. Biotechnol Law Rep. 2005;24(3):344–356.

7. Hauben M, Aronson JK. "Defining 'signal' and its subtypes in pharmacovigilance based on a systematic review of previous definitions. Drug Saf. 2009;32:99–110.

8. Dasic G, Jones T, Frajzyngier V, Rojo R, Madsen A, Valdez H. Safety signal detection and evaluation in clinical development programs: A case study of tofacitinib. Pharmacol Res Perspect. 2018;6(1).

9. Terry RF, Yamey G, Miyazaki-Krause R, Gunn A, Reeder JC. Funding global health product R&D: the portfolio-to-impact model (P2I), a new tool for modelling the impact of different research portfolios." Gates Open Res. 2018;2.

10. Hughes JP, Rees S, Kalindjian SB, Philpott KL. Principles of early drug discovery. Br J Pharmacol. 2011;162(6):1239–1249.

11. Chen J, Luo X, Qiu H, Mackey V, Sun L, Ouyang X. Drug discovery and drug marketing with the critical roles of modern administration. Am J Transl Res. 2018;10(12):4302.

12. Shen J, Swift B, Mamelok R, Pine S, Sinclair J, Attar M. Design and conduct considerations for first-in-human trials. Clin Transl Sci. 2019;12(1):6–19.

13. Van Norman GA. Limitations of animal studies for predicting toxicity in clinical trials: is it time to rethink our current approach? JACC: Basic to Transl Sci. 2019;4(7):845–854.

14. Nair AB., Jacob S. A simple practice guide for dose conversion between animals and human. J Basic Clin Pharm. 2016;7(2):27.

15. Guidance D. Guidance for industry: characterization and qualification of cell substrates and other biological starting materials used in the production of viral vaccines for the prevention and treatment of infectious diseases. Biotechnol Law Rep. 2006;25(6):697–723.

16. S Shin J-W, Seol I-C, Son C-G. Interpretation of animal dose and human equivalent dose for drug development. 대한한의학회지. 2010;31(3):1–7.

17. Zou P, Yu Y, Zheng N, Yang Y, Paholak HJ, Yu LX, Sun D. Applications of human pharmacokinetic prediction in first-in-human dose estimation. AAPS J. 2012;14:262–281.

18. Pugsley MK, Authier S, Curtis MJ. Principles of safety pharmacology. Br J Pharm. 2008;154(7): 1382–1399.

19. Quinn R. Comparing rat's to human's age: how old is my rat in people years? Nutrition. 2005;21(6):775.

20. Londzin P, Brudnowska A, Kurkowska K, Wilk K, Olszewska K, Ziembiński Ł, Janas A, Cegieła U, Folwarczna J. Unfavorable effects of sodium-glucose cotransporter 2 (SGLT2) inhibitors on the skeletal system of nondiabetic rats. Biomed Pharmacother. 2022;155:113679.

21. DeGeorge JJ, Ahn C-H, Andrews PA, Brower ME, Giorgio DW, Goheer MA, Lee-Ham DY. et al. Regulatory considerations for preclinical development of anticancer drugs. Cancer Chemother Pharmacol 1997;41:173–185.

22. Sjöberg Per, Jones David R. Non-clinical safety studies for the conduct of human clinical trials for pharmaceuticals: ICH M3 and M3 (R2). In Global Approach in Safety Testing: ICH Guidelines Explained. New York, NY: Springer New York, 2013. pp. 299–309.

23. International Council for Harmonisation of Technical Requirements for Pharmaceuticals for Human Use. ICH Harmonised Guideline. Addendum to E11: Clinical Investigation of Medicinal Products in the Pediatric Population E11(R1). Ich. 2017;18(August).

24. Andrade EL, Bento AF, Cavalli J, Oliveira SK, Schwanke RC, Siqueira JM, Freitas CS, Marcon R, Calixto JB. Non-clinical studies in the process of new drug development-part II: good laboratory practice, metabolism, pharmacokinetics, safety and dose translation to clinical studies. Braz J Med Biol Res 2016;49.

25. Van Vleet TR, Liguori MJ, Lynch JJ. III, Rao M, Warder S. Screening strategies and methods for better off-target liability prediction and identification of small-molecule pharmaceuticals. SLAS DISCOVERY: Adv Life Sci R&D. 2019;24(1):1–24.

26. Lee H-M, Yu M-S, Kazmi SR, Oh SY, Rhee K-H, Bae M-A, Lee BH et al. Computational determination of hERG-related cardiotoxicity of drug candidates. BMC Bioinform 2019;20:67–73.

27. Sharma V, McNeill JH. To scale or not to scale: the principles of dose extrapolation. Br J Pharmacol. 2009;157(6):907–921.

28. da SilveiraCavalcante L, Acker JP, Holovati JL. Differences in rat and human erythrocytes following blood component manufacturing: the effect of additive solutions. Transfus Med Hemother. 2015;42(3):150–157.

29. Nagpal R, Wang S, Solberg Woods LC, Seshie O, Chung ST, Shively CA, Register TC, Craft S, McClain DA, Yadav H. Comparative microbiome signatures and short-chain fatty acids in mouse, rat, non-human primate, and human feces. Front Microbiol. 2018;9:2897.

30. Agoston DV. How to translate time? The temporal aspect of human and rodent biology. Front Neurol. 2017;8:92.

31. Martignoni M, Groothuis GMM, de Kanter R. Species differences between mouse, rat, dog, monkey and human CYP-mediated drug metabolism, inhibition and induction. Expert Opin Drug Metabol Toxicol. 2006;2(6):875–894.

32. Vargesson N. Thalidomide-induced teratogenesis: history and mechanisms. Birth Defects Res, C: Embryo Today: Rev. 2015;105(2):140–156.

33. Bell AM, Hankison SJ, Laskowski KL. The repeatability of behaviour: a meta-analysis. Anim Behav. 2009;77(4):771–783.

23 Role of Artificial Intelligence in Signal Detection

Pramod Kumar A, Jeesa George, Rethesh Kiran, Priyanka, and Prizvan Lawrence Dsouza

23.1 INTRODUCTION

Pharmacovigilance (PV) is reliant on the extensive collection, management, and analysis of diverse data sources, primarily individual case safety reports (ICSRs). Documentation of these suspected adverse event reports from various channels pose a formidable challenge due to their global origin, linguistic variations, and distinct healthcare systems (1). PV's central challenge is the prompt identification of novel adverse events (AE) that are distinguished by their clinical characteristics, severity, or frequency. Expertise in this field is vital, but there is a growing interest in developing various databases and screening tools in order to assist human reviewers in identifying associations amid a sea of less significant reports deserving further scrutiny, known as 'signals' (2). Data mining algorithms are now under development and are used by health authorities, pharmaceutical organizations, and researchers to aid this process (3). In a comprehensive analysis of postapproval drug safety signal detection, the functioning of current algorithms and critical concerns regarding their validation, comparative effectiveness, and real-world deployment are addressed. They also propose areas for further research and development in this dynamic field, with the ultimate goal of improving PV's ability to safeguard public health by swiftly recognizing emerging adverse drug reactions (ADRs) (2–4).

The advent of artificial intelligence (AI) and machine learning (ML) has prompted speculation about their applicability to PV. Over the last decade, AI and ML have demonstrated remarkable capabilities in multiple scientific and medical domains. Their introduction to PV began in the 1990s but gained traction in the 2000s (5). There are various databases used in PV, such as the United States Food and Drug Administration's (FDA) Adverse Event Reporting System (AERS), World Health Organization's VigiBase, and Europe's EudraVigilance database. Effective case report analysis can be carried out by using various tools developed within these databases (5–7).

The use of AI/ML can effectively reduce case processing costs to improve PV activities. AI is beneficial in the field of life sciences, PV, and medical information (2, 8). ML methods have the potential to enhance various aspects of PV. They can automate the identification and analysis of safety signals, expedite data triage, and improve the detection of emerging safety concerns in medicines and vaccines. By harnessing AI and ML algorithms, PV can enhance signal detection and signal management processes, thus strengthening patient safety (6, 9, 10). Various statistical and data mining techniques are employed, which include proportional reporting ratio (PRR), reporting odds ratio (ROR), Bayesian confidence propagation neural network (BCPNN), and Multi-Item Gamma Poisson Shrinker (MGPS) in order to identify the patterns and associations in PV data. (7, 11, 12) This review underscores the potential for recent AI and ML advancements to revolutionize different facets of PV, providing a pathway towards more proactive and effective drug safety monitoring.

DOI: 10.1201/9781032629940-23

23.2   ARTIFICIAL INTELLIGENCE IN PHARMACOVIGILANCE

The increase in the number of ICSRs result in the complexity of handling and processing of these reports. Since the evaluation process is tedious and time-consuming, regulatory authorities are embracing AI/ML to reduce workforce, cost, and time for processing ICSR cases (13, 14). There are two categories in processing ICSRs using AI:

Insertion of structured and unstructured content:
   The data from the case can be read through XML, DOCX, Image, and PDF in a manner that is compliant with regulatory authorities. Natural language programming (NLP) and ML are used to extract data from ICSR.

Use of AI in decision-making:
   Generally, underreporting or missing information reduces ICSRs quality. Therefore, AI plays an important role in extracting accurate information, categorizing and classifying AEs, and correlating them with the suspected drug. AI tools can also help to automate and facilitate all aspects of PV such as case processing and risk tracking to reduce the processing time.

23.3   BENEFITS OF ARTIFICIAL INTELLIGENCE IN PHARMACOVIGILANCE

The use of AI has been beneficial in the field of PV, which includes the following (13–15):

1. Spontaneous reporting and processing.
2. Reduced PV cycle time.
3. Improvement in data quality and accuracy.
4. Handling and managing different types of incoming data formats.
5. Reduced burden and time for case processing.
6. Identification of ADRs.
7. Accurate and relevant data can be extracted using AI tools in order to evaluate the case validity without the workforce.

23.4   PROMISING NEAR-TERM APPLICATIONS OF MACHINE LEARNING IN PHARMACOVIGILANCE

Translation and multi-language models: These models can help in submitting and collecting case reports and other safety data from any part of the world written in various languages (13–16). Data processing requires translating the available information to a common language prior to signal investigation. In recent years, ML has exceedingly boomed at translating various low-resource languages that do not have a large amount of available training texts. Individual models that are trained on storing and processing large language data into different languages are available. Moreover, many of these models are available in the public domain and can be repurposed for PV. Integration of the translation model directly into PV can open up various capabilities to adapt to a variety of tasks by reducing the auxiliary and separate pre-processing tasks.

Named entity recognition (NER): A frequent task involved in PV data is the automatic extraction of key phrases and nouns, which is referred to as NER in the NLP literature. There has been rapid progress and use of ML in other areas, involving several languages, biomedical applications, and the lack of labelled data.

Text summarization and generation: Large volumes of unstructured text are found in cases that need individual examination. The generation of codified and structured data could enhance easier search as well as reduce time and workforce. In recent years, deep learning has been widely used and easily applied to perform abstractive summarization.

Causal inference: PV has been useful in answering whether a drug-adverse event association that has been reported in the safety reports is real. Causal inference is a statistical method used to provide estimates of treatment effects in real-world data. Recently, the intersection of causal inference and ML has been a topic of interest. There is nearly one-to-one translation of the ideas of causal inference to PV, and this could serve as another tool for signal detection and data analysis.

23.5   SIGNAL DETECTION APPROACHES USING AI

There are seven approaches involved in AI signal detection, which include the following (7, 11, 12):

1. Disproportionality Analysis
2. Traditional Pharmacoepidemiologic Designs
3. Sequence Symmetry Analysis
4. Sequential Statistical Testing
5. Temporal Association Rule
6. Supervised Machine Learning
7. Tree-Based Scan Statistic

23.5.1   DISPROPORTIONALITY ANALYSIS

Disproportionality analysis is largely used to develop a hypothesis between the drug and the adverse effects. These hypotheses are then tested in a clinical setting by using individual case reports. It is generally recommended for large databases with more precise values. A contingency table consisting of all the drugs and events experienced at least once is used to illustrate all spontaneous reports. The ratio of the observed-to-expected report count is then calculated by combining the data into a 2×2 contingency table (Table 23.1).

Longitudinal Gamma Poisson Shrinker (LGPS) and Multi-Item Gamma Poisson Shrinker (MGPS) are better alternative techniques used to assess and analyse longitudinal data. The exposed and non-exposed periods are recorded in patient-days known as longitudinal GPS. To eliminate incorrect signals associated with a protopathic or indication bias, combining the LGPS with the Longitudinal Evaluation of Observational Profiles of Adverse Events Related to Drugs (LEOPARD) technique is very effective. The idea is to examine prescription rates over a set period of time before and after the incident. If the latter rate is higher than the former, LEOPARD interprets the correlation as being associated with the indication (Table 23.2).

TABLE 23.1

The 2×2 Contingency Table for Computing Frequentist Method

	Suspected Drug	All Other Drugs in the Database	Total
Suspected ADRs	N_1	N_2	$N_1 + N_2$
All other database ADRs	N_3	N_4	$N_3 + N_4$
Total	$N_1 + N_3$	$N_2 + N_4$	$N_1 + N_2 + N_3 + N_4$

Notes: N_1 – number of reports that contain the suspected medication with the suspected ADR; N_2 – number of reports that contain other medications with the suspected ADR; N_3 – number of reports that contain the suspected medication with other ADRs; N_4 – number of reports that contain other medications with other ADRs.

TABLE 23.2

Insight on the Computation and Published Threshold Criteria of the Two Methods with Their Advantages and Limitations

DPA	Computation	Published Threshold Criteria	Explanation
Frequentist Methods			
Proportional Reporting Ratio (PRR)	$\dfrac{N_1/(N_1 + N_2)}{N_3/(N_3 + N_4)}$ $95\%CI = e^{\ln(PRR)\pm1.96}$	$PRR \geq 2$ $\chi^2 \geq 4$ $n \geq 3$	PRR calculates the reporting rate of a suspected ADR for a specific drug compared to all drugs in the database. It is easier and more sensitive than Bayesian methods, but cannot be applied to all DECs and is less specific.
Reporting Odds Ratio (ROR)	$\dfrac{N_1 \times N_4}{N_3 \times N_2}$ $95\%CI = e^{\ln(ROR)\pm1.96}$	$95\%CI > 1$ $n \geq 2$	ROR calculates the reporting odds of a suspected ADR for a specific drug compared to all drugs in the database. It is easier and more sensitive than Bayesian methods and logistic regression adjustments, but it is less specific and cannot be calculated if the denominator is zero.
Bayesian Methods			
Multi-Item Gamma Poisson Shrinker (MGPS)	$\dfrac{N_1 (N_1 + N_2 + N_3 + N_4)}{(N_1 + N_3)(N_1 + N_2)}$	$EBGM_{05} > 2$ $n > 0$	Valid for any DECs, highly specific with thresholds. Bayesian statistics are complex and less sensitive.
Bayesian Confidence Propagation Neural Network (BCPNN)	$\log_2 \dfrac{N_1 (N_1 + N_2 + N_3 + N_4)}{(N_1 + N_3)(N_1 + N_2)}$	$IC-2$ $SD > 0$	Applicable to any DECs and highly specific with threshold criteria. Bayesian statistics are complex, less sensitive, but useful for high-dimensional pattern recognition.

Abbreviations: χ^2, Chi-square; CI, confidence interval; DPA, disproportionality analysis; DECs, drug event combinations; EBGM, empirical Bayesian geometric mean; IC, information component; n, reported cases number; SD, standard deviation.

23.5.2 THE TRADITIONAL PHARMACOEPIDEMIOLOGIC APPROACH

In order to decrease the confounding and systematic error associated with observational research, the pharmacoepidemiology research team analyses the drug's use and effects in a larger population or large group of individuals. This is a two-step process, wherein two groups are first identified based on exposures, either prospectively or retrospectively. Then, these groups are compared based on the rate of the drug-event within these groups. With the below designs, statistical methods are applied to regulate potential confounders.

New user cohort design:

   The concept behind this design is to record patient data from the start of a drug exposure in a prospective manner. One cohort would consist of patients who started taking the drug of interest, while the other cohort would consist of patients who started taking another drug with a similar indication. The data from both cohorts will then be compared for the frequency of relevant event(s).

Matched case-control designs:

The fundamental idea behind matched case-control studies is to retrospectively examine any past drug exposure(s) between two groups of patients who are matched on confounders (such as age and sex). The goal is to compare the likelihood of control and intervention with the exposure to the drug(s) of interest.

Self-controlled designs:

The focus of this design is to compare the incident rate of event(s) during exposed time periods to that of unexposed time periods for each patient. On comparing, the impact of a drug on the incidence of an event is quantified. Confounding factors (chronic comorbidities, genetic risk factors, etc.) are implicitly adjusted in self-controlled designs. Self-controlled case series (SCCS) designs are used for safety signal detection and consist of patients who are exposed to the drug and experienced the event of interest at least once.

23.5.3 Sequence Symmetry Analysis (SSA) Method

The SSA method involves choosing drug exposure 'A' as the event of interest and drug exposure 'B' as a substitute event. The comparison of the starting order of the two pre-set drug exposures within a certain time period is the main aim of this approach. The number of patients with drug exposure 'A', followed by drug exposure 'B' is expected to be greater than the number of patients who were started with drug exposure 'B', followed by drug exposure 'A' leading to the prescription of drug exposure 'B' as a result of an ADR.

23.5.4 Sequential Statistical Testing Approach

The event rate is considerably higher among exposed individuals and needs to be tested sequentially (e.g., once a month) using the set of sequential statistical testing techniques on data obtained from prospective cohort studies, as it could be done in a typical signal-detection manner, which is not a requirement in unexposed individuals. Each new analysis in this process considers both the increase in exposure time among the individuals previously studied and the number of additional individuals in both exposed and unexposed groups since the last analysis.

23.5.5 Temporal Association Rule (TAR)

The following two rules are taken into account by TAR algorithms while dealing with signal detection:

1. The occurrence of the event must be after drug exposure.
2. The event must occur in a predetermined timeframe (i.e., the time period when risk exists).

23.5.5.1 Mutara/Hunt

(Mining the unanticipated TARs with given the antecedent)

A third rule algorithm that specifies that the event must happen 'unexpectedly' should be incorporated. A reference period prior to the beginning of drug exposure is established for each user of the subject. It is the researcher's belief that identification of an event is expected following the onset of exposure, given the event under study is seen within that specific time period. The incident is then eliminated from the patient data since it is thought to have a small probability of being an ADR. Thus, using the filtered data, the correlation score is computed.

23.5.5.2 Fuzzy logic rule base

For each case of the drug-event interaction, an individual score of causality is calculated using fuzzy criteria. This rating considers timing, presence of alternative explanations, dechallenge and rechallenge. For instance, based on how much time passes between the beginning of exposure and event occurrence, the fuzzy criteria assesses the temporal link as likely, feasible, or unlikely. Similar to this, dechallenge comes to a conclusion if the patient is alive after stopping the drug causing the event.

23.5.6 SUPERVISED MACHINE LEARNING (SML)

SML involves defining a reference set comprising of drug-event interactions that are a prior known to be linked or not linked to train a classifier. From this sample of trained data, a vector of pre-set parameters corresponding to proxies for each relationship is taken. The classifier uses all of these vectors as input data and is programmed to choose the parameters that will best to identify the true connections in the reference set with the use of impurity criteria and resampling method. It also involves screening the testing sample data to collect the selected parameters for each of the drug-event relationships and using the trained classifier to forecast novel drug-event associations.

23.5.7 TREE-BASED SCAN STATISTICAL METHOD

The core idea behind a tree-based scan statistical approach is to map a tree in accordance with the hierarchical structure of categories that are used to code events. The root of a specific event relates to its broad description, the nodes to its many sublevel definitions, the leaves to its finer definitions, and the branches connect the three elements together.

23.6 CONCLUSION

AI and ML are transforming PV by enhancing signal detection capabilities. These technologies address limitations in traditional methods, such as processing time, data accuracy, and the handling of diverse data formats. AI and ML techniques, including natural language processing, named entity recognition, and causal inference, streamline the identification of ADRs from large databases. Advanced statistical methods like disproportionality analysis and sequence symmetry analysis, combined with AI, improve the detection of drug-event associations. The integration of AI reduces manual workload, enhances data quality, and expedites the detection of potential safety signals. By continuously refining these methodologies, PV can more effectively safeguard public health, ensuring timely identification and response to emerging ADRs. AI's role in PV is pivotal, offering a proactive approach to drug safety monitoring and management.

REFERENCES

1. Beam, Andrew L., and Isaac S. Kohane. "Big data and machine learning in health care." *JAMA* 319, no. 13 (2018): 1317–1318.
2. Murali, Kotni, Sukhmeet Kaur, Ajay Prakash, and Bikash Medhi. "Artificial intelligence in pharmacovigilance: Practical utility." *Indian Journal of Pharmacology* 51, no. 6 (2019): 373.
3. Kompa, Benjamin, Joe B. Hakim, Anil Palepu, Kathryn Grace Kompa, Michael Smith, Paul A. Bain, Stephen Woloszynek, Jeffery L. Painter, Andrew Bate, and Andrew L. Beam. "Artificial intelligence based on machine learning in pharmacovigilance: A scoping review." *Drug Safety* 45, no. 5 (2022): 477–491.
4. Hauben, Manfred, David Madigan, Charles M. Gerrits, Louisa Walsh, and Eugene P. Van Puijenbroek. "The role of data mining in pharmacovigilance." *Expert Opinion on Drug Safety* 4, no. 5 (2005): 929–948.
5. Henegar, Corneliu, Cédric Bousquet, and Patrice Degoulet. "A knowledge based approach for automated signal generation in pharmacovigilance." In *MEDINFO 2004*, pp. 626–630. IOS Press, 2004.
6. Bousquet, Cédric, Corneliu Henegar, Agnès Lillo-Le Louët, Patrice Degoulet, and Marie-Christine Jaulent. "Implementation of automated signal generation in pharmacovigilance using a knowledge-based approach." *International Journal of Medical Informatics* 74, no. 7–8 (2005): 563–571.
7. Caster, Ola, Yasunori Aoki, Lucie M. Gattepaille, and Birgitta Grundmark. "Disproportionality analysis for pharmacovigilance signal detection in small databases or subsets: Recommendations for limiting false-positive associations." *Drug Safety* 43 (2020): 479–487.

8. Gawai, Pushpraj. "Overview of signal detection in pharmacovigilance." *Journal For Innovative Development in Pharmaceutical and Technical Science (JIDPTS)* 4, no. 9 (2021).

9. Jing, Zhang, and Santhosh Kamaraj. "Harnessing machine learning to improve healthcare monitoring with FAERS." *Applied Research in Artificial Intelligence and Cloud Computing* 5, no. 1 (2022): 135–149.

10. Bate, Andrew, and Jens-Ulrich Stegmann. "Artificial intelligence and pharmacovigilance: What is happening, what could happen and what should happen?" *Health Policy and Technology* 12, no. 2 (2023): 100743.

11. Arnaud, Mickael, Bernard Bégaud, Nicolas Thurin, Nicholas Moore, Antoine Pariente, and Francesco Salvo. "Methods for safety signal detection in healthcare databases: A literature review." *Expert Opinion on Drug Safety* 16, no. 6 (2017): 721–732.

12. Jahnavi, Yalamanchili, Kevin James, Prizvan Lawrence Dsouza, Basavaraj BV, and Viswam Subeesh. "Necessity of artificial intelligence to detect signal in the field of pharmacovigilance." *Journal of Dental and Orofacial Research* 16, no. 1 (2020): 71–75.

13. Markey, Jennifer, and Kelly Traverso. "The untapped potential of AI and automation." *Pharmaceutical Engineering* 40, no. 4 (2020): 60–64.

14. Trifirò, Gianluca, and Salvatore Crisafulli. "A new era of pharmacovigilance: Future challenges and opportunities." *Frontiers in Drug Safety and Regulation* 2 (2022): 1.

15. Bae, Ji-Hwan, Yeon-Hee Baek, Jeong-Eun Lee, Inmyung Song, Jee-Hyong Lee, and Ju-Young Shin. "Machine learning for detection of safety signals from spontaneous reporting system data: Example of nivolumab and docetaxel." *Frontiers in Pharmacology* 11 (2021): 602365.

16. VC, Randeep Raj, Divya P, Susmita A, Sushmitha P, Ramya Ch, and Chandni K. "Automation in pharmacovigilance: Artificial intelligence and machine learning for patient safety." *Journal of Innovations in Applied Pharmaceutical Science (JIAPS)* 7 (2022): 118–122.

Index

A

Adverse drug reaction, 6
Anatomic structure in animals, 275
Approaches, 245
Aris G, 37
Artificial intelligence, 5, 280
Asian Pharmacoepidemiology Network (AsPEN), 54
Assisted Model Building with Energy Refinement (AMBER), 186
Available packages, 187

B

Bayesian approaches, 13,138, 162
Behaviour signal, 276
Benefits of artificial intelligence, 280
Biases, 75
    detection, 76
    information, 76
    reporting, 77
    notoriety, 79
    protopathic, 83
    temporal, 78
Biomedical literature, 24
Body compartment, 274

C

Case studies, 129, 163, 172, 198, 210
Causality assessment, 12,160
Chemistry at Harvard Macromolecular Mechanics (CHARMM), 187
Chi-squared – χ^2, 138
Clinical context, 271
Clinical development signal detection and assessment, 111
Clinical translation, 274
Cohort studies, 11, 19
Confounders, 84
COVID-19, 253
Current challenges, 266
Current methods of pharmacovigilance, 9

D

Database of Adverse Event Notifications (DEAN), 48
Data mining, 39
Data, 168
    source, 168
    Base (Cov19VaxKB), 254
Disease classification, 99
Disproportionality analysis, 25, 75, 125, 128, 203
Distributed data networks, 53, 64
Diurnal cycle, 276
Docking, 183
    protocol, 181

Dose extrapolation, 274
Drug–drug interaction (DDI), 169
Drug-induced liver injury (DILI), 148
Drug repurposing, 262

E

Electronic health records (EHRs), 24
Ethical considerations, 249
Eudravigilance, 5, 34
Evolution of pharmacovigilance, 3
Expert judgement, 161

F

FAERS database, 170
Food and Drug Administration (FDA) Adverse (AE) Reporting System (FAERS), 32, 47
Functional enrichment analysis, 177

G

GROMACS, 187

H

Healthcare Systems Research Network (HCSRN), 54
HMG-CoA (3-hydroxy-3-methylglutaryl-CoA), 112, 214

I

Identification and validation, 249
Idiosyncratic Adverse Drug Reactions, 9
Individual case safety reports (ICSRs), 166
Information component, 26
Inhibitors, 214
International Classification of Diseases (ICD), 99
International Organization of Medical Sciences, 2

J

Japan Adverse Drug Event Report Database (JADER), 48

M

Management process, 26
Materials, 199
Measurements of disproportionality, 137
MedDRA, 103
Medical coding, 95
Medical dictionaries, 95
Metabolic variation, 275
Methods of signal detection, 124
Molecular dynamics, 186
Myasthenia gravis, 214

N

National Patient-Centered Clinical Research Network
        (PCORnet), 55
Network pharmacological approaches, 175
Network pharmacology, 177
Neurological events, 255, 257
Non-steroidal anti-inflammatory drug (NSAID), 2

O

Observational Medical Outcomes Partnership (OMOP), 36
Oracle Argus Safety, 36

P

Pharmacogenetics, 6
Pharmacovigilance centres, 3
Pharmacovigilance databases, 32
Pharmacovigilance regulations, 6
Pharmacovigilance system, 12, 14
Pharmacovigilance-based approach, 264
Post-marketing signal detection and assessment, 118
Preclinical data, 270
Promising applications, 280
Propionic acid, 203
Proportional reporting ratio (PRR), 26, 125, 128
Published studies, 130

R

Randomized Controlled Trials, 11
Recent trends, 30

Reporting odds ratio (ROR), 29, 127
Retrospective study, 253

S

Safety signals, 241
Safire, 38
Sensitivity, 275
Sentinel System, 55
Signals, 22, 191
Signal assessment, 28
Signal detection, 24, 230, 281
        qualitative method, 24
        quantitative method, 25
Signal detection process, 28
Signal generation, 45, 136
Signal prioritization, 28
Simulation parameters, 189
Software, 40
Sources of digitalized data for signal generation, 46
Spontaneous reporting systems, 47, 73, 135
Spontaneous reports, 23
Statistical software available, 143
Subgroup and sensitivity analysis, 148

V

Vaccine Adverse Event Reporting System
        (VAERS), 24, 48, 253
Vaccine safety, 227
Vaccine Safety Datalink (VSD), 56
Various databases, 228
VigiBase, 35, 47